Studies in Computational Intelligence

Volume 1183

The series "Studies in Computational Intelligence" (SCI) publishes new developments and advances in the various areas of computational intelligence—quickly and with a high quality. The intent is to cover the theory, applications, and design methods of computational intelligence, as embedded in the fields of engineering, computer science, physics and life sciences, as well as the methodologies behind them. The series contains monographs, lecture notes and edited volumes in computational intelligence spanning the areas of neural networks, connectionist systems, genetic algorithms, evolutionary computation, artificial intelligence, cellular automata, self-organizing systems, soft computing, fuzzy systems, and hybrid intelligent systems. Of particular value to both the contributors and the readership are the short publication timeframe and the world-wide distribution, which enable both wide and rapid dissemination of research output.

Indexed by SCOPUS, DBLP, WTI AG (Switzerland), zbMATH, SCImago.

All books published in the series are submitted for consideration in Web of Science.

Luyi Bai · Lin Zhu

Fuzzy Spatiotemporal XML Data Management

Springer

Luyi Bai
Qinhuangdao, Hebei, China

Lin Zhu
Qinhuangdao, Hebei, China

ISSN 1860-949X ISSN 1860-9503 (electronic)
Studies in Computational Intelligence
ISBN 978-3-031-81032-9 ISBN 978-3-031-81033-6 (eBook)
https://doi.org/10.1007/978-3-031-81033-6

This Springer imprint is published by the registered company Springer Nature Switzerland AG
The registered company address is: Gewerbestrasse 11, 6330 Cham, Switzerland

Preface

As spatiotemporal applications manipulate a considerable amount of spatiotemporal data, there is a growing need for managing spatiotemporal data accurately and efficiently. Spatiotemporal data typically includes both spatial and temporal phenomena. In the meanwhile, much spatiotemporal data is usually fuzzy in the real-world applications since their values are subjective to real applications. Due to its significance to spatiotemporal data, studies on combination of fuzziness and spatiotemporal data need further merits.

On the other hand, with the rapid development of the Internet, eXtensible Markup Language (XML) is rapidly emerging and has been the de-facto standard for representing and exchanging data on the Web due to its simplicity, readability, and portability. As a result, studies on fuzzy spatiotemporal XML data management are attracting an increased attention.

This book goes to concerning the topic of fuzzy spatiotemporal XML data management. The topics of this book include eight chapters: spatiotemporal data and fuzzy spatiotemporal data conceptual model; fuzzy spatiotemporal XML data model supporting dynamic update; transformation of fuzzy spatiotemporal data between XML and object-oriented database; topological relations of fuzzy spatiotemporal XML data over time; algebraic operations on fuzzy spatiotemporal XML data and querying fuzzy spatiotemporal data using XQuery; querying fuzzy spatiotemporal XML data integrated with holistic twig pattern; query relaxation and top-*k* result sorting of fuzzy spatiotemporal XML data; adaptive query relaxation and result categorization of fuzzy spatiotemporal XML data. Concerning spatiotemporal data and fuzzy spatiotemporal data conceptual model, we propose a fuzzy spatiotemporal data conceptual model based on UML by extending object-oriented UML. Concerning fuzzy spatiotemporal XML data model supporting dynamic update, we propose a fuzzy spatiotemporal XML data model by investigating inconsistency conditions and their fixing approaches. Concerning transformation of fuzzy spatiotemporal data between XML and object-oriented database, we study transformation methods of fuzzy spatiotemporal data between XML and object-oriented database by devising a set of transforming rules between them. Concerning topological relations of fuzzy spatiotemporal XML data over time, we extend the 9-Intersection model to establish a model for identifying continuous topological relations over time in XML. Concerning algebraic operations on fuzzy spatiotemporal XML data and querying fuzzy spatiotemporal data using XQuery, we present a general algebraic framework for supporting fuzzy spatiotemporal XML queries, containing set operations and other useful operations. Also, we propose an approach for querying fuzzy spatiotemporal data using XQuery. Concerning querying fuzzy spatiotemporal XML data integrated with holistic twig pattern, we extend Dewey code to mark fuzzy spatiotemporal data for special process and determine structure relationship of fuzzy spatiotemporal nodes in XML documents. Concerning query relaxation and top-*k* result sorting of fuzzy spatiotemporal XML data, we propose a query relaxation method, and propose a top-*k* sorting method for the query results. Concerning adaptive query relaxation and result categorization of fuzzy spatiotemporal XML data, we propose another different efficient algorithm to relax the fuzzy spatiotemporal query and improve the existing algorithms of generating navigation tree and make it rational to categorize the fuzzy spatiotemporal query results.

This book aims to provide a single record of current research in fuzzy spatiotemporal XML data management. The objective of the book is to provide state-of-the-art information to researchers and graduate students of the fuzzy spatiotemporal XML data management. Researchers and graduate students interested in fuzzy spatiotemporal XML data management will find this book a starting point and a reference for their study, research, and development.

We would like to acknowledge the researchers in the area of spatiotemporal databases, XML, and fuzzy databases. Based on both their publications and the many discussions with some of them, their influence on this book is profound. The materials in this book are the outgrowth of research conducted by the authors in recent years. The research work was supported by the National Natural Science Foundation of China (61402087), the Natural Science Foundation of Hebei Province (F2022501015), and the Fundamental Research Funds for the Central Universities (2023GFYD003). Special thanks goes to Thomas Ditzinger, the series editor of *Studies in Computational Intelligence*, and Saranya Sakkarapani, the production editor of Springer Nature, for their advice and help to propose, prepare, and publish this book. This book will not be completed without support from them.

Qinhuangdao, China
2024

Luyi Bai
Lin Zhu

Contents

Chapter 1
Spatiotemporal Data and Fuzzy Spatiotemporal Data Conceptual Model

1.1 Introduction

Space and time are inherent notions in spatiotemporal data which exists widely in many practical applications [1–5]. Spatiotemporal data contains spatial features and temporal features, and can support the change of spatiotemporal objects over time. At the same time, spatiotemporal data is often vague or fuzzy in the real-world applications [6–9], which is influenced by the data randomness and incompleteness, limitations of measuring equipment, delayed data updates and other factors. As a result, fuzzy spatiotemporal data management has received more and more attention.

The management ability of spatiotemporal data is mainly affected by the spatiotemporal data model, and the foundation of establishing spatiotemporal data model is the spatiotemporal data conceptual model. In Semantic Web, UML (Unified Modeling Language) has strong semantic representation and description ability, and can clearly represent the complex semantic relationships between concepts. UML is one of the important means to study the conceptual model. Babar et al. [10] describe a framework which is in the form of an extension to UML because it can be extended to bring in novel elements for different domains. Gonçales et al. [11] suggest a versatile, component-oriented framework to support the creation of comparison methods and provide details on the implementation of the UMLSim tool. Moral et al. [12] propose a conceptual framework for characterizing information visualization systems through UML, leveraging the clarity, precision, versatility, and adaptability that this notation offers. Shaikh et al. [13] provide a comprehensive overview of various approaches to the verification of UML class models, highlighting unresolved issues, prevailing research trends, and potential areas for enhancement. Consistency is identified as a critical metric for assessing the correctness of UML models. For that case, Wen et al. [14] investigate a formal UML model designed to delineate the overall structure of model-based systems in accordance with established consistency rules. Balaban and Maraee [15] conduct a thorough examination of the redundancy associated with multiplicity constraints, which includes a formal analysis of the gaps in redundancy

L. Bai and L. Zhu, *Fuzzy Spatiotemporal XML Data Management*, Studies in Computational Intelligence 1183, https://doi.org/10.1007/978-3-031-81033-6_1

within multiplicity intervals, as well as algorithmic and rule-based strategies for eliminating such redundancies. They also provide a formal assessment of the completeness of these algorithmic procedures in relation to the majority of UML class model constraints. Furthermore, the modeling of spatiotemporal data using UML and the subsequent verification of the correctness of these spatiotemporal UML models have emerged as significant challenges. Zhang et al. [16] introduce a spatiotemporal UML model alongside a Description Logic (DL) methodology for its verification. They first present a UML-based model for spatiotemporal data, followed by an abstract definition and semantic characterization of the spatiotemporal UML models, illustrated through a case study on the cadastral change process. Subsequently, they propose a method for mapping the spatiotemporal UML models to knowledge bases by incorporating specific concepts, roles, and axioms.

In recent years, fuzzy UML models have been applied in various practical contexts, including uncertain enterprise architecture design [17], behavior modeling of uncertain information systems [18], product development process modeling [19], design-level quality analysis [20], the news video domain [21], and meteorological fields [22], etc. The verification of fuzzy UML models is becoming increasingly critical to ensure that effective models can be utilized in the early stages of development. Zhang and Cheng [23] propose a comprehensive and practical methodology for verifying fuzzy UML models using fuzzy description logic (DL). Their approach involves initially transforming a fuzzy UML model into a fuzzy DL knowledge base. Subsequently, the verification tasks associated with the fuzzy UML model are reframed as reasoning problems pertaining to the transformed fuzzy DL knowledge base.

Recent advancements in UML models, particularly in the domains of fuzzy UML models and spatiotemporal UML models, have led to a growing interest in the research of fuzzy spatiotemporal UML models. Cheng et al. [24] introduce a methodology for the representation of fuzzy spatiotemporal knowledge utilizing ontologies grounded in fuzzy description logic. They further explore a formalized approach for the development of fuzzy spatiotemporal ontologies through the application of fuzzy Petri nets. In a related study, Chen et al. [25] present a framework for modeling fuzzy spatiotemporal objects using UML class diagrams. This framework begins with a formal definition of fuzzy spatiotemporal Event-Process, followed by an extension of the UML data model to accommodate fuzzy spatial and temporal data. The practical applicability of the fuzzy spatiotemporal UML object model is illustrated through the example of the Fujiwara Effect. Building upon their analysis of the semantics associated with fuzzy spatiotemporal data, Chen et al. [25] propose an innovative approach to modeling fuzzy spatiotemporal objects by further extending UML to encapsulate fuzzy spatial, fuzzy temporal, and fuzzy spatiotemporal data. Additionally, Wang and Bai [26] develop a fuzzy spatiotemporal data model that expands upon the standard UML modeling language.

This chapter presents an exploration of the various types and characteristics of spatiotemporal data, as well as fuzzy spatiotemporal data. By enhancing the Unified Modeling Language (UML) class diagram, we examine the semantics associated with fuzzy spatiotemporal data, which encompasses fuzzy spatiotemporal entities, attributes, and relationships. Building upon this foundation, we further develop the

UML class diagram by incorporating elements such as transaction time, valid time, and historical topological states, thereby establishing a fuzzy spatiotemporal data model grounded in the UML framework. The chapter also investigates the fundamental operations pertinent to this fuzzy spatiotemporal data model. To demonstrate the practicality of the proposed model, we conclude with a case study.

1.2 Spatiotemporal Data

Spatial databases and temporal databases have historically been regarded as distinct and unrelated research domains. While both have made incremental advancements within their respective areas, traditional spatial and temporal databases exhibit significant limitations in their ability to integrate spatiotemporal characteristics. Specifically, temporal databases often lack robust mechanisms for representing spatial changes, whereas spatial databases do not adequately support the expression and manipulation of temporal semantics. Consequently, researchers in both fields have begun to recognize the inherent challenges and interconnections between spatial and temporal databases. This realization has prompted an exploration of technologies that facilitate the integration of these two types of databases, leading to the emergence of a new research domain known as spatiotemporal databases, which has garnered increasing scholarly attention.

Spatiotemporal data refers to a category of spatial data that incorporates temporal elements and exhibits variations over time. This term encompasses information regarding features within the Earth's environment, which may include spatiotemporal phenomena, natural objects characterized by geometric features, events, states, processes, or properties and relationships that are spatiotemporal in nature. Such data encompasses a diverse array of information, including quantitative measures, shapes, textures, spatial distribution characteristics, and internal relationships pertaining to the Earth's environment, represented through numbers, characters, graphics, and images. These data not only exhibit distinct spatial distribution characteristics but also demonstrate temporal variability. Figure 1.1 illustrates the conceptual framework of spatiotemporal data, depicting how a spatial object A at t_1 transforms into spatial object B at time t_2, which subsequently evolves into spatial object C at time t_3. Thus, spatiotemporal data can be understood as a compilation of the historical changes of spatial objects over time. Consequently, spatiotemporal data represents the information that delineates the transformations of spatial data types in conjunction with temporal data types, with such changes potentially arising from spatial operations, temporal operations, or spatiotemporal operations. This framework also contributes to the multidimensional structure of spatiotemporal data, as illustrated in Fig. 1.2.

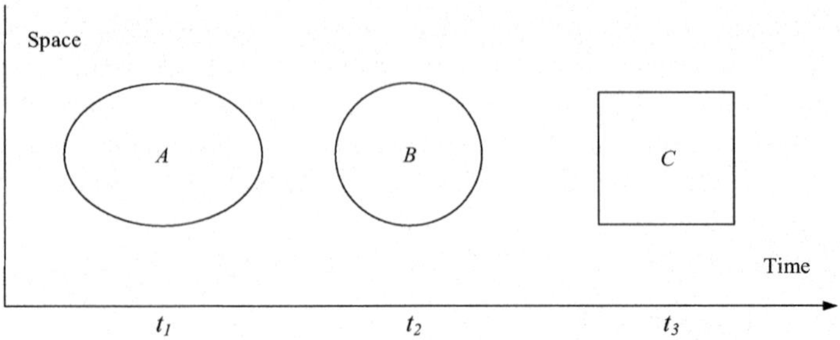

Fig. 1.1 Conceptual implications of spatiotemporal data

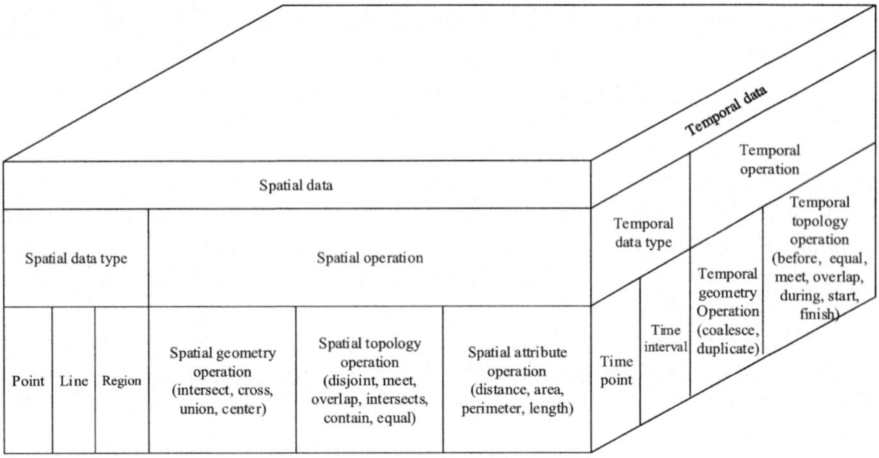

Fig. 1.2 The multidimensional structure of spatiotemporal data

1.3 Fuzziness in Spatiotemporal Data

1.3.1 Fuzzy Entity Set and Fuzzy Theory

A spatiotemporal entity is defined as an entity that undergoes changes in relation to both spatial and temporal dimensions. In various real-world applications, such entities often exhibit fuzzy characteristics. To effectively represent and manage these fuzzy entities, fuzzy theory is extensively employed.

For every element x within the defined universe of study U, there exists a corresponding numerical value associated with it.

$$A(x) \in [0, 1]$$

Here, A denotes a fuzzy set within the universal set U, and $A(x)$ is referred to as the membership degree of the element x in relation to the fuzzy set A. As the element x varies within the universal set U, $A(x)$ serves as the membership function associated with the fuzzy set A.

The degree of membership $A(x)$ is indicative of the extent to which an element x is associated with the set A. Specifically, a membership value approaching 1 signifies a higher degree of belonging, whereas a value approaching 0 indicates a lower degree of belonging. The membership function $A(x)$ quantifies this relationship and is defined within the interval (0, 1).

1.3.2 Fuzzy Time Set and Fuzzy Time Relation

1.3.2.1 Fuzzy Time Set

Fuzzy time is primarily categorized into fuzzy time point and fuzzy time interval.

Fuzzy time point: the time when the spatiotemporal entity occurs.

For instance, the expression "around six o'clock in the morning" does not provide a definitive indication of whether the time is prior to, precisely at, or subsequent to six o'clock. Consequently, this represents a fuzzy time point. The notation (t, μ) is utilized to signify a fuzzy time point, where t represents the fuzzy time itself and μ indicates the degree of membership associated with that fuzzy time point. For example, the notation (12:00, 0.6) suggests that the likelihood of an event occurring at 12:00 is 0.6, with the membership degree μ constrained within the range of 0 to 1 $(0 \leq \mu \leq 1)$.

Fuzzy time interval: consists of two time points that respectively indicate the starting and ending time, wherein at least one of these time points is characterized as a fuzzy time point.

For instance, the phrase "in the morning" lacks precise delineation of both the starting or the ending time of this interval, rendering it a fuzzy time interval. The notation (t_1, t_2, μ_1, μ_2) represents a fuzzy time interval, where t_1 and t_2 correspond to the estimated starting and ending time, respectively, and μ_1 and μ_2 signify the membership degrees associated with the fuzzy time interval t_1, t_2. It is important to note that the values of μ_1 and μ_2 are constrained within the range of 0 to 1 $(0 \leq \mu_1 \leq 1, 0 \leq \mu_2 \leq 1)$.

1.3.2.2 Fuzzy Time Relation

Fuzzy time relation is similar to the exact time relation, which includes fuzzy temporal direction relation, fuzzy temporal metric relation and fuzzy temporal topology relation.

Fuzzy temporal direction relation

Fuzzy temporal direction relation denotes a direct directional association between two temporal entities, which may consist of either two temporal intervals or a combination of one temporal point and one temporal interval. This relation encompasses concepts such as before and after.

Fuzzy temporal metric relation

Fuzzy time metric relation delineates the association between two temporal entities, quantified in units such as hours, minutes, and seconds, among others.

Fuzzy temporal topology relation

Fuzzy temporal topology relation refers to the relation when two events occur simultaneously. These relations can be expressed in terms of *f-Equal*, *f-Contain*, *f-Overlap*, *f-Meet*, *f-Disjoint* and so on.

The relations between two fuzzy time points are *f-Equal*, *f-Meet* and *f-Disjoint*.

The relations between two fuzzy temporal intervals are *f-Equal*, *f-Contain*, *f-Overlap*, *f-Meet* and *f-Disjoint*.

The relations between a fuzzy time point and a fuzzy temporal interval are *f-Contain*, *f-Meet* and *f-Disjoint*.

1.3.3 Fuzzy Spatial Set and Fuzzy Spatial Relation

1.3.3.1 Fuzzy Spatial Set

Fuzzy spatial entities are primarily categorized into three types: fuzzy point, fuzzy line, and fuzzy region.

Fuzzy point

A fuzzy point is defined as a point whose precise location is indeterminate, yet potential positions are identifiable within a specified region. In this context, a fuzzy point is conceptualized as a point within two-dimensional Euclidean space, characterized by a membership degree that reflects the value of the membership function, thereby indicating the possible location of the point. The notation (la, lo, τ) represents a fuzzy point, where la and lo correspond to the longitude and latitude of the fuzzy point, respectively, and τ signifies the membership degree associated with the fuzzy point. For instance, the representation (N: 34°31, E: 69°12, 0.8) suggests that the fuzzy point is approximately located at N: 34°31, E: 69°12, with a membership degree of 0.8, where $0 \leq \tau \leq 1$.

Fuzzy line

A fuzzy line is defined as a line whose precise shape, position, or length remains indeterminate; however, the general area in which the line exists is known. In this context, ones conceptualize a fuzzy line as a continuous entity characterized by a smooth membership degree that varies between adjacent points along the line. The semantics of a line can be understood as a set of points situated between two ending points. The notation (la_1, lo_1, la_2, lo_2, τ_1,τ_2) is employed to represent a fuzzy line, where la_1 and lo_1 correspond to the longitude and latitude of the two endpoints, respectively, while τ_1 and τ_2 signify the membership degrees of these fuzzy endpoints, constrained within the range of 0 to 1 ($0 \leq \tau_1 \leq 1, 0 \leq \tau_2 \leq 1$).

Fuzzy region

A fuzzy region is characterized by its ambiguous boundaries. Drawing parallels with the extension of crisp regions to fuzzy regions, researchers aim to define fuzzy regions utilizing the point set paradigm alongside fuzzy concepts. The notation (la_{min}, lo_{min}, τ_{min}, la_{max}, lo_{min}, τ_{max}) represents a fuzzy region, where (la_{min}, lo_{min}) and (la_{max}, lo_{max}) correspond to the minimum and maximum coordinates of the fuzzy region as mapped to latitude and longitude, respectively. Additionally, τ_{min} and τ_{max} indicate the membership degrees of points situated at the coordinates (la_{min}, lo_{min}) and (la_{max}, lo_{max}), constrained within the range of 0 to 1 ($0 \leq \tau_{min} \leq 1, 0 \leq \tau_{max} \leq 1$).

1.3.3.2 Fuzzy Spatial Relation

Fuzzy spatial relation is similar to the exact spatial relation and encompass fuzzy spatial direction relation, fuzzy spatial metric relation and fuzzy spatial topology relation.

Fuzzy spatial direction relation

Fuzzy spatial direction relation denotes a direct directional relationship between two points, two lines, or two regions. The relative directions of spatial objects are categorized as east, south, west, north, northeast, southeast, northwest and southwest.

Fuzzy spatial metric relation

Fuzzy spatial metric relation refers to the distance between two fuzzy spatial objects, quantified in units such as centimeters, meters, kilometers, and so on.

Fuzzy spatial topology relation

Fuzzy spatial topology relation denotes the relation of fuzzy point, fuzzy line, fuzzy area. These relations can be expressed in terms of *f-Equal, f-Contain, f-Overlap, f-Meet, f-Disjoint* and so on.

The relations between two fuzzy points are *f-Equal, f-Meet* and *f-Disjoint.*

The relations between two fuzzy lines are *f-Equal, f-Contain, f-Overlap, f-Intersect, f-Meet* and *f-Disjoint.*

The relations between two fuzzy areas are *f-Equal, f-Contain, f-Overlap, f-Meet* and *f-Disjoint.*

The relations between a fuzzy point and a fuzzy line are *f-Contain, f-Meet* and *f-Disjoint.*

The relations between a fuzzy point and a fuzzy area are *f-Contain*, *f-Meet* and *f-Disjoint*.

The relations between a fuzzy line and a fuzzy area are *f-Intersect*, *f-Contain*, *f-Meet* and *f-Disjoint*.

1.4 Spatiotemporal Class Relation

The UML class diagram delineates several fundamental relations, which include Generalization, Association, Aggregation, Composition, and Dependency.

Generalization: This exemplifies an inheritance relationship as well as a generalized-specialized relationship. For instance, there exists a generalization between the terms "typhoon" and "tropical cyclone," wherein "typhoon" is a specific instance of a "tropical cyclone," while the latter term does not exclusively refer to typhoons.

Association: It reflects a strong dependency between two classes, or a semantic relationship between a class and an interface. This encompasses one-way associations, two-way associations, and self-associations. For instance, the relationship between subtropical high and tropical cyclone exemplifies such an association.

Aggregation: This illustrates the ownership relationship that exists between a whole and its constituent parts. For instance, there exists an aggregation relationship between the eye of a typhoon and the typhoon itself, as the eye is a component of the typhoon.

Composition: Composition refers to the interdependent relationship between a whole and its constituent parts, wherein the individual parts cannot exist independently of the whole.

Dependency: Dependency refers to a relationship characterized by reliance, wherein the functionality of one class necessitates the support of another class. Modifications made to one class can have repercussions on the other class; therefore, it is advisable to minimize two-way interdependence. An illustrative example of this concept can be observed in the relationship between typhoons and cyclones.

1.5 Representation of Fuzzy Spatiotemporal Data

A fuzzy spatiotemporal data model is conceptualized as a 5-tuple (P, S, T, O, A) for the purpose of representing fuzzy spatiotemporal data. In this representation, P denotes an entity, S signifies a spatial attribute associated with the spatiotemporal data, T indicates a temporal attribute, O refers to an attribute that is independent of both spatial and temporal dimensions, and A represents the relationships among various fuzzy spatiotemporal entities. In instances where fuzziness is not involved,

Fig. 1.3 Fuzzy spatiotemporal entity and real entity

the data is managed according to the standard format established by the Unified Modeling Language (UML).

1.5.1 Fuzzy Spatiotemporal Entity

Regardless of whether an entity is classified as fuzzy or crisp, the presence of attributes is essential. In the context of the extended spatiotemporal UML class diagram, fuzzy entities are depicted using a dashed box, while solid frames are utilized to represent crisp entities. For instance, as illustrated in Fig. 1.3, desertification is categorized as a fuzzy entity, whereas a computer is identified as a crisp entity.

1.5.2 Fuzzy Spatiotemporal Attribute

Fuzzy spatiotemporal entities may encompass fuzzy spatial, temporal, and other attributes, as delineated by the attribute category inherent in the spatiotemporal entity. Furthermore, within the UML model for fuzzy spatiotemporal data, attributes and methods possess visibility, which is represented by three distinct symbols: "+" (for public), "–" (for private), "#" (for protected).

1.5.2.1 Fuzzy Temporal Attribute Set

Fuzzy temporal attributes encompass fuzzy time points, fuzzy time intervals, and fuzzy temporal relationships. A designated set, referred to as "$<T>$", can be utilized to denote collections of fuzzy temporal properties. The degree of fuzzy membership associated with an attribute is contingent upon the attribute itself and is denoted within brackets. For instance, as illustrated in Fig. 1.4, the notation "Beginning time (0.8)" signifies that the event possesses a membership degree of 0.8 at this specific time point, indicating a probability of 0.8 for its occurrence at that moment.

Fuzzy time point: "$<t_p>$" is used to describe the fuzzy time point of the fuzzy temporal attribute.

Fuzzy time interval: "$<t_i>$" is used to represent the fuzzy time interval of fuzzy temporal attribute.

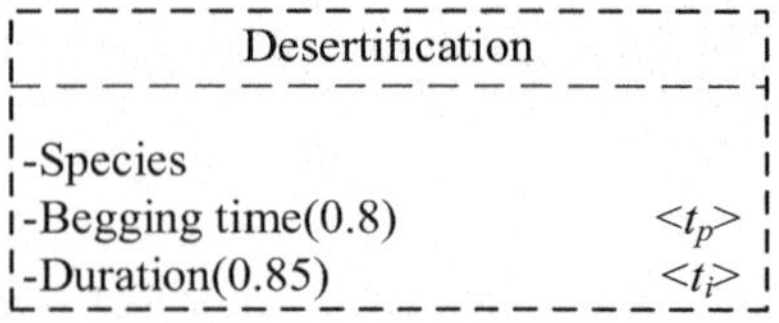

Fig. 1.4 Representation of a fuzzy temporal object

Fuzzy time relation: “$<t_r>$” is used to represent the fuzzy time relation of fuzzy temporal attribute, where “t_d” represents fuzzy temporal direction relations, “t_m” represents fuzzy temporal metric relations, and “t_o” represents fuzzy temporal topological relations.

1.5.2.2 Fuzzy Spatial Attribute Set

Fuzzy spatial attributes, as distinguished from ordinary spatial attributes, can be categorized into fuzzy points, fuzzy lines, fuzzy regions, and fuzzy spatial relationships. To denote these fuzzy spatial property sets, we utilize a notation represented by the symbol “$<S>$”. The degree of fuzzy membership associated with a particular attribute is denoted within brackets. For instance, as illustrated in Fig. 1.5, the notation “Starting place (0.9)” signifies that the event possesses a membership degree of 0.9 at this specific location, indicating a 90% probability that this location serves as the starting point.

Fuzzy point: “$<s_p>$” is used to represent the fuzzy spatial point of fuzzy spatial attribute.

Fuzzy line: “$<s_l>$” is used to represent the fuzzy spatial line of fuzzy spatial attribute.

Fuzzy region: “$<s_a>$” is used to represent the fuzzy spatial area of fuzzy spatial attribute. The classification of a region as either a fuzzy point or a fuzzy region is contingent upon the spatial context in which it is evaluated. When the region in question is diminutive in comparison to the contrasting space, and its area is deemed negligible, it is characterized as a fuzzy point. Conversely, if attributes such as the outline are needed, the region is classified as a fuzzy surface.

Fuzzy spatial relation: “$<s_r>$” is used to represent the fuzzy time relation of fuzzy temporal attribute, where “s_d” represents fuzzy temporal direction relations, “s_m” represents fuzzy temporal metric relations, and “s_o” represents fuzzy temporal topological relations.

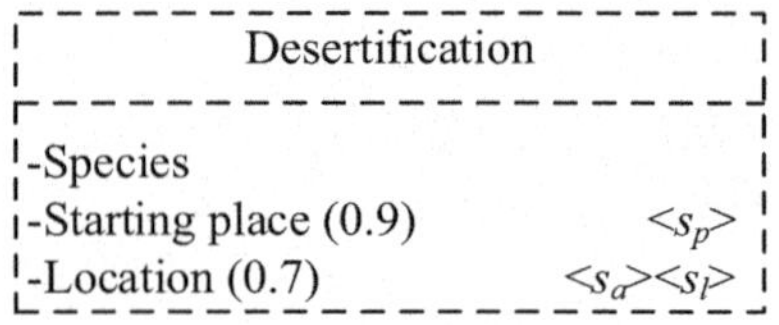

Fig. 1.5 Representation of a fuzzy spatial object

1.5.2.3 Other Fuzzy Attribute

Certain attributes do not fall within the categories of spatial, temporal, or spatiotemporal. To denote fuzzy properties that do not conform to these classifications, we can utilize a set referred to as "<*o*>". In essence, this set encompasses properties that are associated with temporal, spatial, or spatiotemporal dimensions. Furthermore, if additional entities or attributes are defined, their format will adhere to the established standards of UML.

In accordance with the established UML framework, the relationships between common fuzzy spatiotemporal entities are Generalization, Association, Aggregation, Composition, Dependency, and Realization.

Generalization

In the extended spatiotemporal UML class diagram, dotted lines and triangle arrows are employed to illustrate a fuzzy spatiotemporal generalization relationship. For instance, land desertification can be broadly categorized into four types: wind erosion, water erosion, freeze–thaw desertification, and soil salinization. As depicted in Fig. 1.6, the relationship among these four categories of desertification and the overarching concept of desertification is characterized as a generalization.

Association

In the extended spatiotemporal UML class diagram, fuzzy spatiotemporal association relationships are depicted using dotted lines and arrows, with the direction of the arrow denoting the nature of the association. As illustrated in Fig. 1.7, the primary factor contributing to land desertification is the prolonged destruction of extensive vegetation, thereby establishing an associative relationship between these elements.

Dependency

In the extended spatiotemporal UML class diagram, fuzzy spatiotemporal dependency relationships are denoted by dotted lines and solid circles, with the circles

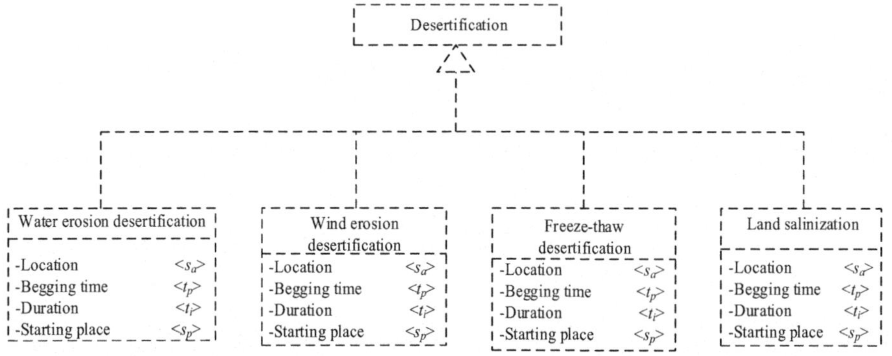

Fig. 1.6 Representation of fuzzy generalization

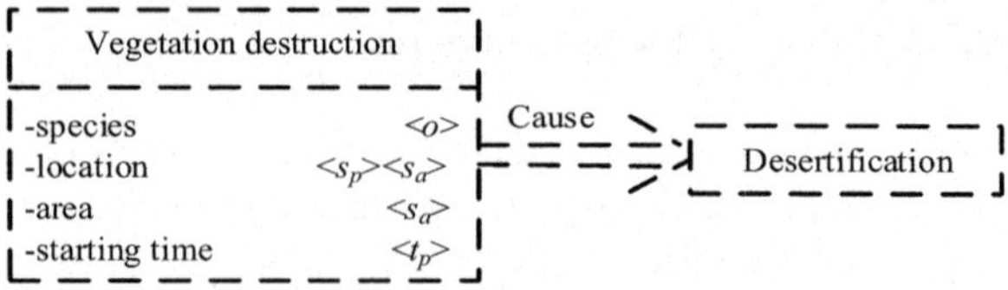

Fig. 1.7 Representation of fuzzy association

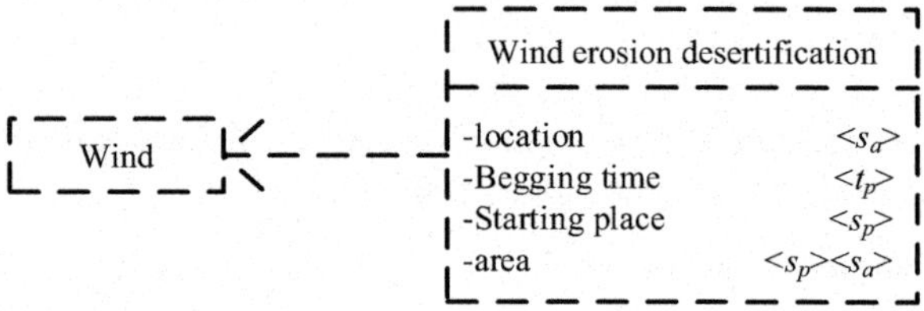

Fig. 1.8 Representation of fuzzy dependency

oriented towards the user. As illustrated in Fig. 1.8, the relationship between wind and wind erosion desertification is characterized as a dependency, as the occurrence of wind erosion desertification cannot transpire in the absence of wind.

Aggregation

In the extended spatiotemporal UML class diagram, a dotted line accompanied by a hollow diamond (with the diamond oriented towards the entirety) is employed to denote a fuzzy spatiotemporal aggregation relationship. As illustrated in Fig. 1.9, soil salinization is primarily characterized by reduced crop yields, water scarcity, and arid climatic conditions; thus, the interrelationship among these factors can be conceptualized as an aggregation.

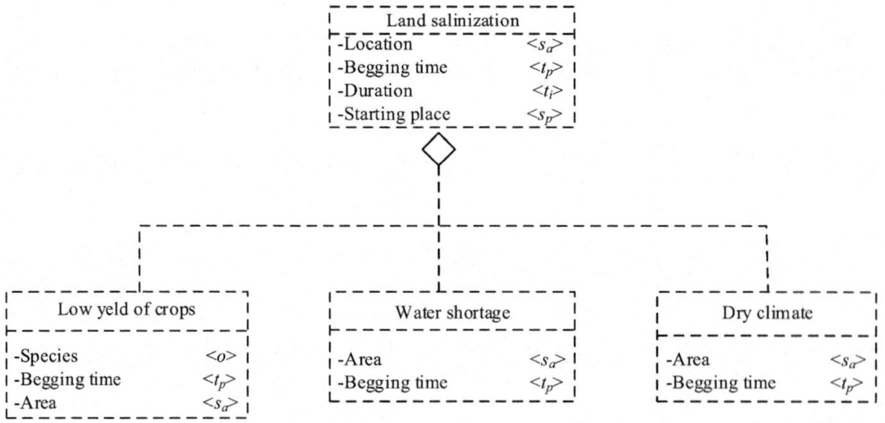

Fig. 1.9 Representation of fuzzy aggregation

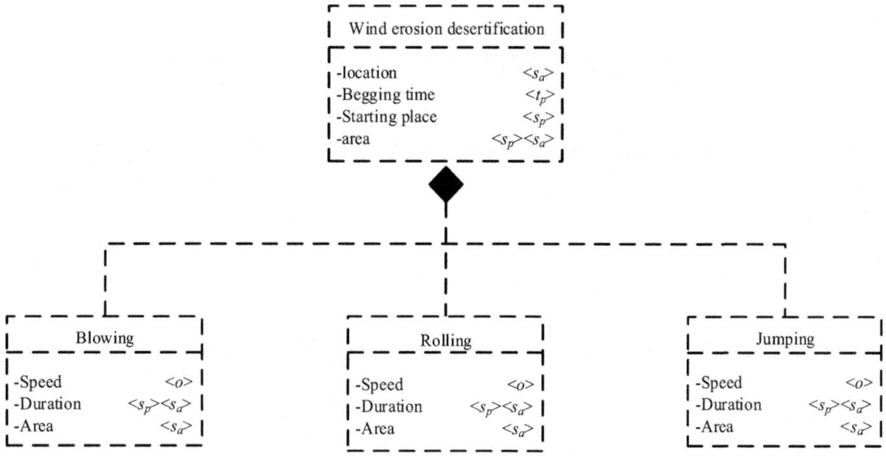

Fig. 1.10 Representation of fuzzy composition

Composition

In the extended spatiotemporal UML class diagram, a dashed line accompanied by a solid diamond (with the diamond oriented towards the entirety) is employed to illustrate a fuzzy spatiotemporal composition relationship. The predominant manifestations of wind erosion desertification include migration, suspension, and rolling. As depicted in Fig. 1.10, these forms are interrelated in a composite manner; none of these forms can exist independently of wind erosion desertification, indicating their inherent interdependence.

1.6 Fuzzy Spatiotemporal Data Conceptual Model Based on UML

Definition 1.1 (*Historical topological state*). A 3-tuple (U_p, U_s, T_a) is characterized as a historical topology state. It denotes the dynamic temporal evolution of an object, U_p denotes the object identifier set corresponding to the preceding object, U_s signifies the object identifier set associated with the succeeding object, and T_a indicates the valid time or the transaction time for this state.

The historical topological state integrates the states of various moments to construct a timeline that illustrates the developmental trajectory of an entity. Valid time refers to the duration during which a fact within a database is deemed accurate in the context of the modeled reality, whereas transaction time pertains to the period during which a fact recorded in the database is regarded as true. For instance, the historical topological state of an entity can be represented as (U_p, U_s, 2017). If this time is interpreted as valid time, it indicates that the fuzzy spatiotemporal data model representing the entity is applicable during this period. Conversely, if this time is understood as transaction time, it signifies that the information was entered into the database in 2017.

A fuzzy spatiotemporal data model, denoted as (U_p, U_s, T_a), is developed for an entity characterized by its historical topological state. In this model, U_p and U_s signify a fuzzy spatiotemporal data quintuple (P, S, T, O, A) pertaining to the preceding and succeeding objects, respectively. Here, P denotes the fuzzy entity; S is defined as (s_p, s_l, s_a, s_r), where s_r is further specified as (s_d, s_m, s_{to}); T is represented as (t_p, t_i, t_r), with t_r being detailed as (t_d, t_m, t_{to}); O signifies an attribute that is independent of both spatial and temporal dimensions; and A encompasses the components of Generalization, Association, Aggregation, Composition, and Dependency.

The current model under consideration is a dynamic framework that captures the spatiotemporal data representation of fuzzy entities across various time intervals. A conceptual model for fuzzy spatiotemporal data, grounded in the Unified Modeling Language (UML), can be developed in accordance with the aforementioned model and the specific requirements identified.

The fuzzy spatiotemporal UML data model illustrated in Fig. 1.11 provides a more accurate representation of the fuzzy spatiotemporal data associated with land desertification types, including land desertification due to salinization and wind erosion, particularly in the context of prolonged vegetation absence. As depicted in Fig. 1.11, there exists an associative relationship between vegetation destruction and land desertification, as extensive and sustained destruction of vegetation is a contributing factor to the phenomenon of land desertification. Land desertification can be broadly categorized into four types: wind erosion, water erosion, freeze–thaw processes, and soil salinization, indicating that the relationship among these four categories and the overarching concept of desertification is one of generalization. The primary manifestations of wind erosion desertification include processes such as migration, suspension, and rolling, suggesting that their relationship is one of combination; these processes cannot exist independently of wind erosion desertification. Lastly, the relationship between wind and wind erosion desertification is characterized as dependency, as the occurrence of wind erosion desertification is contingent upon the presence of wind.

1.7 Operations Based on UML

1.7.1 Common Operations of UML Class Diagrams

In a UML class diagram, attributes serve to delineate the data that is subject to processing, whereas operations specify the methods employed for data manipulation. A class may encompass multiple operations or, conversely, may not include any operations at all. Operations elucidate the various ways in which data within a class can be managed, including accessing and modifying property values, as well as

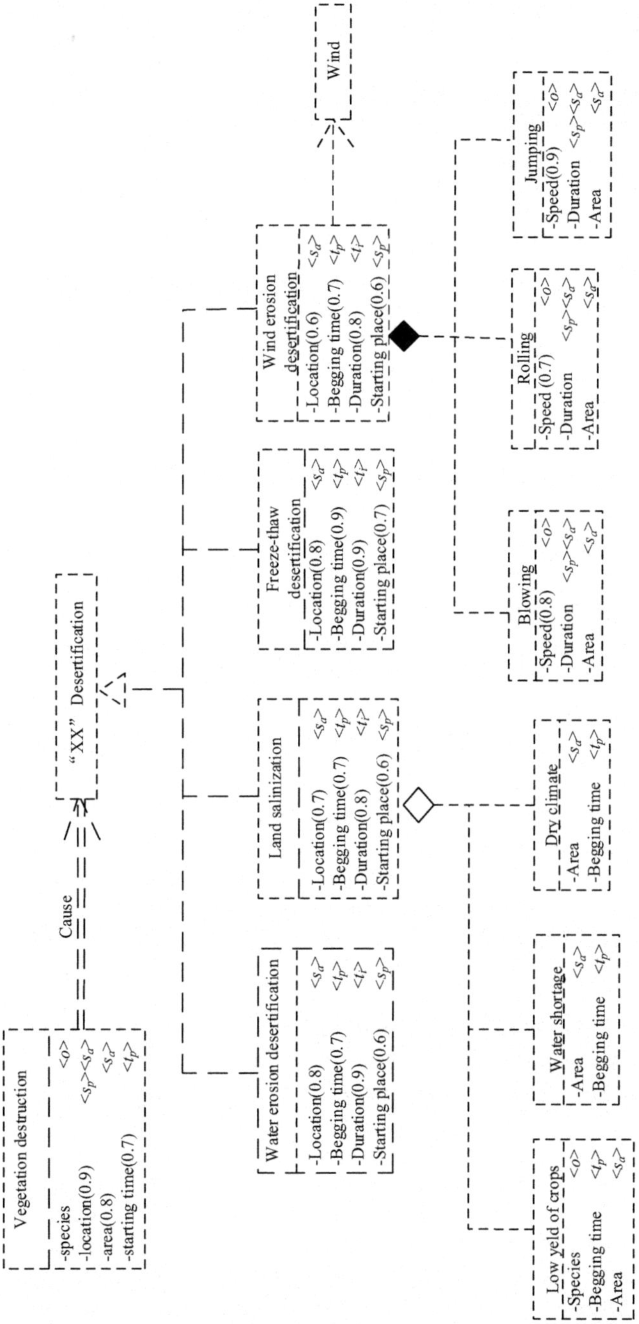

Fig. 1.11 Fuzzy spatiotemporal data model based on UML

executing specific actions. It is important to note that these operations are confined within the class and are limited to interacting with instances of that particular class.

The operation of a class offers two distinct types of services. The first type involves the alteration of the object's state as a result of the operation, which may also encompass the manifestation of associated dynamic behaviors. The second type pertains to the provision of return results to the entity that requested the service.

1.7.2 Operations of Fuzzy Spatiotemporal Data Model Based on UML

1.7.2.1 Visibility of the Operation

The fuzzy spatiotemporal data model, as delineated by UML, encompasses three distinct types of visibility. These types correspond to the visibility of fuzzy spatiotemporal data attributes and are represented by the symbols "+" (for public), "-" (for private), and "#" (for protected).

1.7.2.2 Basic Operation

Theory

In the context of a UML-based fuzzy spatiotemporal data model, operations are employed to alter or retrieve attributes of a class, as well as to execute specific actions. Each spatiotemporal object's operations must be documented within the appropriate class symbol, detailing both the operation name and a comprehensive description of the operation. Furthermore, when delineating the process of message transmission, it is essential to specify which object operations are necessary for the execution of the operation. This entails identifying the class to which the spatiotemporal object that receives the message belongs, along with the name of the operation that facilitates the message.

Definition 1.2 In the UML-based spatiotemporal data model, the syntax for describing operations is: [Visibility] Operation Name [(Property Table): (Parameter Table)] [: Return Type] [{Attribute string}].

In Definition 1.2, visibility includes "+" (for public), "–" (for private) and "#" (for protected). Operation names typically employ verbs or verb phrases to articulate the functionality of a class. The property table delineates the specific table associated with the data that is to be returned. A parameter table refers to a variable that accepts the parameter values provided during the invocation of the operation, with multiple variables being delineated by a "," (and potentially none). The return type indicates the data type that the operation yields.

For instance, "+Get (int part, int part [ ]): int" indicates that the public method "Get" receives parameters of types integer, "int [ ]" and returns a value of type int.

Basic operation

The following is a summary of fundamental operations associated with several widely utilized spatiotemporal data models:

Get ( ): this operation retrieves information from the property sheet. By specifying the name of the desired parameter table, the information is returned in the specified return type. For instance, the notation "-Get (s_p): void" indicates the retrieval of the fuzzy time point attribute from the fuzzy time attribute set.

Add ( ): this operation facilitates the creation of a new attribute, which is subsequently incorporated into the attribute table.

Delete ( ): this operation removes an attribute from the attribute table.

Alert ( ): this operation allows for the modification of an attribute within the attribute table. For example, the notation "-Alert (t_p: string): void" signifies the modification of the fuzzy space point attribute "t_p" within the fuzzy space attribute set.

Extended operation

The fundamental operations outlined previously facilitate the acquisition of certain necessary attribute values, thereby enabling the execution of supplementary operations, including comparative analyses and the formulation of summary conclusions.

The primary focus is on comparative analysis. By acquiring relevant data, we are able to conduct a comparative analysis of the same fuzzy entity across different time periods or of different fuzzy entities at a single time point. This approach enables us to derive meaningful conclusions.

In the UML-based fuzzy spatiotemporal data model, solid lines serve to link two opposing fuzzy entities, while small circles indicate the fuzzy entities under comparison. Additionally, dashed lines and ellipses are employed to represent the fuzzy spatiotemporal attributes that exist between these entities.

As illustrated in Fig. 1.12, the extent of desertification at a specific point in time can be distinctly observed in two separate regions. For instance, there is a 36-day interval between the initiation of desertification in these two locations, with "'xx' wind erosion desertification" occurring to the southeast of "'xx' soil salinization".

1.8 Case Study

The Alashan region of Inner Mongolia, which was historically abundant in water and vegetation, has now become one of the most arid areas in China due to a multitude of factors. In recent years, various initiatives have been implemented to mitigate the effects of desertification. Nevertheless, the extent of desertification in the region exhibits a dynamic pattern of both increase and decrease over time.

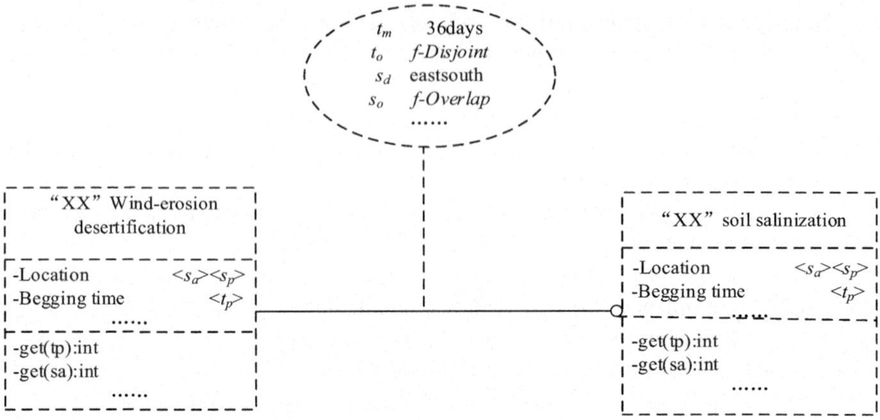

Fig. 1.12 Comparison of two fuzzy spatiotemporal entities

Utilizing the data gathered from prior studies [27, 28], a cartographic representation of the desert and desertification distribution within the Alashan region for the year 2010 has been constructed, as illustrated in Fig. 1.13. The map indicates that the delineation of boundaries within the region exhibits a degree of ambiguity, resulting in the data possessing fuzzy characteristics. Consequently, this data can be conceptualized as a fuzzy spatiotemporal data object.

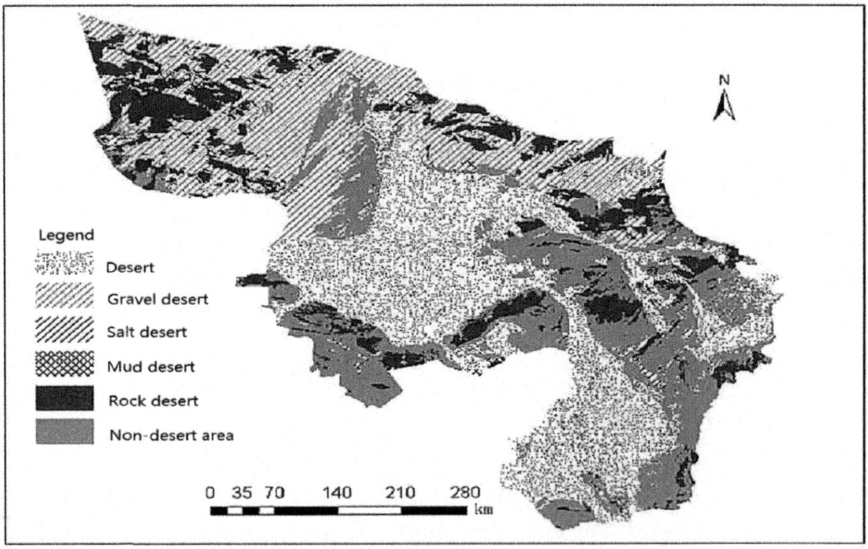

Fig. 1.13 The distribution of sandification and desertification in Alashan region

The alterations in desertified land across the three regions of Alashan from the 1970s to 2010, along with the dynamics of various forms of desertification over the past four decades, can be ascertained from sources [27, 28].

In conclusion, land desertification in the Alashan region can be characterized as a complex spatiotemporal phenomenon that encompasses several key aspects. Firstly, the delineation of desertified areas within Alashan is not well-defined and has exhibited fluctuations over time, particularly from the 1970s to 2010. Secondly, the distinctions between different types of desertification are ambiguous, and these classifications are subject to temporal variability. Lastly, this spatiotemporal phenomenon also incorporates metrics such as the rates of expansion and contraction of desertified land, as well as the initiation period and the initial extent of the affected areas.

The proposed fuzzy spatiotemporal data conceptual model, which is based on UML, facilitates the construction of fuzzy spatiotemporal data models as illustrated in Fig. 1.14. For instance, if the model representing the 1970s is denoted as P_{1970s} (U_s, U_p, 1970s) and the model for the year 2018 is represented as P_{2018} (U_s, U_p, 1970s)), then the model pertaining to land desertification in Alashan for the year 2010 can be formulated as ($P_{1970s,}$ P_{2018}, 2010). Each of these models is characterized by a fuzzy spatiotemporal data quintuple (P, S, T, O, A).

As illustrated in Fig. 1.14, significant environmental degradation has led to desertification in the Alashan region, establishing a "Cause" relationship between these phenomena. The model encompasses the fuzzy spatiotemporal object of Alashan land desertification, along with three additional fuzzy spatiotemporal objects: Alashan Zuoqi desertification, Alashan Youqi desertification, and Eqina Qi desertification. The interrelations among these objects are characterized as compositional. Each object possesses distinct attributes that delineate specific characteristics, which can be categorized as fuzzy spatiotemporal objects denoted as "*P*". Furthermore, the fuzzy spatiotemporal class "Alashan Zuoqi desertification" comprises numerous fuzzy attributes. The extent of each type of desertification is inherently ambiguous, lacking precise boundaries and subject to temporal variation, thus categorizing them as fuzzy spatiotemporal attributes marked with "$<s_a>$" (primarily indicating the fuzzy area). For instance, the designation "-Desert area: about 30000 km^2 (0.9)" signifies that the fuzzy membership degree of the desert area measuring 30,000 km^2 in Alashan Zuoqi in 2010 is 0.9. The attribute "-Begging time" is labeled "$<t_p>$", indicating that the exact initiation time is indeterminate, with only an approximate fuzzy timeframe available. Similarly, the attribute "speed" is also vague and classified as fuzzy data, although it is not related to spatiotemporal characteristics, hence it is labeled "$<o>$". The attributes associated with the fuzzy spatiotemporal class "Alashan Youqi desertification" mirror those of "Alashan Zuoqi desertification." The relationship between the fuzzy spatiotemporal class "Eqina Qi desertification" and the fuzzy spatiotemporal classes "Rock desert," "Desert," and "Gravel desert" is characterized as Aggregation. The primary factor contributing to the formation of "Rock desert" is

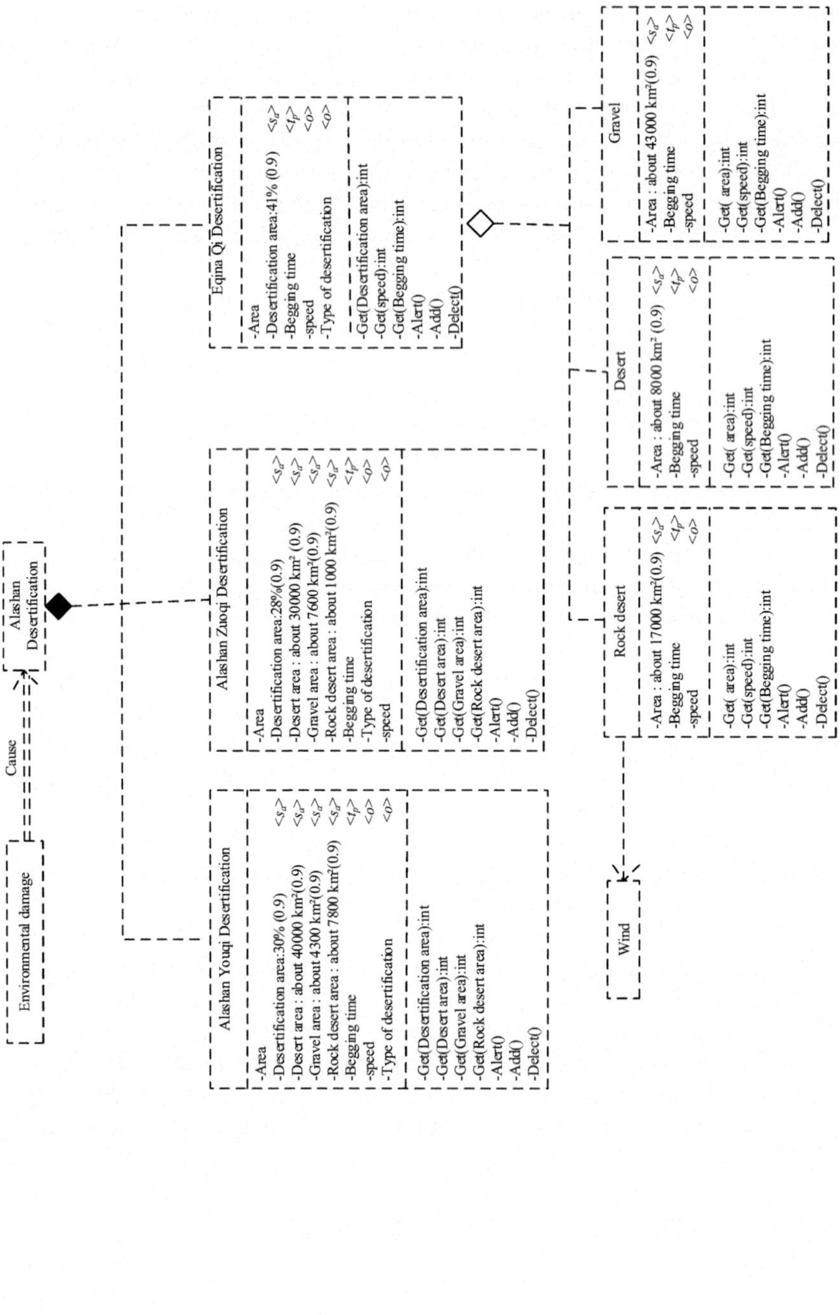

Fig. 1.14 Fuzzy spatiotemporal data modeling based on UML of land desertification in Alshan in 2010

the influence of strong winds, establishing a dependency relationship between "Rock desert" and wind.

Fuzzy attributes can be retrieved, modified, deleted, or added through operations defined in UML. The "Get" operation facilitates the retrieval of specific attributes within the fuzzy spatiotemporal class, allowing for subsequent comparisons with other classes or various temporal states of the same class. Figure 1.15 presents a comparative analysis of the fuzzy spatiotemporal classes "Alashan Zuoqi desertification" and "Eqina Qi desertification" for the year 2010, while Fig. 1.16 illustrates a comparison of the fuzzy spatiotemporal class "Alashan desertification" between 2010 and the 1970s. Analyzing Figs. 1.15 and 1.16 enable a clearer understanding of the semantic relationships and distinctions among different fuzzy spatiotemporal entities.

1.9 Summary

This chapter commences with an introduction to spatiotemporal data and the inherent fuzziness associated with it. Following this, it delineates the spatiotemporal class relations and subsequently emphasizes the semantics of fuzzy spatiotemporal data, encompassing fuzzy spatiotemporal entities, attributes, and relationships. A conceptual model for fuzzy spatiotemporal data is proposed, which extends the object-oriented Unified Modeling Language (UML). This model incorporates a historical topological state, thereby rendering it dynamic; it integrates multiple temporal states to construct a timeline that illustrates the developmental trajectory of an entity. The model effectively accommodates the fuzziness of data and the dynamic evolution of objects, thereby enhancing the representation of fuzzy spatiotemporal objects. Additionally, the operational aspects of the UML class diagram are incorporated, allowing for the model's data to be manipulated and compared according to specific requirements, thus augmenting its practical applicability.

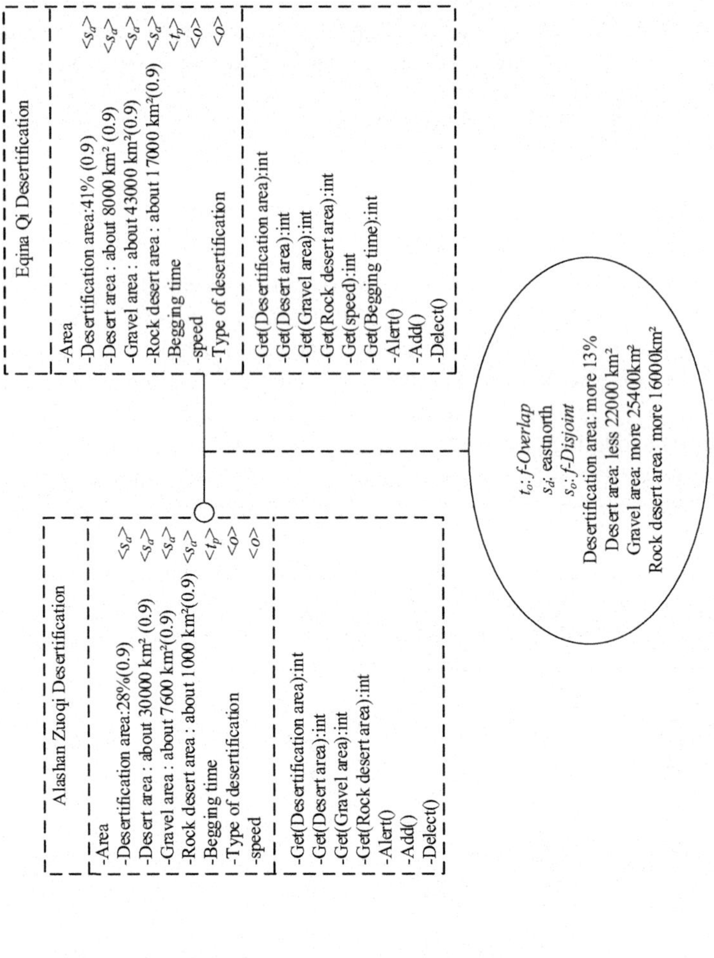

Fig. 1.15 Comparison of “Alashan Zuoqi desertification” and “Eqina Qi desertification” in 2010

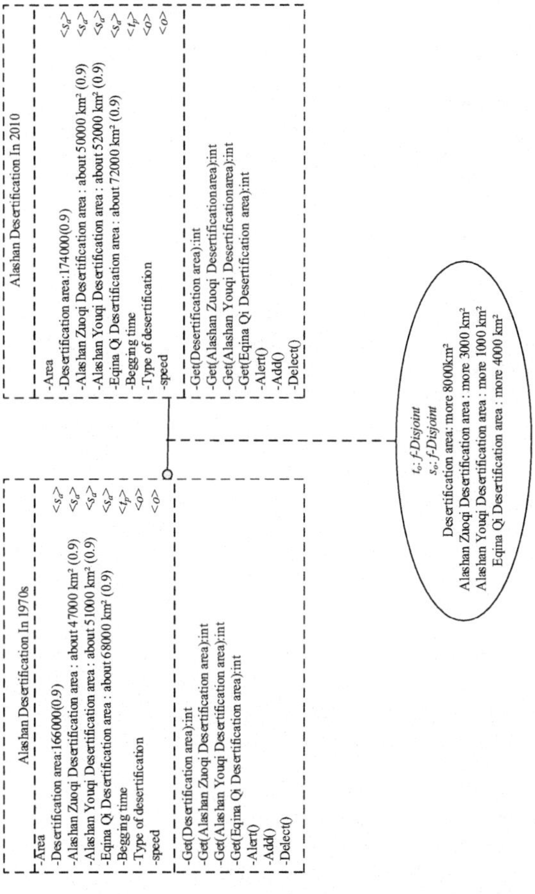

Fig. 1.16 Comparison of "Alashan desertification" in 2010 and the 1970s

References

1. Z. Fang, Y. Wang, C.J. van Westen, L. Lombardo, Landslide hazard spatiotemporal prediction based on data-driven models: estimating where, when and how large landslide may be. Int. J. Appl. Earth Obs. Geoinf.Geoinf. **126**, 1–13 (2024)
2. Z. Li, Y. Wei, G. Chen, K. Lu, X. Zheng, Learning dynamics of multi-level spatiotemporal graph data for traffic flow prediction. Comput. Commun.. Commun. **223**, 26–35 (2024)
3. X. Chen, C. Zhang, X. Chen, N. Saunier, L. Sun, Discovering dynamic patterns from spatiotemporal data with time-varying low-rank autoregression. IEEE Trans. Knowl. Data Eng.Knowl. Data Eng. **36**(2), 504–517 (2024)
4. R. Song, T. Li, X. Dong, Z. Ding, An effective points of interest recommendation approach based on embedded meta-path of spatiotemporal data. Expert Syst.—J. Knowl. Eng. **40**(2), 1–19 (2023)
5. B. Cardone, F.D. Martino, A novel spatiotemporal prediction method based on fuzzy transform: application to demographic balance data. Inf. Sci. **634**, 677–695 (2023)
6. Y. Yin, Y. Sheng, Y. He, J. Qin, Modeling vague spatiotemporal objects based on interval type-2 fuzzy sets. Int. J. Geogr. Inf. Sci.Geogr. Inf. Sci. **36**(6), 1258–1273 (2022)
7. H. Ji, L. Yan, Z. Ma, FSTRE: Fuzzy spatiotemporal RDF knowledge graph embedding using uncertain dynamic vector projection and rotation. IEEE Trans. Fuzzy Syst. **32**(2), 435–444 (2024)
8. H. Ji, L. Yan, Z. Ma, Multihop fuzzy spatiotemporal RDF knowledge graph query via quaternion embedding. IEEE Trans. Fuzzy Syst. **32**(5), 3026–3036 (2024)
9. H. Cheng, Z. Ma, A tractable temporal description logic for reasoning fuzzy spatiotemporal knowledge. World Wide Web J. **26**(5), 3155–3182 (2023)
10. M. Babar, A.S. Khattak, F. Arif, M.U. Tariq, An improved framework for modelling data warehouse systems using UML profile. Int. Arab J. Inf. Technol. **17**(4), 562–571 (2020)
11. L. Gonçales, K. Farias, V. Bischoff, M. Segalotto, Toward an architecture for comparing UML design models. J. Softw. **12**(7), 559–569 (2017)
12. C. Moral, A. de Antonio, X. Ferré, J. Ramírez, A proposed UML-based common model for information visualization systems. Multimedia Tools Appl. **80**(8), 12541–12579 (2021)
13. A. Shaikh, A. Hafeez, A.A. Wagan, M. Alrizq, A. Alghamdi, M.A. Reshan, More than two decades of research on verification of UML class models: a systematic literature review. IEEE Access **9**, 142461–142474 (2021)
14. H. Wen, J. Wu, J. Jiang, G. Tang, Z. Hong, A formal approach for consistency management in UML models. Int. J. Softw. Eng. Knowl. Eng.Softw. Eng. Knowl. Eng. **33**(5), 733–763 (2023)
15. M. Balaban, A. Maraee, Removing redundant multiplicity constraints in UML class models. Softw. Syst. Model.. Syst. Model. **18**(1), 557–583 (2019)
16. F. Zhang, D. Cao, J. Cheng, Spatio-temporal UML modeling and verifying with description logic. J. Inf. Sci. Eng. **35**(6), 1279–1298 (2019)
17. J. Afshani, A. Harounabadi, M. Dezfouli, A new model for designing uncertain enterprise architecture. Manage. Sci. Lett. **2**(2), 689–696 (2012)
18. A. Haroonabadi, M. Teshnehlab, Behavior modeling in uncertain information systems by fuzzy-UML. Int. J. Soft Comput.Comput. **4**(1), 32–38 (2009)
19. J. Xiao, P. Pinel, L. Pi, V. Aranega, C. Baron, Modeling uncertain and imprecise information in process modeling with UML, in *Proceedings of the 14th International Conference on Management of Data*, (2008), pp. 237–240
20. M. Vimaladevi, G. Zayaraz, Design level quality analysis using fuzzy UML models and weighted complex networks, in: *Proceedings of the 2017 International Conference on Communication and Signal Processing*, (2017), pp. 270–275
21. D. Kucuk, N.B. Ozgur, A. Yazici, M. Koyuncu, A fuzzy conceptual model for multimedia data with application to news video domain, in *Proceedings of the 2008 International Conference on Information Processing and Management of Uncertainty in Knowledge-Based Systems*, (2008), pp. 1741–1748

22. E.R. Tudoroiu, A. Astilean, T. Letia, G. Neacsu, M. Zoltan, N. Tudoroiu, Fuzzy UML and petri nets modeling investigations on the pollution impact on the air quality in the vicinity of the Black Sea Constanta Romanian resort, in: *Proceedings of the 2011 Federated Conference on Computer Science and Information Systems*, (2011), pp. 763–766
23. F. Zhang, J. Cheng, Verification of fuzzy UML models with fuzzy description logic. Appl. Soft Comput.Comput. **73**, 134–152 (2018)
24. H. Cheng, L. Yan, Z. Ma, S. Ribaric, Fuzzy spatio-temporal ontologies and formal construction based on fuzzy petri nets. Comput. Intell.. Intell. **35**(1), 204–239 (2019)
25. X. Chen, H. Cheng, H. Wang, W. Li, Fuzzy spatiotemporal object modeling based on UML class diagram. J. Intell. Fuzzy Syst. **33**(5), 2727–2736 (2017)
26. Y. Wang, L. Bai, Fuzzy spatiotemporal data modeling based on UML. IEEE Access **7**, 45405–45416 (2019)
27. Y. Tian, Desertification land dynamic analysis and restoration countermeasures in Alashan area, in *Proceedings of the National Symposium on Ecological Restoration of Soil and Water Conservation*, (2009), pp. 173–183
28. J. Zhou, Y. Gu, L. Da, Z. Liu, Remote sensing monitoring and dynamic change of desertification in Alashan area. Resour. Environ. Arid Zone **28**(01), 126–130 (2014)

Chapter 2
Fuzzy Spatiotemporal XML Data Model Supporting Dynamic Update

2.1 Introduction

A considerable amount of spatiotemporal data emerges in spatiotemporal applications [1, 2]. At the same time, spatiotemporal data is usually fuzzy in the real world applications since their values are subjective in real applications. Due to its significance to spatiotemporal data, researches on combination of fuzziness and spatiotemporal data have been extensively proposed [3–7].

On the other hand, XML (eXtensible Markup Language) has the characteristics of strong operability, uniform standard and good scalability [8]. Hence, XML has attracted a great deal of attention as a format for exchanging data over the Web as well as an important language for exchanging structural documents. In that case, the advent of XML seems to provide an opportunity for modeling fuzzy spatiotemporal data. Therefore, studies on fuzzy spatiotemporal data based on XML have been investigated by many researchers [9–13]. Bai et al. [9] present a methodology for modeling fuzzy spatiotemporal data utilizing XML. Their research further explores the representation of fuzzy spatiotemporal data within XML documents and the enhancement of XML Schema to facilitate the description of such data while capturing the structural characteristics of fuzzy spatiotemporal XML documents. In a subsequent study [10], Bai et al. employ XML twig pattern to ascertain the topological relationships inherent in fuzzy spatiotemporal XML data, extending the region coding scheme to ensure compatibility with this type of data. Zhu et al. [11] propose an integration model specifically designed for fuzzy spatiotemporal XML data. This model involves the transformation of the XQuery query language into a compatible format, thereby addressing semantic heterogeneity through the implementation of a field mapping table. The results of the queries are then uniformly represented as data source views utilizing XML Schema. Additionally, Bai et al. [12] present an adaptive query relaxation and result categorization strategy for fuzzy spatiotemporal XML data. To efficiently manage complex queries, they suggest classifying the results into two distinct sets following query relaxation [13]. Furthermore, they propose a result

L. Bai and L. Zhu, *Fuzzy Spatiotemporal XML Data Management*, Studies in Computational Intelligence 1183, https://doi.org/10.1007/978-3-031-81033-6_2

sorting algorithm aimed at mitigating the challenges associated with information overload.

The management of fuzzy spatiotemporal XML data is predominantly influenced by the underlying fuzzy spatiotemporal XML data model. However, the inherent characteristics of fuzzy spatiotemporal XML data can lead to inconsistencies among the nodes and edges within these documents. Such inconsistencies manifest as temporal discrepancies, inconsistencies in the conversion from time to spatial dimensions, and spatiotemporal inconsistencies. Temporal inconsistency arises from the synchronization issues between fuzzy spatiotemporal XML nodes and the delays in information transmission, resulting in nodes exhibiting distinct temporal attributes. For instance, applications A and B operate with their respective clocks, leading to differing recorded times for event E, denoted as t_A and t_B. This temporal inconsistency can subsequently cause variations in the spatial observations of the same entity, a phenomenon referred to as time-to-space conversion inconsistency. Furthermore, discrepancies in the observations of the same entity by different nodes are categorized as spatiotemporal inconsistency. Given these inconsistencies, it is imperative to investigate methodologies for rectifying these discrepancies within fuzzy spatiotemporal XML documents and to propose a fuzzy spatiotemporal XML data model that accommodates dynamic updates.

In this chapter, we examine the challenges associated with the modeling and implementation of fuzzy temporal labels, fuzzy positional labels, and attribute nodes within XML documents. We initiate our discussion by formulating a model for fuzzy spatiotemporal XML documents, conceptualized as a graph characterized by annotated edges that include both containment edges and reference edges. Containment edges serve to illustrate the nesting of elements and the values of attributes, whereas reference edges pertain to ID references. Subsequently, we investigate the consistency conditions pertinent to fuzzy spatiotemporal XML documents in accordance with our proposed data model. We propose algorithms designed to verify consistency, analyze their time complexity, and propose methodologies for rectifying inconsistencies.

2.2 Related Work

Consistency in general database. Concerning data consistencies of general database, two main areas of researches are proposed [14–16]. A reconciliation model has been introduced to uphold database consistency within distributed database systems [16]. This model integrates Neighbor Replication with Grid (NRG), wherein data is replicated to neighboring grid nodes utilizing an update ordering methodology. The proposed mechanism significantly enhances the performance of the replicated database, achieving improvements of up to two orders of magnitude in terms of data consistency. However, it operates under the assumption of a reliable replicated environment, without accounting for potential node or network failures. Conversely, Bertossi et al. [14] and Greco et al. [15] tackle the challenge of deriving consistent information from databases that exhibit inconsistencies and violate specified integrity constraints. Bertossi et al. [14] employ a technique that involves rewriting constraints into extended disjunctive rules, utilizing two distinct forms of negation: negation as failure and classical negation, to address inconsistencies. Building on the identified inconsistencies, Greco et al. [15] propose a disjunctive programming approach that serves dual purposes: computing "repairs" for the database and generating consistent answers. This technique has been demonstrated to be both sound and complete, and it is more comprehensive than previously established methods. In the realm of fuzzy logic, Wang and Chen [17] extend the issue of inconsistency to the fuzzy Analytic Hierarchy Process (AHP), achieving a reduction in the number of comparisons by a factor of 15. They present a series of mechanisms aimed at identifying consistent fuzzy rankings.

Consistency in temporal database. The periodic update model is predominantly utilized in hard real-time systems, which may yield erroneous outcomes [18–20]. It is essential to ensure temporal consistency, as user transactions or queries may access invalid data objects. In contrast to conventional databases, the primary issues to be addressed include: (1) the determination of the period and deadline for each update task to uphold the temporal consistency of each real-time data object; and (2) the formulation of a scheduling strategy that guarantees the deadlines of all update tasks. One of the earliest methodologies introduced is the Half-Half (HH) method [19], wherein both the period and relative deadline of an update task are established as half of the validity interval of the data object being updated. To further alleviate the update workload, the More-Less (ML) approach [21] has been proposed, which employs deadline monotonic scheduling to prioritize various update tasks. In contrast to HH and ML, the deferrable scheduling algorithm for fixed-priority transactions (DS-FP) [22] adopts a sporadic update model. DS-FP capitalizes on the semantics of the temporal validity constraints of real-time data objects by strategically deferring the sampling times of update jobs to the latest possible moment, thereby significantly reducing processor workload compared to ML. To mitigate the online scheduling overhead associated with DS-FP, Xiong et al. [20] have introduced two extensions aimed at generating a hyperperiod for DS-FP, ensuring that the schedule produced by the infinite repetition of the hyperperiod adheres to the temporal validity constraints of real-time data objects. Additionally, Falda et al. [18] have expanded the scope of inconsistent problems to encompass fuzzy temporal fields. To facilitate the mapping from an alternative temporal formalism that addresses temporal uncertainty and preferences, the Conditional Temporal Problems with Fuzzy Preferences (CTPPs) [23] have been proposed.

Consistency in spatial database. In contemporary applications, Geographic Information Systems (GIS) are increasingly integrated into processes that rely on the spatial characteristics of objects [24]. However, existing geographical datasets often contain inaccuracies that, while not affecting visualization, impede or obstruct spatial reasoning [25]. To address this issue, Servigne et al. [26] have introduced a novel concept known as topo-semantic consistency, which is a subset of logical consistency. This concept pertains to the accuracy of the topological relationships between two objects in relation to their semantics. For example, a scenario in which a building is situated within another building is unequivocally an error, whereas a building located within a parcel does not constitute an error. The authors propose a comprehensive methodology aimed at enhancing the spatial consistency of existing geographical datasets formatted as vectors. Nonetheless, this methodology is not without its limitations; certain errors can only be rectified if underlying structural errors are addressed first, necessitating a specific order in the correction processes. In the context of fuzzy spatial fields, Ferrari-Trecate and Rovatti [27] have proposed a regression algorithm designed to resolve inconsistencies within fuzzy spatial databases.

Consistency in spatiotemporal database. Del et al. [23] examine and contrast the spatial and temporal consistencies of various models, focusing on their data structures and semantics. The challenge of addressing multidimensional data, compounded by the absence of real datasets with established ground truth and the ambiguous relationship between optimal spatial and temporal processing for correspondence matching, has deterred many researchers from proposing effective solutions to the video disparity estimation problem. Building on previous successful methodologies, Williams et al. [28] enhance the matching cost representation for video by developing a three-dimensional Markov Random Field. Their primary strategy involves minimizing the use of spatial and aggregation operations that lead to resolution reduction, while implementing a series of multiple-match consistencies that assess both spatial and temporal dimensions to eliminate errors. To address spatiotemporal consistency, Khoshabeh et al. [29] introduce a two-stage algorithm that improves consistency by solving a novel three-dimensional total variation minimization problem. In the context of fuzzy spatiotemporal fields, Bejaoui et al. [30] propose a framework for defining integrity constraints that accommodate fuzzy spatiotemporal data. However, their work primarily categorizes inconsistencies without offering corrective measures.

Consistency in fuzzy spatiotemporal XML documents. The storage of fuzzy spatiotemporal data within XML documents can lead to inconsistencies in the nodes and edges of the corresponding fuzzy spatiotemporal XML data tree, primarily due to the discontinuity or overlap of temporal labels associated with certain incoming edges. However, the majority of existing research addressing consistency issues tends to focus on general data, temporal data, spatial data, or spatiotemporal data, thereby lacking the capacity to effectively manage inconsistencies specific to fuzzy spatiotemporal XML data. While these prior studies do not directly tackle the inconsistency challenges associated with fuzzy spatiotemporal XML data, they provide a foundational framework for addressing such inconsistencies in XML documents containing fuzzy spatiotemporal data, thus enabling further research based on their findings. Therefore, it is essential to examine the consistency of fuzzy spatiotemporal data within XML documents. Ma et al. [31] investigate the conditions necessary for maintaining consistency in fuzzy spatiotemporal data within XML documents, illustrating the impact of update, insertion, and deletion operations on the consistency of such data. Additionally, Bai et al. [32] propose methodologies for identifying and rectifying inconsistencies in fuzzy spatiotemporal XML documents based on established consistency states.

2.3 Preliminaries

2.3.1 Fuzzy Set and Possibility

In real-world applications, information is frequently characterized by imperfections. Various models have been developed to address distinct categories of data quality deficiencies, which are inconsistency, imprecision, vagueness, uncertainty, and ambiguity. Inconsistency refers to a semantic conflict wherein the same aspect of reality is irreconcilably represented multiple times within a single database or across different databases. Imprecision and vagueness pertain to the content of an attribute value, necessitating a selection from a specified range (interval or set) of values without definitive knowledge of the appropriate choice. Typically, vague information is expressed through linguistic values. Uncertainty denotes a situation in which one can assign partial belief to a particular value or set of values, although this paper does not address random uncertainty, which is typically modeled using probability theory. Ambiguity arises when certain elements of a model lack comprehensive semantics, resulting in multiple potential interpretations. It is important to note that various forms of imperfection can coexist concerning the same piece of information. Among these, imprecision, uncertainty, and vagueness are three primary types of imperfect information that can be effectively modeled using fuzzy sets and possibility theory [33, 34].

Since its introduction in 1965 by Zadeh, the theory of fuzzy sets has been presented in various forms across multiple disciplines. Fuzzy sets are characterized by elements that possess varying degrees of membership. This concept serves as a foundational basis for developing a conceptual framework that, while analogous to that of traditional sets, is more encompassing and may offer broader applicability, especially in the domains of pattern classification and information processing. Fundamentally, this framework facilitates the analysis of issues where imprecision arises from the lack of clearly defined criteria for class membership, rather than from the influence of random variables.

The concept of fuzzy sets has led to a significant number of practical applications across a wide array of fields and disciplines. Throughout the years, numerous methodologies addressing imprecision and uncertainty have been developed, largely grounded in the principles of fuzzy set theory.

Let U be a universe of discourse. A fuzzy value on U is characterized by a fuzzy set F in U. A membership function.

$$\mu_F{:}U \rightarrow [0, 1]$$

is defined for the fuzzy set F, where $\mu_F(u)$, for each $u \in U$, denotes the degree of membership of u in the fuzzy set F. For instance, $\mu_F(u) = 0.9$ means that u is "likely" to be an element of F by a degree of 0.9. For the purpose of facilitating representation, a fuzzy set F over universe U is organized into a set of ordered pairs:

$$F = \{\mu_F(u_1)/u_1, \mu_F(u_2)/u_2, \ldots, \mu_F(u_\mathrm{n})/u_\mathrm{n}\}.$$

When the membership function $\mu_F(u)$ above is interpreted as an indicator of the likelihood that a variable X has the value u, with X being defined over U, a fuzzy value is characterized by a possibility distribution π_X:

$$\pi_X = \{\pi_X(u_1)/u_1, \pi_X(u_2)/u_2, \cdots, \pi_X(u_n)/u_n\}$$

Here, $\pi_X(u_i)$, $u_i \in U$, represents the possibility that u_i is true.

Furthermore, fuzzy data is characterized by similarity relations among elements within a given domain. The inherent fuzziness arises from the similarity relations between pairs of values within a particular universe of discourse, rather than from the intrinsic properties of the objects themselves. Consequently, similarity relations serve to quantify the degree of similarity between two values belonging to the same universe of discourse. A similarity relation, denoted as Sim, on the universe of discourse U is defined as a mapping: $U \times U \rightarrow [0, 1]$, such that:

(i) for $\forall x \in U$, $Sim(x, x) = 1$, (reflexivity);
(ii) for $\forall x, y \in U$, $Sim(x, y) = Sim_\mathrm{i}(y, x)$, (symmetry); and
(iii) for $\forall x, y, z \in U$, $Sim(x, z) \geq \max_y(\min(Sim(x, y), Sim(y, z)))$, (transitivity).

Furthermore, the subsequent notions related to fuzzy sets can be defined.

Support. The set of the elements that have non-zero degrees of membership in F is called the support of F, denoted by

$$\mathrm{supp}(F) = \{u | \mathrm{u} \in \mathrm{U} \text{ and } \mu_F(u) > 0\}$$

Kernel. The set of the elements that completely belong to F is called the kernel of F, denoted by

$$\ker(F) = \{u | u \in U \text{ and } \mu_F(u) = 1\}.$$

α-Cut. The set of the elements which degrees of membership in F are greater than (greater than or equal to) α, where $0 \leq \alpha < 1$ ($0 < \alpha \leq 1$), is called the strong (weak) α-cut of F, respectively denoted by

$$F_{\alpha+} = \{u | \mathrm{u} \in \mathrm{U} \text{ and } \mu_F(u) > \alpha\} \text{ and}$$

$$F_\alpha = \{u | \mathrm{u} \in \mathrm{U} \text{ and } \mu_F(u) \geq \alpha\}$$

Additionally, to facilitate the manipulation of fuzzy sets and possibility distributions, a number of standard *set* operations have been established. The conventional *set operations*, including union, intersection, and complementation, have been adapted to accommodate fuzzy sets. Let A and B represent fuzzy sets defined over the same universe of discourse U, with their respective membership functions denoted as μ_A and μ_B. Consequently, we can express the following:

Union. The union of fuzzy sets A and B, denoted $A \cup B$, is a fuzzy set on U with the membership function $\mu_{A\cup B}:U \rightarrow [0, 1]$, where

$$\forall u \in U, \mu_{A\cup B}(u) = \max(\mu_A(u), \mu_B(u)).$$

Intersection. The intersection of fuzzy sets A and B, denoted $A \cap B$, is a fuzzy set on U with the membership function $\mu_{A\cap B}:U \rightarrow [0, 1]$, where

$$\forall u \in U, \mu_{A\cap B}(u) = \min(\mu_A(u), \mu_B(u)).$$

Complementation. The complementation of fuzzy set $\bar{A}$, denoted by $\bar{A}$, is a fuzzy set on U with the membership function $\mu_{\bar{A}}$: $U \rightarrow [0, 1]$, where

$$\forall u \in U, \mu_{\overline{A}}(u) = 1 - \mu_A(u).$$

On the basis of those definitions, the *difference* of the fuzzy sets B and A can be defined as:

$$B - A = B \cap \overline{A}.$$

Furthermore, the majority of properties applicable to classical set operations, including DeMorgan's Laws, have been demonstrated to be valid for fuzzy sets as well. The sole principle from traditional set theory that does not retain its validity in this context is the law of the excluded middle, which states that:

$$A \cap \overline{A} \neq \emptyset \text{ and } A \cup \overline{A} \neq U.$$

Let A, B and C be fuzzy sets in a universe of discourse U. Then the operations on fuzzy sets satisfy the following conditions:

- Commutativity laws: $A \cup B = B \cup A, A \cap B = B \cap A$
- Associativity laws: $(A \cup B) \cup C = A \cup (B \cup C)(A \cap B) \cap C = A \cap (B \cap C)$
- Distribution laws: $(A\cup B)\cap C = (A\cap C)\cup(B\cap C), (A\cap B)\cup C = (A\cup C)\cap(B\cup C)$
- Absorption laws: $(A \cup B) \cap A = A, (A \cap B) \cup A = A$
- Idempotency laws: $A \cup A = A, A \cap A = A$
- de Morgan laws: $\overline{A \cup B} = \overline{A} \cap \overline{B}, \overline{A \cap B} = \overline{A} \cup \overline{B}$

Given two fuzzy sets A and B in U, B is a fuzzy subset of A, denoted by $B \subseteq A$, if

$$\mu_B(u) \leq \mu_A(u) \text{for all } u \in U.$$

Two fuzzy sets, denoted as A and B, are considered to be equivalent if $A \subseteq B$ and $B \subseteq A$.

In order to define Cartesian product of fuzzy sets, let $U = U_1 \times U_2 \times \ldots \times U_n$ be the Cartesian product of n universes and $A_1, A_2, \ldots, A_n$ be fuzzy sets in $U_1, U_2, \ldots, U_n$, respectively. The Cartesian product $A_1 \times A_2 \times \ldots \times A_n$ is defined to be a fuzzy subset of $U_1 \times U_2 \times \ldots \times U_n$, where

$$\mu_{A1\times\ldots\times\ldots An}(u_1 \ldots u_n) = \min(\mu_{A1}(u_1), \ldots, \mu_{An}(u_n))$$

and $u_i \in U_i$, $i = 1, \ldots, n$.

In classical set theory, the membership of elements within a set is evaluated using a binary criterion, wherein an element is classified as either belonging to the set or not. In contrast, fuzzy set theory allows for a more nuanced evaluation of membership, employing a membership function that assigns values within the real unit interval [0, 1]. Fuzzy sets can be viewed as a generalization of classical sets, as the indicator functions of classical sets represent specific instances of the membership functions found in fuzzy sets, particularly when the latter are restricted to values of 0 or 1. Within the framework of fuzzy set theory, classical bivalent sets are typically referred to as crisp sets. Furthermore, fuzzy set theory conceptualizes set membership as a possibility distribution, a key notion in possibility theory, which serves as a mathematical framework for addressing certain forms of uncertainty and provides an alternative to probability theory. A possibility distribution functions as a measure that delineates the likelihood of a variable assuming specific values, thereby associating each outcome with its corresponding probability of occurrence. In conclusion, a fuzzy set serves as a conceptual representation, while a possibility distribution pertains to the likelihood of a value manifesting within a given distribution.

2.3.2 *Fuzziness in XML Documents*

XML is classified as a meta-markup language utilized for the purpose of data description. The term "markup" pertains to the arrangement of characters or symbols inserted at specific locations within a text or word processing document. XML data is characterized as self-describing or self-defining, indicating that the data's structure is inherently included within the data itself. Consequently, there is no requirement to establish a predefined structure for data storage upon its arrival. XML serves as a straightforward and adaptable descriptive text markup format, enabling the definition of data structures, facilitating comprehension of their meaning and context, and offering methods to articulate the semantics and organization of the information. Structured information encompasses both content and an indication of the role that content fulfills. A markup language functions as a mechanism for identifying structures within a document. The XML specification delineates a standardized approach for incorporating markup into documents.

The XML standard provides a versatile framework for the creation of information formats and the electronic exchange of structured data over the public Internet. XML is essentially a more straightforward and user-friendly subset of the Standard Generalized Markup Language (SGML), designed specifically for Internet applications. The fundamental component of an XML document is the element, which is delineated by tags. Each element comprises a start tag and an end tag, and all elements within an XML document are encapsulated within a singular outermost element referred to as the root element. Furthermore, XML accommodates nested elements, allowing for the inclusion of elements within other elements, thereby facilitating the representation of hierarchical structures. The names of elements serve to describe their content, while the overall structure elucidates the interrelationships among the elements. Additionally, XML permits the definition of attributes for elements, which convey specific characteristics within the start tag of an element. The efficacy of XML lies in its inherent simplicity, enabling the aggregation of substantial amounts of information into a cohesive XML document.

An XML document comprises a collection of elements, wherein elements may be nested within one another and can also possess attributes. It is important to note that XML imposes a restriction whereby attributes can only hold a unique single value. Consequently, two distinct levels of fuzziness can be discerned within an XML document. The first level of fuzziness pertains to the elements themselves, and the second level of fuzziness relates to the attribute values associated with these elements.

Focusing exclusively on the second level of fuzziness within XML documents, specifically the fuzziness present in attribute values, There are three distinct types of fuzzy data into XML structures. These include discrete fuzzy sets, which can effectively represent possibility distributions; continuous fuzzy sets; and linguistic variables, which can be linked to fuzzy sets within a continuous universe of discourse.

There are two levels of fuzziness within XML documents. The initial level of fuzziness is represented by fuzzy data representation. To denote the degree of existence of a specific element within the XML document, a novel attribute termed "*Poss*" is introduced. This attribute is utilized in conjunction with a new construct referred to as "*Val*." The combination of the pairs <*Val Poss*> and </*Val*> serves to indicate the degree to which the element is present in the XML document. Furthermore, to articulate all degrees of an element's existence, an additional construct called "*Dist*" is proposed. Within the pair < *Dist* > and < /*Dist* > , multiple *Val* elements may be included as child elements.

The data model encompasses two distinct forms of fuzziness. The first form involves the assignment of membership degrees to each tuple, which can be understood as the likelihood of the tuple being a member of the associated relation. The second form pertains to the representation of attribute values through possibility distributions, indicating that the precise value of the attribute in terms of time and position is uncertain; instead, it delineates a range of potential values that the attribute may assume.

Now we examine the implications of membership degrees associated with an element, particularly in the context of hierarchical structures where elements may be nested and multiple elements may possess corresponding membership degrees. The existential membership degree attributed to an element reflects the likelihood that the state of the element encompasses both the element itself and the sub-tree that it anchors. Within this sub-tree, each node is not independent; rather, it is contingent upon the chain connecting it to its root node. The potentialities derived from the source XML documents are predicated on the certainty of the existence of the parent element. In essence, this potentiality is relative, based on the assumption that the probability of the parent element is precisely 1.0. To ascertain the absolute possibility of an element, it is necessary to incorporate the relative possibility associated with the parent element. Generally, the absolute possibility of an element *e* can be determined by multiplying the relative possibilities identified in the source XML along the path from the root. It is important to note that each of these relative possibilities is accessible within the source XML documents.

In our data model, we employ the concept of membership degree to characterize the fuzzy information inherent in fuzzy spatiotemporal XML documents. For illustrative purposes, consider a path from A to C in fuzzy spatiotemporal XML documents, represented as $A \rightarrow B \rightarrow C$. The relative probability of C, B, A can be indicated as $Poss(C|B)$, $Poss(B|A)$ and $Poss(A)$ (the possibilities of node A) respectively [35]. Then there is

$$Poss(B) = Poss(B|A) \times Poss(A)$$
$$Poss(C) = Poss(C|B) \times Poss(B|A) \times Poss(A)$$

Here, $Poss(C|B)$, $Poss(B|A)$ and $Poss(A)$ can be directly obtained from the source fuzzy spatiotemporal XML documents.

In XML, attribute values are constrained to possess a unique singular value. However, this limitation does not consistently apply in practice. It is frequently observed that certain data items may be associated with multiple values, which may be entirely unknown and can be represented through a possibility distribution. For instance, an individual's email address may consist of multiple character strings, as they may have several active email addresses concurrently. In situations where complete information regarding Jaren Jackson's email address is lacking, one might express the potential email addresses as follows: "JJack@hotmail.com" with a possibility of 0.75, "Jaren_Jackson@hotmail.com" with a possibility of 0.85, "JJack@yahoo.com" with a possibility of 0.90, "Jaren_Jackson@yahoo.com" with a possibility of 0.65, "JJack@msn.com" with a possibility of 0.55, and "Jaren_Jackson@msn.com" with a possibility of 0.70.

In contrast, certain data items are characterized by a single unique value. For example, an individual's age, expressed in years, is represented as a unique non-negative integer. In instances where this value remains unknown, one may employ the following possibility distribution: {0.55/24, 0.70/25, 0.85/26, 1.0/27, 1.0/28, 1.0/29, 0.8/30, 0.65/32, 0.45/35}. The preceding discussion elucidates that there are two distinct interpretations of fuzzy data as represented by a possibility distribution: fuzzy disjunctive data and fuzzy conjunctive data.

In conclusion, there are two distinct forms of fuzziness present in XML: the fuzziness associated with elements, which is represented through membership degrees, and the fuzziness related to the attribute values of these elements, which is depicted using possibility distributions. It is important to note that there are two categories of possibility distributions (disjunctive and conjunctive) which may manifest in child elements, regardless of whether they possess additional child elements within the ancestor–descendant hierarchy.

2.3.3 *Fuzzy Sets in Database Modeling*

An uncertain value can be intuitively represented by a fuzzy set. A fuzzy set, denoted as $\{(u_1, \mu(u_1)), (u_2, \mu(u_2)), \ldots, (u_n, \mu(u_n))\}$, consists of pairs where each pair $(u_i, \mu(u_i))$ includes an element from the fuzzy set and its corresponding membership degree, with the membership function μ defined over the universe of discourse U (where $u_i \in U$ and $1 \leq i \leq n$). It is important to note that establishing the membership function for a fuzzy set can be a complex endeavor. In some instances, it may be impractical or unnecessary to precisely define a formal membership function. Instead of calculating the membership degrees through a membership function, these degrees may be assigned directly. For instance, consider the uncertain value "around 30 years old," which can be represented by the fuzzy set {(28, 0.75), (29, 0.90), (30, 1.0), (31, 0.90), (32, 0.75)}. In this case, the membership degrees are provided based on subjective judgment, and the membership function is implicitly defined. It is crucial to emphasize that a fuzzy set can be defined in this manner only if all elements of the fuzzy set are explicitly enumerated, indicating that the universe of discourse is discrete, and their membership degrees can be directly assigned. Conversely, a fuzzy set may also be defined over a continuous universe of discourse. For example, a fuzzy set representing the uncertain value "around 30.67 m" for length cannot enumerate all elements, necessitating the use of a formally defined membership function.

Membership functions can theoretically take on various shapes; however, in practice, trapezoidal and triangular membership functions are the most commonly utilized. A trapezoidal membership function is typically characterized by the parameters (*a*, *b*, *c*, *d*), where '*a*' and '*d*' denote the minimum and maximum values within the membership interval, respectively, while '*b*' and '*c*' define an interval where the membership degree is equal to 1. Conversely, a triangular membership function is represented by the parameters (*a*, *b*, *d*), with '*a*' and '*d*' indicating the minimum and maximum values of the membership interval, and '*b*' signifying the point at which the membership degree is equal to 1. Notably, the triangular membership function (*a*, *b*, *d*) can be considered a specific instance of the trapezoidal membership function (*a*, *b*, *c*, *d*). In other words, the trapezoidal membership function (*a*, *b*, *c*, *d*) simplifies to a triangular membership function (*a*, *b*, *d*) when the values of '*b*' and '*c*' are equal.

A fuzzy set is characterized by its membership function, which is defined within a specific universe of discourse. Consequently, the representation of a fuzzy set is largely determined by its corresponding universe of discourse. Generally, four distinct categories of universes of discourse can be identified: finite countable universes of discourse, finite uncountable universes of discourse, infinite countable universes of discourse, and infinite uncountable universes of discourse. To differentiate among these four categories, several illustrative examples can be considered. For instance, the universe of discourse pertaining to "the twentieth century" consists of a finite set of countable dates, specifically from January 1, 1901, to December 31, 2000. In contrast, the universe of discourse for "water temperature" encompasses a finite set of uncountable real numbers, ranging from 0 to 100 °C. Furthermore, the universe of discourse for "age" is represented by an infinite set of countable natural numbers, whereas the universe of discourse for "body weight" is characterized by an infinite set of uncountable real numbers.

Fuzzy sets can be utilized to characterize fuzzy values. In accordance with the aforementioned categorization of the universes of discourse upon which fuzzy sets are established, we delineate two distinct types of fuzzy values, which are fuzzy values within a discrete (countable) domain of discourse and fuzzy values within a continuous (uncountable) domain of discourse.

In order to accurately represent a fuzzy value within a continuous universe of discourse, it is essential to precisely define a formal membership function for the corresponding fuzzy set. Conversely, when dealing with a fuzzy value in a discrete universe of discourse, one has the option to either explicitly define a formal membership function for the fuzzy set or to present the fuzzy set directly, thereby obviating the necessity for a formal definition of the membership function.

In the context of defining fuzzy sets, it is essential to consider not only the discrete or continuous nature of the universe of discourse but also the granularity of the data involved. A fuzzy value can be classified as an atomic fuzzy value when all elements of the corresponding fuzzy set consist of atomic values, or as a non-atomic fuzzy value when all elements comprise non-atomic values. Generally, an element within a fuzzy set representing an atomic fuzzy value corresponds to a specific, concrete value (either numerical or categorical) while an element within a fuzzy set representing a non-atomic fuzzy value corresponds to an entity (object). For instance, the term "around 30 years old" exemplifies an atomic fuzzy value, as the elements of its corresponding fuzzy set (such as 28, 29, 30, 31, and 32) are atomic values. Conversely, the term "old men" represents a non-atomic fuzzy value, as the elements of its corresponding fuzzy set (individuals such as Chris, Daniel, Tonny, Bill, and George) are non-atomic values that require further elaboration through a collection of atomic values (e.g., ID, name, age, and phone number). Typically, a fuzzy relational instance characterized by membership degrees (i.e., a set of tuples with associated membership degrees) is classified as a non-atomic fuzzy value, whereas a fuzzy attribute value within a tuple is regarded as an atomic fuzzy value. Thus, from the perspective of data granularity in data representation, two distinct categories of fuzzy values can be identified, which are atomic fuzzy values and non-atomic fuzzy values.

A non-atomic fuzzy value is formally defined as a fuzzy set represented by the collection $\{(o_1, \mu(o_1)), (o_2, \mu(o_2)), \ldots, (o_n, \mu(o_n))\}$, where the membership function $\mu(o_i)$ operates within the universe of discourse U ($o_i \in U$, $1 \leq i \leq n$). Each element o_i is expressed as a vector $< u_{i1}, u_{i2}, \ldots, u_{im} >$, indicating that o_i is non-atomic and is characterized by this vector representation. It is important to note that the components u_{ij} ($1 \leq j \leq m$) may consist of atomic values, non-atomic values, atomic fuzzy values, or non-atomic fuzzy values. For instance, an atomic fuzzy value u_{ij} is further delineated by a fuzzy set of the form $\{(v_1, \eta(v_1)), (v_2, \eta(v_2)), \ldots, (v_p, \eta(v_p))\}$, where the membership function $\eta(v_k)$ is defined over the universe of discourse V ($v_k \in V$, $1 \leq k \leq p$), with v_k representing either a numerical or categorical value. Consequently, while an atomic fuzzy value can be directly represented by a fuzzy set, a non-atomic fuzzy value necessitates representation that adheres to its data granularity until such time as no non-atomic fuzzy values remain. Fuzzy values are categorized into atomic and non-atomic types based on their data granularity, as this distinction influences their representation within data models, particularly in the context of XML tree structures.

A non-atomic fuzzy value is typically characterized within a discrete universe of discourse. The associated fuzzy set is often presented in a direct manner, lacking a formally defined membership function. In the context of a non-atomic fuzzy value, attention is primarily directed towards each pair represented as $(o_i, \mu(o_i))$, where o_i denotes an entity in the real world and $\mu(o_i)$ signifies a degree within the interval [0, 1]. This degree is frequently interpreted as a measure of confidence or truth. In contrast to non-atomic fuzzy values, which are predominantly defined within a discrete universe of discourse, atomic fuzzy values can be defined in either a discrete or a continuous universe of discourse. For atomic fuzzy values associated with a continuous universe of discourse, a formal membership function is essential for their description. Conversely, for atomic fuzzy values within a discrete universe of discourse, the necessity of a formal membership function may vary. It is important to note that, in practical applications, fuzzy notations often correspond to linguistic values (terms or labels) found in natural language, such as "tall," "good," or "cold." From the perspective of processing linguistic values, an atomic fuzzy value can be directly represented by a linguistic value that is linked to a fuzzy set defined within either a discrete or continuous universe of discourse.

In the context of atomic fuzzy values within a discrete universe of discourse, an alternative representation known as similarity-based representation can be identified. This type of fuzzy value is characterized by a set of values that exhibit a degree of similarity to one another, thereby rendering them indistinct. Consequently, these values collectively constitute an uncertain representation of a fuzzy value as it pertains to real-world scenarios.

In summary, we have the following fuzzy data representations:

- *Type-1*: direct representation with fuzzy set for a non-atomic fuzzy value;
- *Type-2*: linguistic label representation for an atomic fuzzy value on a discrete universe of discourse;
- *Type-3*: linguistic label representation for an atomic fuzzy value on a continuous universe of discourse;
- *Type-4*: indirect representation with membership function for an atomic fuzzy value on a discrete universe of discourse;

- *Type-5*: indirect representation with membership function for an atomic fuzzy value on a continuous universe of discourse;
- *Type-6*: direct representation with fuzzy set for an atomic fuzzy value on a discrete universe of discourse;
- *Type-7*: similarity-based representation for a non-atomic fuzzy value on a discrete universe of discourse.

A significant number of data models have been developed to address the challenges of uncertainty and vagueness. The majority of these models are grounded in similar theoretical frameworks. Typically, vagueness and uncertainty are represented through fuzzy sets and possibility theory [33, 34]. Numerous existing methodologies that tackle imprecision and uncertainty are rooted in the principles of fuzzy set theory. The application of fuzzy information has been thoroughly examined within the context of the relational model. Recent advancements have sought to extend these findings to object-oriented databases by incorporating relevant concepts. Furthermore, the modeling of fuzzy data has been explored within conceptual data models, such as Entity-Relationship (ER) [36] and Enhanced Entity-Relationship (EER) models [37]. In contemporary research, the management of XML data has garnered increasing attention, particularly due to its widespread utilization on the Internet. Consequently, the modeling of fuzzy information within XML represents a foundational aspect of the implementation of web-based intelligent information processing.

2.3.4 *Principles of Fuzzy Spatiotemporal Data Modeling*

Data models exhibit various characteristics, including reality-oriented, computer-oriented, and application-oriented aspects. Distinct spatiotemporal databases possess unique requirements regarding the structure and operations of spatiotemporal data. To develop a comprehensive fuzzy spatiotemporal data model, it is essential to consider several guiding principles. A set of spatiotemporal requirements is established for crisp spatiotemporal data at the modeling level. When addressing fuzzy spatiotemporal data modeling utilizing XML, it is important to acknowledge that certain requirements, akin to those for crisp data, must also be taken into account.

- Need for the representation of spatiotemporal objects incorporating fuzzy spatial and fuzzy temporal information utilizing XML is evident.
- Need for representation for structure of fuzzy spatiotemporal data in XML document.
- Need to capture the changes of spatiotemporal position over time. In this context, both spatiotemporal position and time may be fuzzy, leading to the classification of four distinct scenarios: (a) crisp spatiotemporal position over crisp time; (b) fuzzy spatiotemporal position over crisp time; (c) crisp spatiotemporal position over fuzzy time; (d) fuzzy spatiotemporal poison over fuzzy time. Furthermore, spatiotemporal position changes can be categorized into continuous changes and discrete changes depending on the specific applications of interest.

- Need to capture the changes of spatiotemporal attributes over time. Here, these changes of spatiotemporal attributes can also be categorized into four distinct cases and two types similar as changes of spatiotemporal position previously discussed.
- Need for the establishment of definitions pertaining to fuzzy attributes that encompass fuzzy time, particularly in the context of spatiotemporal data.
- Need for representation for changing history of spatiotemporal data.
- Need for representation for fuzzy topological relations of two fuzzy spatiotemporal objects.
- Need for representation for motion of fuzzy spatiotemporal data. The motion may exhibit fuzzy due to the potential fuzzy spatiotemporal position.
- Need for checking inconsistencies of fuzzy spatiotemporal data in XML documents when change operations (update, insert, and delete) occur.
- Need for fixing inconsistencies of fuzzy spatiotemporal data in XML documents when change operations (update, insert, and delete) occur.

2.4 Fuzzy Spatiotemporal XML Data Model

2.4.1 Classification of Nodes in Fuzzy Spatiotemporal XML Documents

In this subsection, we categorize nodes within fuzzy spatiotemporal XML documents. We enhance the XPath data model to incorporate historical data pertinent to these documents. XPath, or XML Path Language, serves as a query language designed for the selection of nodes from XML documents. Furthermore, XPath can be utilized to derive values from the content contained within an XML document. The XPath language, which operates on a tree structure representing the XML document, facilitates navigation through the tree and allows for the selection of nodes based on various criteria.

The data model comprises four distinct categories of nodes: root nodes, element nodes, attribute nodes, and text nodes. Each category of nodes is associated with a specific string value. The subsequent subsections will provide a comprehensive overview of each type of node within the data model.

- Root node *r*: the root node is the root of the XML data tree. A root node *r* is unique which does not have an incident edge. Furthermore, each node within the XML data tree is capable of establishing a path to the root node.
- Element nodes *e*: the element nodes are child nodes of the root node *r*. The child nodes of an element node are element and/or text nodes. Each element node has an associated set of attribute nodes.
- Attribute nodes *a*: the attribute nodes are child nodes of an element node or text node. They are labeled with the names of attributes or other attributes of the mark.

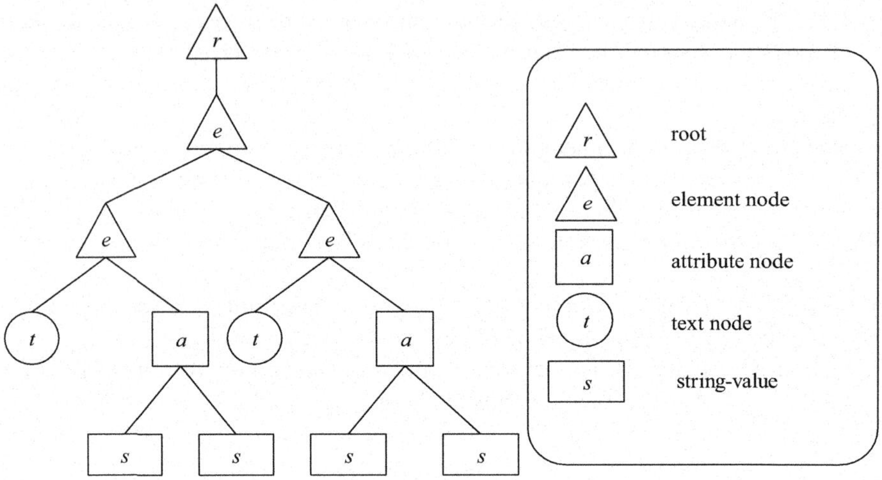

Fig. 2.1 Classification of nodes in fuzzy spatiotemporal XML documents

- Text nodes *t*: the text nodes denote name or key value of element nodes. They have only one incoming edge and one outgoing edge from element nodes (or from the root node).

Figure 2.1 illustrates the four distinct types of nodes within the fuzzy spatiotemporal XML data model, which serves as an abstract data model.

Definition 2.1 (*Node*). Let D be an XML document, $V(D)$ is a set of nodes of D. The four kinds of nodes are referred to r, $V_e(D)$, $V_a(D)$, and $V_t(D)$, respectively. These sets are pairwise disjoint and $V(D) = \{r\} \cup V_e(D) \cup V_t(D) \cup V_a(D)$.

Definition 2.2 (*Edge*). $E(D)$ is a set of edges (p, c) in an fuzzy spatiotemporal XML document D, where each pair is one of the followings: $p = r$, $c \in V_e(D)$; or $p \in V_e(D)$, $c \in V_e(D)$; or $p \in V_e(D)$, $c \in V_t(D)$; or $p \in V_e(D)$, $c \in V_a(D)$; or $p \in V_a(D)$, $c \in V_t(D)$. These types are referred to as *r-e*, *e-e*, *e-t*, *e-a*, and *a-t*, respectively.

Node types and edge types may exhibit mixed content, meaning that an element node can contain various types of child nodes.

2.4.2 Fuzzy Temporal Labels

A fuzzy temporal label is defined as an interval T_e (where natural numbers are employed to denote T_e) along with a membership degree associated with a node, which may label either a containment edge or an element edge. In the context of an edge e connecting nodes n_i and n_j, $Poss(n_i)$ signifies the relative possibilities pertaining to node n_i. This chapter focuses on the temporal aspects of nodes and edges within fuzzy spatiotemporal XML documents. For example, T_{er} denotes the transaction time associated with the reference edge r. Edges that are assigned temporal labels are referred to as temporal edges. Generally, if an edge e is designated with a temporal label T_e, we utilize $T_{e \cdot from}$ and $T_{e \cdot to}$ to indicate the endpoints of the interval T_e. We consider two temporal labels T_{ei} and T_{ej} to be consecutive if $T_{ej \cdot from} = T_{ei \cdot to} + 1$. As previously noted, our analysis is confined to a singular time interval rather than a time element.

In accordance with standard practices in fuzzy spatiotemporal databases, the current time point is represented by the term '*Now*'. The timestamp corresponding to the creation of the document is referred to as t_0. Subsequently, we will provide definitions for the concepts of current node and edge.

Definition 2.3 (*Current node and edge*). A fuzzy spatiotemporal containment (reference) edge characterized by the condition $T_{e \cdot to} = Now$ is referred to as a current containment (reference) edge. A node is designated as a current node if at least one of its incoming containment edges is classified as current.

2.4.3 Fuzzy Position Labels

A fuzzy position label represents a potential location, with multiple fuzzy position labels collectively encompassing the complete probability distributions associated with a given node. In the context of an edge e_c connecting an element node n_i and an attribute node n_j, p_{ec} is designated as the fuzzy position label. Edges that are annotated with fuzzy position labels are referred to as fuzzy position edges. Unlike fuzzy temporal labels, which may encompass broader temporal ranges, fuzzy position labels are characterized solely as discrete fuzzy points rather than as fuzzy regions.

2.4.4 Attribute Nodes

In the ordinary XML data model, attributes are expected to maintain uniqueness. There are two potential approaches to address this: (a) prohibiting attributes from exhibiting temporal variability, or (b) conceptualizing them as a distinct category of elements. We opt for the latter approach to facilitate the representation of multiple node possibilities. From a formal modeling perspective, we do not differentiate between attribute nodes and element nodes, with the exception that attribute nodes are not permitted to contain other elements. We introduce a specialized element, referred to as < ATTRIBUTES >. For instance, if an element < cloud > possesses an attribute, the value associated with this attribute will be subject to change in accordance with the cloud's positional alterations.

```
<clouds>
    <cloud number = "1501">
        <ATTRIBUTES>
            <Val Poss = 0.8></Val>
            <cloud name Time: From = "0" Time: To = "Now">
                Mekkhala
            </cloud>
            <position>
                <Dist type = "disjunctive">
                    <Val Poss = 0.8 Sorsogon> </val>
                    <Val Poss = 0.2 Eastern samar> </val>
                    <Val Poss = 0.7 Camarines Norte> </val>
                </Dist>
            </position>
        </ATTRIBUTES>
</clouds>
```

After this cloud changes the position at time t_1, the element will contain:

```
<clouds>
    <cloud number = "1501">
        <ATTRIBUTES>
            <Val Poss = 0.7></Val>
            <cloud name Time: From = "0" Time: To = "t1">
                Mekkhala
            </cloud>
            <Val Poss = 0.9>
            <cloud name Time: From = "t1" Time: To = "Now">
                Mekkhala
            </cloud>
            <position>
                <Dist type = "disjunctive">
                        <Val Poss = 0.9 Eastern samar> </val>
                        <Val Poss = 0.7 Camarines Norte> </val>
                </Dist>
            </position>
        </ATTRIBUTES>
</clouds>
```

It is evident from the aforementioned example that the definition of a possibility attribute, referred to as *Poss*, is a prerequisite. This attribute assumes values within the interval [0, 1]. The possibility attribute is utilized in conjunction with a fuzzy structure known as Val to delineate the likelihood of a particular element's presence within fuzzy spatiotemporal XML documents.

2.4.5 Semantics of Fuzzy Spatiotemporal Data

Spatiotemporal data possess a unique set of characteristics that distinguish them from the more conventional alphanumeric data typically utilized in traditional business applications. When addressing fuzzy spatiotemporal data, we measure spatiotemporal entities based on five dimensions. The first dimension facilitates the reference to various forms of fuzzy spatiotemporal data and delineates the evolving history of such data in relation to both its antecedents and successors. The antecedent elucidates the origin and formation process of the object, while the successor illustrates the eventual transformation of the object and the underlying reasons for this change. Furthermore, the dynamic categories of fuzzy spatiotemporal data generally encompass four classifications: *creation*, *splitting*, *merging*, and *elimination*. For example, due to the complex and irregular shape of the land, a land may be created at time t_1 and subsequently divides into two lands at time t_2. At time t_3, one of the land may further divides into two lands. At time t_4, two lands may merge to form one land. At time t_5, one land may be eliminated. The second dimension is employed to characterize the static characteristics of fuzzy spatiotemporal data, which may include variables such as land ownership, area measurements, typhoon intensity, and air pressure, among others. A spatiotemporal dataset may encompass one or multiple attributes, which can exhibit fuzziness. The specific nature of this dimension is significantly influenced by the particular application domain in question. The third dimension encompasses fuzzy points, fuzzy lines, and fuzzy regions. A fuzzy point refers to a location whose precise coordinates are indeterminate, yet potential locations are identifiable within a specified area. A fuzzy line denotes a line for which the exact configuration, location, and length remain unknown, although the region it occupies is established. A fuzzy region is characterized by its ambiguous boundaries. From the framework of fuzzy position, one can derive: (a) the fuzzy position of fuzzy spatiotemporal data; and (b) the relationships between two fuzzy spatiotemporal datasets, including topological, directional, and distance relationships. The fourth dimension encompasses both the direction and motion value. The direction aspect indicates that the trajectory of motion is indeterminate and can occur in any direction. For instance, a cloud may be displaced by the wind in various directions, illustrating the uncertainty in its motion direction. The magnitude aspect refers to the velocity associated with spatiotemporal data. Using clouds as an example, the velocity may vary depending on the strength of the wind, highlighting the indeterminate nature of the motion value. The last dimension delineates the concept of fuzzy time within the context of fuzzy spatiotemporal data, which encompasses both fuzzy time points and fuzzy time intervals. The fuzzy time point is understood as a representation of a collection of potential time points, each associated with varying degrees of likelihood regarding the occurrence of an event. Conversely, the fuzzy time interval is characterized as a representation of a range of possible time intervals, also accompanied by degrees of possibility. Fuzzy time is utilized to describe: (a) the temporal dynamics of changes; and (b) the temporal relationships, including aspects such as temporal topology.

Consider the behaviors of fire in real-world scenarios: (a) each fire event is assigned a unique identifier for the purpose of tracking and identification, which reflects the evolving history of the fire. For example, if fire *A* transforms into fires *B* and *C*, the nature of fires *B* and *C* can be delineated, with fire *A* serving as their predecessor and both *B* and *C* having no successors at that moment. (b) Fires may possess various attributes, such as their designated names and affected areas, which can exhibit multiple values, leading to potential fuzzy. (c) The spatial information of the fire can be represented using Minimum Bounding Rectangle (MBR). (d) It is capable of documenting the movement dynamics associated with the fire's spread. (e) The temporal aspect of the fire indicates its state, which may remain stable over a period or occur at a specific moment in time; furthermore, this temporal information can also be fuzzy. For instance, a fire may be located in a particular area at a given moment but is expected to leave that area subsequently.

The fuzzy spatiotemporal XML data model serves as the foundation for examining the consistencies of fuzzy spatiotemporal data within XML documents. According to the semantics of fuzzy spatiotemporal data mentioned above, fuzzy spatiotemporal data contains five parts: (a) for the first part, we assume that the predecessor and successor of spatiotemporal objects are not characterized by fuzziness when they are known. This is predicated on the notion that the identification of fuzzy predecessors and successors is unnecessary and potentially superfluous. Consequently, we assign a membership degree of 1.0 to all values within this part. (b) For the sake of simplicity, we utilize the second part to refer to the static attributes associated with spatiotemporal data. It is acknowledged that a spatiotemporal dataset may encompass one or more attributes, which may themselves exhibit fuzziness. To represent this fuzziness, we employ a possibility distribution. (c) The fuzzy position and its corresponding membership degree can be derived from the fuzzy position component. In this study, we adopt MBR to depict two-dimensional fuzzy spatiotemporal data. (d) The fuzzy motion part may exhibit fuzziness, as it encompasses both the direction and magnitude of motion; the former suggests that motion can occur in any direction, while the latter signifies the velocity associated with the spatiotemporal data. (e) The fuzzy time part articulates the temporal aspects of spatiotemporal data, which includes fuzzy time points and fuzzy time intervals. We simplify our analysis by assuming that time is isomorphic to natural numbers, given that time is inherently linearly ordered.

XML, due to its flexible structure and self-descriptive nature, is well-suited for representing fuzzy spatiotemporal information. In the context of XML, elements may be associated with membership degrees, while attribute values can be linked to possibility distributions when these elements or attributes exhibit fuzziness. To elucidate the concept of a membership degree associated with an element, it can be understood as the likelihood that the current state of the world encompasses this element along with the sub-tree that it anchors. For any given element, the nodes within its sub-tree are contingent upon the chain connecting the root to each node. Each possibility within the source fuzzy spatiotemporal XML document is assigned based on the condition that the parent element is present. This implies that the assigned possibility is relative, predicated on the assumption that the existence of the parent element is certain (i.e., with a probability of 1.0). To derive the absolute possibility of an element, it is necessary to account for the relative possibility associated with its parent element. Generally, the absolute possibility of an element can be calculated by multiplying the relative possibilities encountered along the path in the source fuzzy spatiotemporal XML. By default, relative possibilities are assumed to be 1.0.

In relation to fuzzy spatiotemporal XML data, as previously discussed, our focus is on the membership degree associated with a given element. The membership functions employed are typically defined by users based on specific practical applications. Various methods for determining these functions exist, including triangular functions, trapezoidal functions (both R-functions and L-functions), Gaussian functions, and similarity relations, among others.

2.4.6 Fuzzy Spatiotemporal Data in XML Documents

In this subsection, we propose consistent conditions that a fuzzy spatiotemporal XML document should follow. Prior to outlining the conditions that may lead to inconsistencies, we give the definition of lifespan of a node.

Definition 2.4 (*Lifespan of a node*). The lifespan of a node in a fuzzy spatiotemporal XML document, denoted as *lifespan*(n), is the temporal element of constraint edge incoming to the node. The lifespan of the root node is the temporal interval [t_O, *Now*]. *lifespan*(n) $= T_i$, where T_i represents the incident edge of node n in the fuzzy spatiotemporal XML documents. Besides, *Now* is the current time point of node n according to Definition 2.3. In order to facilitate discussion, we consider that each node has only one incident edge.

According to Definition 2.4, we can summarize a set of consistent conditions that a graph must fulfill to qualify as a fuzzy spatiotemporal XML document. The subsequent definition ourlines these conditions.

Definition 2.5 (*Fuzzy spatiotemporal XML document*). Fuzzy spatiotemporal XML documents should adhere to specific conditions to ensure the consistencies of time and position.

- Node n_i is the child node of node n_j in fuzzy spatiotemporal XML document tree. Then, *lifespan*(n_i) $\subset$ *lifespan*(n_j).
- The value of a possibility attribute denoted as *Poss*(n) is an element of [0, 1]. Certain nodes may possess multiple potential values, including position information.
- The time labels and position labels associated with containment edges ought to be incorporated into the lifespan and position labels of the node, respectively.
- Let $e_j(n_i, n_j, T_{ej})$, $e_k(n_i, n_k, T_{ek})$ be two outgoing edges from node n. Then, $T_{ej} + 1 = T_{ek}$.

- For any time instant t, the sub-graph is composed by all containment edges e_c where $t \in T_{ec}$ is a tree with root r. We call this sub-graph a snapshot of fuzzy spatiotemporal XML documents at time t, denoted as $D(t)$.
- For any containment edge $e_c(n_i, n_j, T_{ec})$, if n_j is a node of type ID (identification), the time label of e_c is the same as the lifespan of n_i; moreover, in fuzzy spatiotemporal XML documents, if two elements possess identical values for an ID attribute, it can be concluded that both elements are equivalent. This indicates that the ID of a node is invariant across all snapshots of the fuzzy spatiotemporal XML document.
- For a containment edge $e_c(n_i, n_j, T_{ec})$, if n_j is identified as an attribute node possessing certain attributes such as name or position, then there is a reference edge $e_r(n_j, n_k, T_{er})$. Besides, $T_{ec} = T_{er}$ and $P_{ec} = P_{er}$ holds.
- Let $e_r(n_i; n_j, T_{er})$ be a reference edge, then we have $T_{er} \subseteq \mathit{lifespan}(n_j)$.

We develop a fuzzy spatiotemporal XML data model utilizing the information derived from the cloud database, as illustrated in Fig. 2.2. Figure 2.2 serves as a specific representation of a fuzzy spatiotemporal XML document pertaining to a segment of the cloud database. The cloud is characterized by its cloud wall and spiral band, both of which are subject to temporal variations. Several considerations must be addressed in this context. Firstly, for the fuzzy nodes, we employ possibilities to represent their membership degrees, with the node values and their corresponding possibilities functioning as child nodes. Secondly, the labels associated with edges emanating from time and position nodes signify the possibilities. Lastly, the child nodes of the time and position nodes denote the lifespan and spatial location of the cloud, respectively. This example will be referenced throughout this chapter.

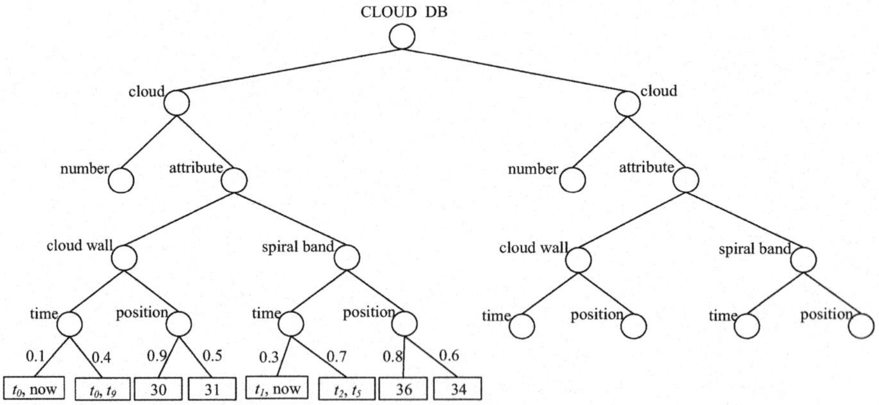

Fig. 2.2 Fuzzy spatiotemporal XML data model with the information of cloud database

2.4.7 Discussion

This subsection discuss several characteristics of the fuzzy spatiotemporal XML data model, along with the foundational assumptions that will be employed throughout the remainder of this chapter.

We will elucidate the conditions of inconsistency conditions as outlined in Definition 2.5.

- The initial inconsistency condition stipulates that containment edges may solely denote containment relations of a uniform nature at any given moment. However, this containment relationship may vary at different points in time. To represent any alternative relationships that occur concurrently, it is necessary to employ reference edges.
- The second inconsistency condition indicates that a node may possess multiple potential lifespans and locations.
- In the third condition of inconsistency, the possibility attribute is utilized within a fuzzy framework referred to as *Val*, which is employed to denote the likelihood of a particular element's presence within fuzzy spatiotemporal XML documents.
- The fourth condition of inconsistency indicates that the duration of existence for sibling nodes must be continuous.
- The fifth inconsistency condition indicates that the focus is on a discrete time interval rather than a continuous time element. This assumption streamlines the examination of the inconsistency conditions and enhances the efficiency of the implementations. Furthermore, our definitions and theorems can be generalized to encompass scenarios involving both time and space elements. Consequently, this assumption will yield both semantic and practical implications for the fuzzy spatiotemporal XML data model.
- The identifier (ID) specified in the sixth inconsistency condition serves as a unique constraint across the entirety of fuzzy spatiotemporal XML documents, thereby imposing various restrictions. However, this unique constraint may also give rise to additional complications. Specifically, the ID attribute fails to consistently identify a cloud node, as it may possess different IDs for two instances of the same cloud within a single snapshot. While it is possible to address fuzzy spatiotemporal queries, this approach compromises the desirable characteristic of maintaining coherent information regarding a cloud within the same node. It is posited that the temporal key for a cloud should be represented by the value node that contains its name. In the subsequent subsections of this chapter, we will operate under the assumption that fuzzy spatiotemporal XML documents adhere to the constraint that containment relationships are many-to-one from child nodes to parent nodes (as illustrated in Fig. 2.2) within each snapshot. In essence, all nodes along a path of containment edges are considered relative keys within a given snapshot. Coupled with the ID constraint, this framework facilitates the identification of a node throughout the historical context of the fuzzy spatiotemporal document, represented as a fuzzy spatiotemporal XML data tree.

- The seventh inconsistency condition stipulates that the temporal and spatial data pertaining to the containment edge and the reference edge must be identical.
- The eighth inconsistency condition indicates that the time interval associated with the reference edge is encompassed within the lifespan of the connection point's existence.

In accordance with the fifth inconsistency condition outlined in Definition 2.5, every fuzzy spatiotemporal XML document is associated with a current tree. Likewise, a subtree is also required to possess a current tree, which we designate as a current subtree.

Definition 2.6 (*Current subtree*). We define D_c as the subgraph of the fuzzy spatiotemporal XML document *D* composed without the referencing edges. Given a fuzzy spatiotemporal XML document *D* and a current node *n*, the current subtree of *n* is the subtree of $D_c(Now)$ with current node *n*.

In the subsequent subsections of this chapter, we will focus exclusively on containment edges for the sake of simplicity. It is important to note that all the concepts discussed can be generalized to include reference edges as well.

In accordance with the seventh consistency condition outlined in Definition 2.5, a containment edge is defined as a pathway connecting one node to another. Consequently, there exists a continuous path from the root node, denoted as *r*, to the element nodes or attribute nodes.

Definition 2.7 (*Continuous path and maximal continuous path*). For a continuous path (*cp*) with interval *T* from node n_1 to node n_k in a sequence of *k* nodes (n_1, ..., n_k, *T*), and an interval *T* with a sequence of containment edges of the form $e_1(n_1, n_2, T_1)$, $e_2(n_2, n_3, T_2)$, ..., $n_k(n_{k-1}, n_k, T_{k-1})$ such that $T = \prod_{i=1,k} T_i$, then we have that there is a maximal continuous path (*mcp*) with interval *T* from node n_1 to node n_k if *T* is the union of a maximal set of consecutive intervals T_i such that there is a continuous path from n_1 to n_k with interval T_i.

The *mcp* traverses each node within the query path a single time. This characteristic can be leveraged to enhance query processing. Consider two nodes n_1 and n_k, and let *N* represent the set of nodes n_1, where $i \neq 1, i \neq k$. For example, there exists a continuous path from n_1 to n_i characterized by the interval T_{ni}, and a containment edge from n_i to n_k with a temporal label T_{ei}. Consequently, each continuous path from n_1 to n_k will be associated with the interval $T_i = T_{ni} \cap T_{ei}$. The union of the intervals corresponding to these continuous paths will constitute the interval of the *mcp* between n_1 and n_k, provided that the intervals are consecutive. This indicates that all *mcps* within a graph can visit each node only once, starting from the root node.

2.5 Fixing Consistencies of Fuzzy Spatiotemporal XML Data

Fuzzy spatiotemporal XML documents are subject to alterations and modifications. It is essential to ensure the consistency of these documents during various operations, including updates, insertions, and deletions. Consequently, it is necessary to check the consistency of fuzzy spatiotemporal XML documents. This assessment entails evaluating the associated costs and developing efficient algorithms. The algorithms proposed herein are designed not only to check the consistency of fuzzy spatiotemporal XML documents but also to fix any identified inconsistencies.

In the following, we give inconsistency conditions in the fuzzy spatiotemporal XML documents.

Definition 2.8 (*Inconsistency conditions in fuzzy spatiotemporal XML documents*). Fuzzy spatiotemporal XML documents need to satisfy the following consistent conditions in order to accommodate operations that could potentially result in inconsistencies:

1. Type *I*: let *i* be a child node of *n*, *lifespan(i)* $\not\subset$ *lifespan(n)*.
2. Type *II*: let e_i and e_j be adjacent edges from node *n*, $T_{ej.from} - T_{ei.to} > 1$ or $T_{ej.from} - T_{ei.to} < 1$.
3. Type *III*: there is a cycle in fuzzy spatiotemporal data XML documents.

Given that each inconsistency condition encompasses multiple inconsistent intervals, we provide a definition for the inconsistency interval within the context of a fuzzy spatiotemporal XML document.

Definition 2.9 (*The inconsistency interval*). Let *I* be one of the inconsistencies in Definition 2.5 denoted as I_I, which is a closed interval where the inconsistency appears, the inconsistency interval is partial to *I*. This indicates the potential existence of multiple instances of I_I within fuzzy spatiotemporal XML documents that exhibit inconsistencies.

Example 2.1 Figure 2.3a–c show examples of inconsistency intervals for type *I*, type *II*, and type *III*. In Fig. 2.3a, II = $[T_6, Now]$ (the temporal label of edge outgoing from n_2 is not included in the lifespan of its parent node); in Fig. 2.3b, II = $[T_4, T_5]$ (a gap in node n_2) and $[T_{13}, T_{15}]$ (an intersection in node n_3); in Fig. 2.3c, there is a cycle in every snapshot within the interval $I_I = [T_5, T_7]$.

Upon identifying the type of inconsistencies as outlined in Definitions 2.8 and Definition 2.9, it is imperative to check and fix these inconsistencies. The procedure for checking and fixing inconsistencies in fuzzy spatiotemporal XML documents can be categorized into two distinct phases:

- Check whether fuzzy spatiotemporal XML documents are inconsistent or not.
- If there are inconsistencies, it is necessary to fix them, which necessitates the calculation of the inconsistency interval.

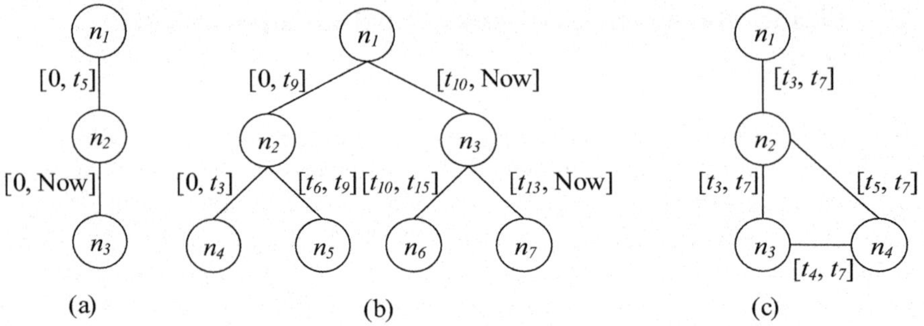

Fig. 2.3 The inconsistent cases

Given that the correction of fuzzy spatiotemporal documents may lead to alterations in semantics, users have the autonomy to determine whether to engage in the second step. Furthermore, we will examine the intricacies involved in checking and fixing inconsistencies in fuzzy spatiotemporal XML documents following the proposed algorithms.

2.5.1 *Checking Inconsistencies of Fuzzy Spatiotemporal XML Documents*

In this subsection, we investigate the complexities associated with checking inconsistencies within fuzzy spatiotemporal XML documents and propose algorithms designed to address these tasks.

We use the following notions throughout this subsection: given two intervals T_1 and T_2, if $T_{1\cdot to} > T_{2\cdot to}$, then we have: T_1 is early than T_2, denoted as $T_1 > T_2$. Analogously, if $T_{1\cdot to} < T_{2\cdot to}$, then we have: T_1 is later than T_2, denoted as $T_1 < T_2$.

Corollary *If there is a cycle in the fuzzy spatiotemporal XML graph during an interval* I_1, *then we have: there will be a node i, where* $T_{mcp}(n_i) \neq$ *lifespan*(n_i) *and* $T_{mcp}(n_i)$ *is the temporal interval of the path(mcp) from the root node* n_i.

Proof According to the definition of fuzzy spatiotemporal XML documents, it is established that each node has a maximum of one incoming containment edge at any given time instant *t*. In the event that a cycle exists within a temporal interval I_1, let n_i be a node belonging to this cycle. Based on the definition, it follows that *lifespan*(n_i) $\supset I_1$ and $T_{mcp} \cap I_1 = \emptyset$. However, if there is a cycle, it will appear that $T_{mcp} \cap I_1 \neq \emptyset$. Consequently, this leads to the conclusion that there exists a node with multiple parent nodes at time instant *t*. □

We will employ the corollary to check inconsistencies of type *III*. If there is a cycle in fuzzy spatiotemporal XML documents we will fix it. The initial step to check inconsistencies involves changing the fuzzy possibility distributions into a normal kind. Algorithm 2.1 (CI) checks inconsistencies of type I that the time period of child node is not contained within the time period of its parent node. In Algorithm 2.1, δ represents the possibility distribution associated with the outgoing containment edge (according to Definition 2.7) of node n. For the purpose of simplifying the discussion, we choose the greater possibility distribution as the *lifespan*(n). L denotes a time interval set that can contain the time intervals of all nodes in a fuzzy spatiotemporal XML document. L is considered null if the fuzzy spatiotemporal XML document is empty. L[i].to + 1 = L[i + 1] indicates that time intervals of two adjacent nodes are continuous. Subsequently, these intervals are put into L. In the rest of this chapter, all algorithms will utilize the time interval contained within L.

Algorithm 2.1 Check Inconsistencies of Type I

Input: A fuzzy spatiotemporal XML root node n
Output: $I = [T_{from}; T_{to}]$; True if D has no inconsistencies of type I; False otherwise
boolean checkNodeConsistencies (document D){
1. Initialize a list of temporal stack L to null
2. for each edge e incident to n with label T_e do
3.   for each $poss(n_i)$ of node i
4.   if $\delta_{ik} > \delta_{i(k+1)}$ //δ_{ik} indicates the possibility distribution of node i_k
5. $T_i = T_{ik}$
6.   read lifespan(e) into L;
7.   End if
8.   End for
9. $I = L[1]$;
10. for each i between 1 and $length(L) - 1$ do
11. if L[i].to + 1 $\neq$ L[i + 1].from then
12.   Return False;
13. else
14. $I = I \cup L$[i + 1];
15. end for;
16. Return True;}

In Algorithm 2.1, δ represents the possibility distribution associated with the outgoing containment edge of node n, as defined in Definition 2.7. For the purpose of simplifying the discussion, we choose the greater possibility distribution as *lifespan*(n). L denotes a time interval set that can contain the time interval of all nodes in a fuzzy spatiotemporal XML document. It is important to note that L is null if the fuzzy spatiotemporal XML document is empty. L[i].to + 1 = L[i + 1] indicates that time intervals of two adjacent nodes are continuous. Subsequently, these intervals are put into L. In the rest of the chapter, all algorithms will utilize the time intervals in L.

According to Algorithm 2.1, lines 4 to 7 are responsible for populating a temporal stack L with time intervals based on the $poss(e)$ of the element node. Subsequently, lines 11 to 12 are utilized to check inconsistencies of type I. Should the temporal stack be found to be NULL, the algorithm will return a value of false, indicating the inconsistency of type I occurs.

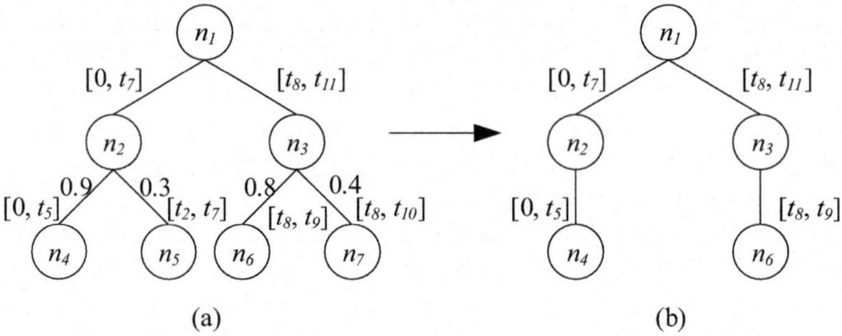

Fig. 2.4 Checking the time intervals between child nodes and parent nodes

It can be observed that the main loop of Algorithm 2.1 iterates a maximum of $|V|$ times (the number of nodes in fuzzy spatiotemporal XML documents). The time complexity is $O(deg_{in}(n) \times deg_{in}(n) + deg_{in}(n) \times \log(deg_{in}(n)))$, where $deg_{in}(n)$ is the number of edges incident to node n. In the worst scenario (this considers that all edges are incident to the node), the order of the algorithm is $O(|E| \times |E|)$ ($|E|$ is the number of edges incoming to the node n). This, in the average case (all nodes have the same number of incoming edges, which is $\frac{|E|}{|V|}$), results in complexity $O\left(\frac{|E|}{|V|} \times \frac{|E|}{|V|}\right)$. In the best cases (when each node has only a single incoming edge), the computation of lifespan occurs in constant time.

Example 2.2 Figure 2.4 will be utilized to demonstrate the operational steps of Algorithm 2.1. The fuzzy spatiotemporal XML data tree can change from Fig. 2.4a to b in accordance with the procedures outlined in Algorithm 2.1. In Fig. 2.4, it is observed that $\delta_{n4} > \delta_{n5}$ (δ_{ik} denotes the possibility distribution of node i_k), leading to the selection of node n_4 as the child node of node n_2. Similarly, node n_6 is chosen following the same criteria as node n_4. Furthermore, in Fig. 2.4, the time intervals associated with child nodes n_4 and n_6 are contained within the time intervals of their parent node. Consequently, Algorithm 2.1 will yield a result of truth.

After checking the time intervals between child nodes and their corresponding parent nodes, we will proceed to analyze the consistency of the time intervals among sibling nodes. The detailed procedural steps for this checking process are outlined in Algorithm 2.2 (C*II*). In Algorithm 2.2, line 3 checks inconsistencies of type *II*, specifically determining whether the time intervals of the incident edges are consecutive; if they are not, the algorithm will return a null value. As illustrated in Fig. 2.5, the time intervals associated with nodes n_4 and n_5 overlap, while there exists a gap in the time intervals of nodes n_6 and n_7, specifically within the interval $[t_9, t_{10}]$.

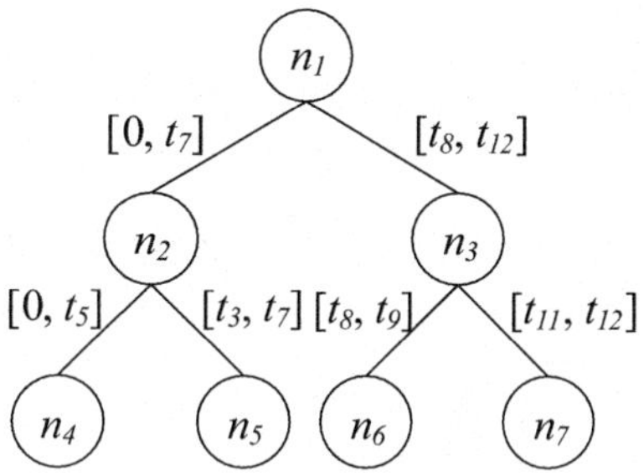

Fig. 2.5 Checking the time intervals between sibling nodes

Algorithm 2.2 Check Inconsistencies of Type *II*

Input: A fuzzy spatiotemporal XML document *D*
Output: True if *D* has no inconsistencies of type *II*; False otherwise
boolean checkNodeConsistencies(document *D*) {
01 for each node *n* in *D* do
02 $I = lifespan(n)$;
03 for each child node *e* of *n* do
04   if $T_e \not\subset I$ then
05   Return False;
06   end if;
07 end for;
08 end for;
09 Return True;}

It can be observed that the main loop of Algorithm 2.2 iterates a maximum number of $|V|$ times (the number of nodes in fuzzy spatiotemporal XML documents). Based on Algorithm 2.2, the time complexity of the internal cycle is $O(deg_{in}(\mathrm{n}) \times \log(deg_{in}(n)))$. Accordingly, the overall time complexity of Algorithm 2.2 is $\sum_{n\in|V|} deg_{in}^2(\mathrm{n}) \times \log(deg_{in}(n))$. In the average scenario, $deg_{in}(n)$ is equal to $|E|$. This leads to time complexity $O\big(|V| \times |E|^2 \times \log(|E|)\big)$.

Example 2.3 As delineated in Algorithm 2.2, Fig. 2.5 serves to exemplify the procedural steps involved in its execution. Specifically, lines 4 to 7 of Algorithm 2.2 are dedicated to the examination of inconsistency intervals within fuzzy spatiotemporal XML documents. In Fig. 2.5, the identified inconsistency intervals are denoted as $[t_3, t_7]$ and $[t_9, t_{11}]$. Should any inconsistencies be detected, the algorithm will yield a false outcome. Conversely, if the temporal intervals between sibling nodes are continuous, the algorithm will return a true result.

After checking inconsistencies of types *I* and *II*, we will proceed to check inconsistency of type *III* involving cycles occurring within specific intervals of fuzzy spatiotemporal XML documents. Algorithm 2.3 (C*III*) is designed to detect cycles by identifying nodes within the designated inconsistency intervals. In Algorithm 2.3, lines 3 to 6 are dedicated to the checking of inconsistencies of type *III*. A node is ended when all of its outgoing edges have been traversed. If the queue is empty and all nodes are ended, it can be inferred that the fuzzy spatiotemporal XML documents are free of cycles. Conversely, if nodes are not ended, indicating that there is a cycle.

Algorithm 2.3 Check Inconsistencies of Type *III*

Input: A fuzzy spatiotemporal XML document D, such that checkNodeConsistencies(D) = true
Output: True if D has no inconsistencies of type *III*; False otherwise

```
boolean checkCycles(document D) {
01  n = first(nodes);
02  for each e outgoing from n do
03      if !traversed(e) then
04          if T_e ∩ T_e' ≠ Ø then
05              Return False;
06          end if;
07      end if;
08  end for;
09  if !empty(nodes) then
10      Return False;
11  end if;
12  Return True;}
```

We will now compute the time complexity of Algorithm 2.3. Each node may be visited multiple times, contingent upon the number of incident edges. The best scenario occurs in the absence of a temporal cycle, wherein line 5 remains unexecuted. The loop delineated in line 2 is executed $deg_{in}(n)$ times. Conversely, in the worst case, line 4 is executed $|V|$ times. Finally, the overall time complexity of Algorithm 2.3 is $O(|E| + |V|)$.

Example 2.4 We will utilize Fig. 2.6 to demonstrate the operational steps of Algorithm 2.3. In Fig. 2.6, $T_{n3} \cap T_{n4} \neq \emptyset$, resulting in a return value of false.

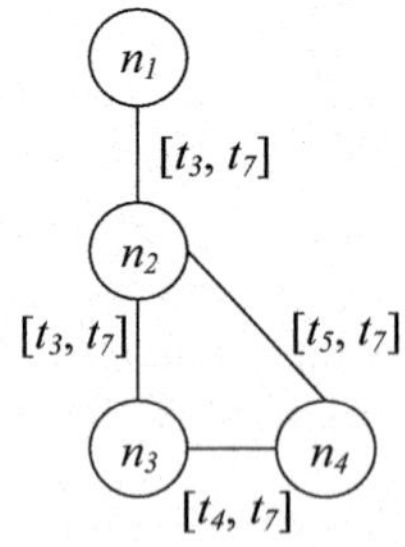

Fig. 2.6 Checking cycle in fuzzy spatiotemporal XML documents

2.5.2 *Fixing Inconsistencies of Fuzzy Spatiotemporal XML Documents*

In the previous subsection, we presented valid algorithms for identifying inconsistencies within fuzzy spatiotemporal XML documents. In this subsection, we will explore methods for fixing these inconsistencies. We will examine potential fixing strategies for each category of inconsistency (types *I*, *II* and *III*). Should these proposed solutions be deemed unsatisfactory, individuals may opt to discard the fuzzy spatiotemporal XML documents rather than attempt to fix them. Our analysis will focus on isolated inconsistencies, operating under the assumption that each inconsistency is treated as if it were the sole issue present within the fuzzy spatiotemporal XML documents.

2.5.2.1 Fixing Inconsistencies of Types *I* and *II*

In the cases of type *I* and *II*, an edge *e* of node *i* is inconsistent if $T_{ei} \not\subset lifespan(i)$. For inconsistencies classified as type *I*, the inconsistency interval of *I* is the maximum interval that does not fall within the lifespan of the inconsistent node. It is important to note that type *I* can be a set of intervals actually. For example, the lifespan of the inconsistent node may be entirely contained within the lifespan of the inconsistent edge. This subsection will study the issues arising from the presence of an inconsistent edge.

This research investigates a method for addressing the issue of inconsistency through a process termed correction by reduction, which involves the modification of the fuzzy spatiotemporal label associated with the inconsistent edge. In this approach, we decrease the lifespan of the inconsistent nodes until it aligns with the lifespan of the parent node, thereby encompassing the interval of inconsistency. We examine the incoming edge of the child node and adjust its fuzzy spatiotemporal label to ensure that the lifespan of the parent node fully encompasses the entire label of the inconsistent edge.

The initial step in addressing inconsistent states involves fixing the inconsistencies classified as types *I* and *II*. This approach is warranted due to the analogous characteristics exhibited by these two types of inconsistencies during temporal intervals. The procedure for the fixing methods is elaborated in Algorithm 2.4 (F*I*&&*II*). Algorithm 2.4 examines all nodes in search of gaps or overlaps in fuzzy spatiotemporal XML documents. In instances where an overlap is detected, one of the edges implicated in the overlap will be removed during the corresponding time interval. Whenever a node *n* is visited, the calling to the node duplication algorithm is performed $deg_{in}(n)$ times, which gives Algorithm 2.4 an order of $O(|E|^2)$.

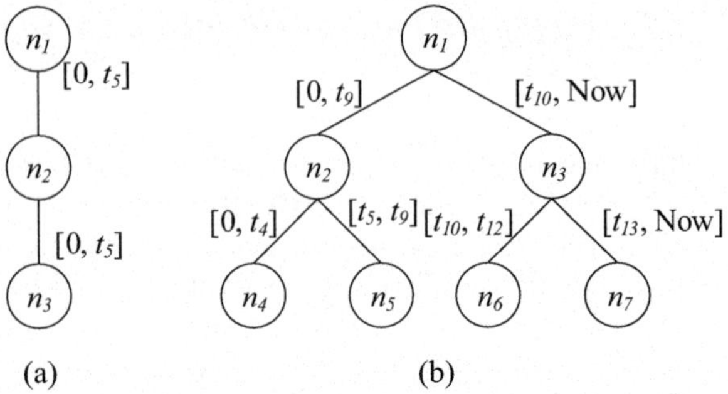

Fig. 2.7 Fixing inconsistent time intervals between nodes

Algorithm 2.4 Fixing Inconsistencies of Types *I* and *II*

Input: A fuzzy spatiotemporal XML document D
Output: a legal fuzzy spatiotemporal XML document
Fixinconsistencies(document *D*) {
for each child node e of *n* do
    read lifespan(*e*) into *L*;
    if lifespan(*e*) $\not\subset I$ then{
        if $t_{e \cdot to} < t_{n \cdot from}$ or $t_{e \cdot from} > t_{n \cdot to}$ then delete *e*;
        else if $t_{e \cdot from} < t_{n \cdot from}$ then $t_{e \cdot from} = t_{n \cdot from}$;
        else if $t_e.to > t_n.to$ then $t_e.to = t_n.to$;}
end for;
for each *i* between 0 and length(*L*)-1 do
    if $L[i]._{from} = L[i+1]._{from}$ && $L[i]._{to} = L[i+1]._{to}$ then delete *i*
if $L[i]._{to} < L[i+1]._{from} - 1$ then $L[i]._{to} = L[i+1]._{from} - 1$;
if $L[i] \cap L[i+1] \neq null$ then $L[i]._{to} = L[i+1]._{from} - 1$;
end for;
Return *D*;}

Example 2.5 Figure 2.7a and b illustrate instances of fixing inconsistencies of types *I* and *II*, respectively. Figure 2.7a is an inconsistency example of type *I*, wherein the time interval is constrained to [0, t_5], corresponding to the time interval of the edge (n_2, n_3). Figure 2.7b depicts the outcome of an inconsistent scenario, with the time interval being limited to [0, t_4], which aligns with the time interval of the edge (n_2, n_4).

2.5.2.2 Fixing Inconsistencies of Type *III*

The inconsistencies of type *III* are characterized by the presence of cycles. In such instances, a potential resolution to these inconsistencies is to remove one of the nodes that exists within the time interval of the inconsistency in the cycle.

Algorithm 2.5 (F*III*) executes cycle elimination by deleting the node that falls within the inconsistency time interval. Let $deg_{in}(n)$ be the number of incident edges

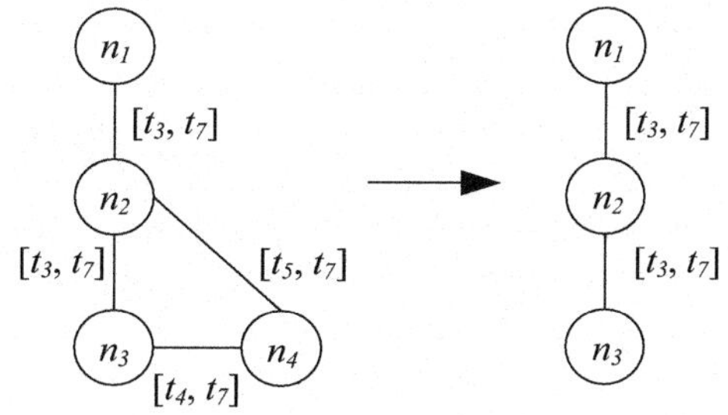

Fig. 2.8 Fixing the cycle in fuzzy spatiotemporal XML documents

of node n, then $|E|$ is equal to $deg_{in}(n)$. The computational complexity of Algorithm 2.5 is $O(deg_{in}(in) \times \log(deg_{in}(n)))$, that is $O(|E| \times \log(|E|))$.

Algorithm 2.5 Fixing Inconsistencies of Type *III*

Input: A fuzzy spatiotemporal document D, with a cycle C in an interval T_c.
Output: a legal fuzzy temporal XML document
Fixinconsistencies(D, C, T_c) {
01 n_c = a node in C;
02 $nodes\_stack = [n_c]$;
03 while node stack is not empty do
04 $n = nodes\_stack.pop()$;
05 visited(n) = True;
06 for each edge $e = (n; n_d; T_e)$ outgoing from n, $T_c \cap T_e \neq \emptyset$
07 delete e in T_c;
08 end for
09 end while;
10 Return D;}

Example 2.6 Figure 2.8 illustrates the procedure of fixing inconsistencies of type *III*. In Algorithm 2.5, lines 6 to 7 fix the inconsistencies of type *III*. As depicted in Fig. 2.8, the node n_4 involved in the cycle will be removed during the interval $I_I = [t_4, t_7]$. Consequently, upon identifying the cycle, the inconsistent outgoing edges of node n will be eliminated.

2.6 Experiments

To evaluate the efficiency of algorithms designed for checking and fixing inconsistencies in fuzzy spatiotemporal XML documents, we conduct a comparative analysis utilizing two datasets: a tropical storm dataset, specifically the *Ingrid* dataset from the North Atlantic Ocean [38], and a synthetic virtual dataset. Our investigation comprises six distinct experimental groups. In the first group, we present the CPU time required for checking and fixing type *I* inconsistencies within virtual fuzzy spatiotemporal documents. The second group focuses on the CPU time associated with the checking and fixing of type *II* inconsistencies. The third group details the CPU time for addressing type *III* inconsistencies. In the fourth group, we analyze the memory consumption of five algorithms across both datasets. The fifth group offers comparative results against two additional algorithms aimed at fixing inconsistencies. Finally, the sixth group provides a comparative analysis of the performance of inconsistency checking and fixing algorithms implemented in DOM versus those in dom4j.

2.6.1 Experimental Setup

The experiments were conducted using Eclipse Mars.1 (version 4.5.1) on a Pentium desktop computer operating at 3.20 GHz, equipped with 4 GB of RAM and a 500 GB hard drive, and running the Windows 7 operating system. The programming tool employed for this research is dom4j, an open-source Java library designed for handling XML, XPath, and XSLT. Dom4j is recognized as a robust Java XML API that is not only powerful but also user-friendly, exhibiting exceptional performance characteristics.

For our experimental evaluations, we utilize both a real storm dataset and a virtual dataset, selected for their distinct characteristics. The storm dataset contains certain inconsistent states, whereas the virtual dataset encompasses a broader range of inconsistencies. We have chosen the tropical storm as a representative system, grounded in spatiotemporal path summaries. This selection is made for practical reasons, as the tropical storm data is readily available and exhibits similarities to the fuzzy and spatiotemporal properties under investigation. We anticipate that the results derived from the real storm dataset will not significantly differ from those obtained using the virtual dataset. The details of the real storm dataset and the virtual dataset are presented in Tables 2.1 and 2.2, respectively.

Storm dataset: The tropical storm dataset serves as a focal point for our analysis, encompassing a fuzzy spatiotemporal dataset related to tropical storms. While the utilization of this dataset simplifies the complexities associated with algorithms designed to check and fix inconsistencies in fuzzy spatiotemporal XML documents, it nonetheless facilitates a deeper understanding of fuzzy spatiotemporal data in practical spatiotemporal applications. The structure of the tropical storm dataset encapsulates key characteristics of fuzzy spatiotemporal XML data, including aspects of fuzziness, spatial information, and temporal information as delineated in XML documents. However, akin to other conventional fuzzy spatiotemporal structural summaries, it is inherently limited in its capacity to address all inconsistencies that may arise during the processes of updating, inserting, or deleting fuzzy spatiotemporal XML documents. Given these limitations, we also incorporate a virtual dataset to enhance our analysis.

Virtual dataset: The virtual dataset adheres to the stipulations outlined in Definition 2.5. It encompasses elements of fuzziness, spatial data, and temporal data. Overall, it is capable of representing all inconsistent states as defined in Definition 2.8, thereby

Table 2.1 Real storm dataset

Date	Time	Lat	Lon	Wind (mph)	Pressure	Poss
Sep/12/2013	2100GMT	19.7	−93.6	35	1003	0.7
Sep/13/2013	0000GMT	19.7	−93.7	35	1003	0.5
Sep/13/2013	0300GMT	19.7	−94	35	1003	0.9
Sep/13/2013	0600GMT	19.8	−94.2	35	1003	0.2
Sep/13/2013	0900GMT	19.7	−94.5	35	1002	0.2
Sep/13/2013	1200GMT	19.5	−95	35	1000	0.3
Sep/13/2013	1500GMT	19.4	−95.3	45	1000	0.6
Sep/13/2013	1800GMT	19.4	−95.4	45	999	0.8
Sep/13/2013	2100GMT	19.2	−95.4	45	999	0.1
Sep/14/2013	0000GMT	19.2	−95.3	60	993	0.7
Sep/14/2013	0300GMT	19.3	−95.2	60	993	0.2
Sep/14/2013	0600GMT	19.7	−95.2	60	993	0.9
Sep/14/2013	0900GMT	19.8	−95	60	991	0.7
Sep/14/2013	1200GMT	20.3	−94.5	65	989	0.8
Sep/14/2013	1500GMT	20.6	−94.5	70	988	0.1
Sep/14/2013	1800GMT	20.9	−94.4	70	988	0.2
Sep/14/2013	2100GMT	21.3	−94.4	75	987	0.7
Sep/15/2013	0000GMT	21.6	−94.7	80	983	0.9
Sep/15/2013	0300GMT	22	−95	85	986	0.6
Sep/15/2013	0600GMT	22.2	−95.2	85	986	0.6
Sep/15/2013	0900GMT	22.4	−95.4	85	986	0.7
Sep/15/2013	1200GMT	22.5	−95.6	85	986	0.8
Sep/15/2013	1500GMT	22.5	−95.8	75	990	0.6
Sep/15/2013	1800GMT	22.6	−96.1	75	990	0.1
Sep/15/2013	2100GMT	22.7	−96.3	75	990	0.3
Sep/16/2013	0000GMT	22.9	−96.1	75	987	0.8
Sep/16/2013	0300GMT	23.1	−96.5	75	989	0.2
Sep/16/2013	0600GMT	23.2	−96.9	75	989	0.9
Sep/16/2013	0900GMT	23.4	−97.1	75	989	0.9
Sep/16/2013	1200GMT	23.8	−97.8	65	991	0.8
Sep/16/2013	1500GMT	23.8	−98.2	60	993	0.9
Sep/16/2013	1800GMT	23.7	−98.6	45	998	0.7
Sep/16/2013	2100GMT	23.7	−99	35	1002	0.8
Sep/17/2013	0300GMT	23.7	−99.4	30	1006	0.7
Sep/17/2013	0900GMT	23.7	−99.9	25	1008	0.7

Table 2.2 Virtual dataset

Node	Lifespan	Position	Poss
n_1	$[t_0-t_{20}]$	P1	0.9
n_2	$[t_0-t_{10}]$	P2	0.8
n_3	$[t_{11}-t_{20}]$	P3	0.3
n_4	$[t_0-t_6]$	P2	0.7
n_5	$[t_7-t_{10}]$	P2	0.2
n_6	$[t_{11}-t_{15}]$	P3	0.8
n_7	$[t_{16}-t_{20}]$	P3	0.8
n_8	$[t_0-t_4]$	P4	0.6
n_9	$[t_5-t_6]$	P5	0.9
n_{10}	$[t_7-t_9]$	P2	0.6
n_{11}	$[t_8-t_{10}]$	P2	0.8
n_{12}	$[t_{11}-t_{15}]$	P3	0.7
n_{13}	$[t_{11}-t_{15}]$	P3	0.8
n_{14}	$[t_{16}-t_{17}]$	P3	0.7
n_{15}	$[t_{19}-t_{20}]$	P3	0.6
n_{16}	$[t_0-t_1]$	P4	0.4
n_{17}	$[t_3-t_4]$	P4	0.3
n_{18}	$[t_5-t_6]$	P5	0.6
n_{19}	$[t_7-t_{10}]$	P5	0.7
n_{20}	$[t_7-t_{10}]$	P2	0.8
n_{21}	$[t_8-t_9]$	P2	0.7

facilitating the checking ad fixing of inconsistencies in fuzzy spatiotemporal XML documents. This capability is advantageous in the management of actual data.

2.6.2 *Performance Analysis*

All experiments were conducted utilizing the checking and fixing algorithms. Furthermore, the methodology for each algorithm involves the creation and manipulation of fuzzy spatiotemporal XML documents. In the present program implementations, all data structures are established prior to the execution of the checking and fixing processes. Additionally, we have optimized the internal representation of time intervals and attributes, resulting in a reduction of the size of fuzzy spatiotemporal XML documents. The objective of these efforts is to develop an effective approach for checking and fixing inconsistencies in fuzzy spatiotemporal XML documents.

2.6.2.1 Performance of Checking and Fixing Algorithms

In the examination of the three types of inconsistencies in fuzzy spatiotemporal XML documents, it is essential to establish an order among the relevant fuzzy nodes. The fuzzy spatiotemporal XML data model presented indicates that the spatial and temporal attributes of the nodes may exhibit fuzziness. Consequently, it is imperative to eliminate the effects of fuzzy attributes prior to initiating the checking process. Subsequently, we utilize CPU time as a metric to evaluate the performance of the checking algorithms.

CPU (central processing unit) time refers to the duration utilized for executing the instructions of a computer program or operating system. This time is quantified in terms of clock ticks or seconds. Measuring CPU time as a percentage of the CPU's total capacity, known as CPU usage, provides valuable insights into the efficiency and performance of the processing unit.

Tables 2.3, 2.4, 2.5 present CPU time (*ms*) of checking and fixing inconsistencies of type *I*, *II* and *III* on virtual dataset respectively.

In Tables 2.3, 2.4, 2.5, it is evident that checking algorithms demonstrate greater efficiency in addressing inconsistency steps. This efficiency can be attributed to the fact that checking algorithms solely focus on identifying inconsistencies without

Table 2.3 CPU time (*ms*) of checking and fixing inconsistency of type *I*

Algorithm	The number of fuzzy spatiotemporal nodes			
	5	10	15	20
C*I*	54	118	266	532
F*I*	69	145	326	697
CF*I*	105	231	543	1040

Table 2.4 CPU time (*ms*) of checking and fixing inconsistency of type *II*

Algorithm	The number of fuzzy spatiotemporal nodes			
	5	10	15	20
C*II*	65	140	310	620
F*II*	75	158	347	711
CF*II*	116	277	598	1138

Table 2.5 CPU time (*ms*) of checking and fixing inconsistency of type *III*

Algorithm	The number of fuzzy spatiotemporal nodes			
	5	10	15	20
C*III*	49	108	246	492
F*III*	55	120	270	540
CF*III*	74	178	472	853

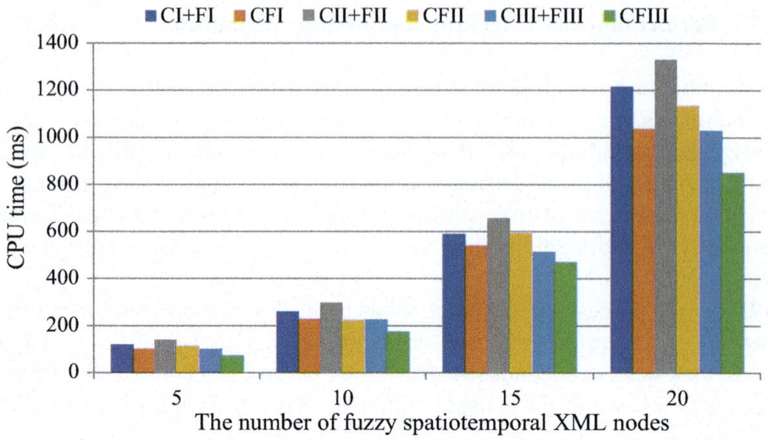

Fig. 2.9 CPU time (*ms*) of checking and fixing inconsistency

engaging in the fixing actions. In contrast, the fixing steps involve the deletion and replacement of fuzzy spatiotemporal nodes and their associated properties. Furthermore, the CPU time associated with the CF (checking and fixing algorithm) does not simply represent the sum of the CPU times for the C*I* (checking algorithm) and F*I* (fixing algorithm); rather, it reflects an integrated CPU time that is less than the cumulative time of the individual sub-steps. This reduction is due to the CF's ability to minimize the number of calls made to fuzzy spatiotemporal XML documents. Additionally, it is important to note that CPU time is significantly influenced by the number of fuzzy spatiotemporal nodes.

We conducted an evaluation of the CPU time (measured in *ms*) required for the processes of checking and fixing inconsistencies, as illustrated in Fig. 2.9. The analysis reveals that an increase in the number of fuzzy spatiotemporal nodes correlates with a longer CPU time for the checking and fixing algorithms. This phenomenon can be attributed to the fact that a larger dataset necessitates more time for the generation of a fuzzy spatiotemporal XML document, consequently leading to an increase in the time required for checking and resolving inconsistencies. Furthermore, it is noteworthy that the CPU time associated with the algorithm $(C + F)$ exceeds that of CF. This observation suggests that the fixing process is not merely an additive combination of the checking and fixing procedures, as the CF approach effectively reduces the time needed to read the fuzzy spatiotemporal XML documents.

2.6.2.2 Memory Cost

Memory cost pertains to the overhead associated with the processes of checking and fixing inconsistencies within algorithms. The memory costs associated with the checking and fixing algorithms are detailed in Table 2.6. Furthermore, a comparative analysis of memory costs between the virtual dataset and the storm dataset is

Table 2.6 The memory cost (*kb*) of algorithms

Algorithm	Virtual data	Storm data	Numerical difference
CI	2502	2611	109
CII	2501	2599	98
CIII	2525	2594	69
FI&II	2506	2607	101
FIII	2525	2633	108

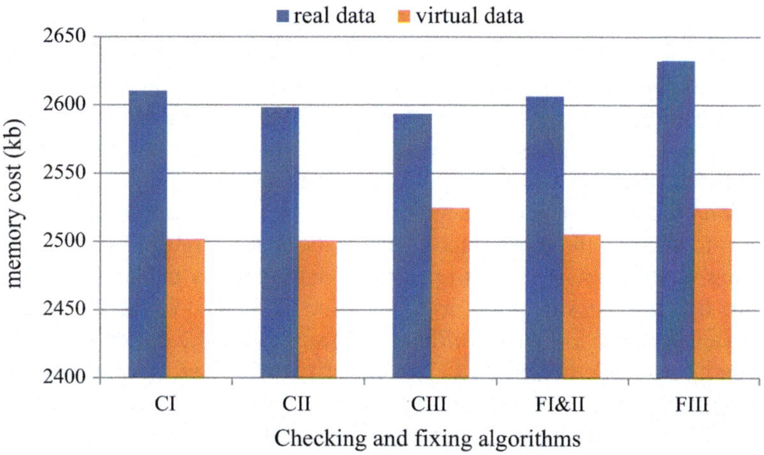

Fig. 2.10 The different memory cost (*kb*) between two datasets

conducted. As indicated in Table 2.6, the memory cost associated with the storm dataset exceeds that of the virtual dataset. This discrepancy can be attributed to the storm dataset's abundance of attributes related to fuzziness, spatial information, and temporal information. Conversely, the virtual dataset encompasses all potential inconsistency scenarios present in fuzzy spatiotemporal XML documents. Additionally, it is noteworthy that the memory costs associated with the checking and fixing algorithms exhibit variability. This variation can be elucidated by the observation that the real dataset contains a greater volume of information compared to the virtual dataset, as illustrated in Fig. 2.10.

2.6.2.3 Experimental Comprehensive Comparison Results

In this subsection, we conduct a comparative analysis of the algorithms for checking and fixing inconsistencies in relation to the methodologies proposed by Hamrouni et al. [39] and Rizzolo et al. [40]. Hamrouni et al. [39] introduce a detection and repair method (*DR*) aimed at addressing inconsistencies arising from the insertion, deletion, or updating of fuzzy spatiotemporal XML documents. Conversely, Rizzolo et al.

[40] present a classification framework for inconsistencies and propose algorithms (*TXCF*) designed to identify and resolve these inconsistencies. Our investigation encompasses three distinct experimental groups: the first compares the *DR* method when operations (insertion, deletion, and updating) result in a singular type of inconsistency; the second compares the *DR* method when these operations lead to multiple types of inconsistencies; and the third evaluates the *TXCF* algorithm for each specific type of inconsistency.

The initial set of experiments examines the performance of fixing inconsistency algorithms, specifically *CF* and *DR*, in scenarios characterized by a singular type of inconsistency. As illustrated in Fig. 2.11, an increase in the number of temporal nodes correlates with an extended CPU processing time for both *DR* (in terms of insert, delete, and update operations) and *CF*. The data presented in Fig. 2.11 indicates that our algorithm (*CF*) incurs a longer CPU processing time when applied to XML documents. This phenomenon can be attributed to the fact that *DR* provides a robust mechanism for checking and fixing fuzzy spatiotemporal XML documents that exhibit only one instance of inconsistency of type *I*. *DR* employs targeted repair strategies to address the emergence of inconsistencies. Figure 2.12 presents CPU processing times associated with operations that result in inconsistencies of type *II*. In this figure, it is evident that as the volume of XML documents increases, the CPU processing time for our algorithms (*CF*) surpasses that of *DR*, consistent with the observations made in Fig. 2.11. The disparity in CPU processing time becomes increasingly pronounced with the escalation of temporal XML data. Figure 2.13 contrasts CPU processing times when operations lead to inconsistencies of type *III*. A notable finding from Fig. 2.13 is that *DR* demonstrates superior efficiency compared to *CF*. This can be elucidated by the fact that *CF* necessitates a comprehensive traversal of all XML documents when addressing fuzzy spatiotemporal inconsistencies, which becomes particularly cumbersome when the temporal XML documents exhibit only a single type of inconsistency.

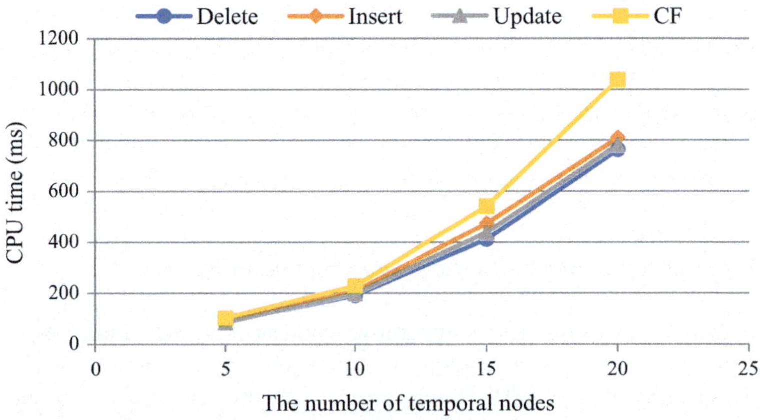

Fig. 2.11 CPU time of fixing inconsistency of type *I*

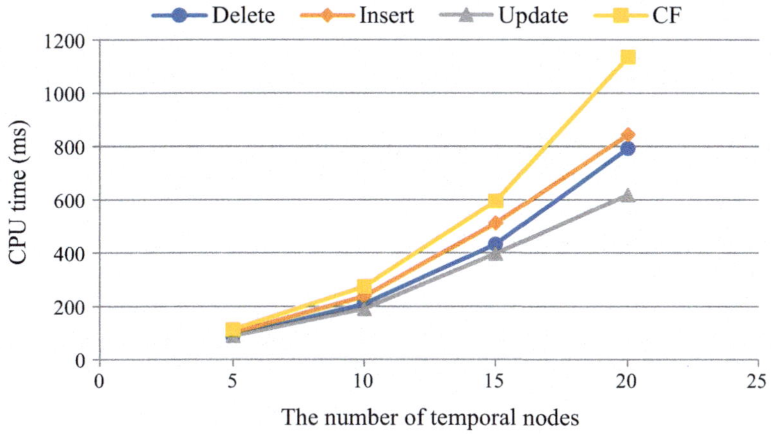

Fig. 2.12 CPU time of fixing inconsistency of type *II*

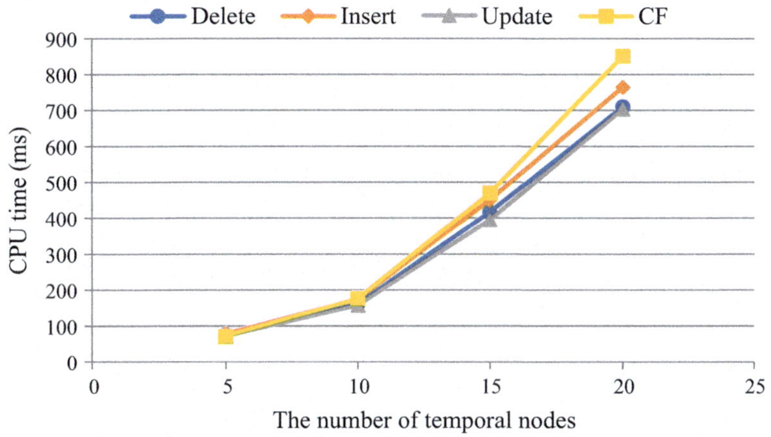

Fig. 2.13 CPU time of fixing inconsistency of type *III*

The second set of experiments examines the performance of fixing inconsistency algorithms, specifically *CF* and *DR*, in scenarios involving multiple types of inconsistencies. Figure 2.14 illustrates the CPU time required to fix inconsistencies of types *I* and *II* when operations modify temporal XML documents. It is evident that the CPU time associated with the *CF* algorithm is lower than that of the *DR* algorithm. Furthermore, as the volume of temporal XML data escalates, the advantages of *CF* become increasingly pronounced. This phenomenon can be attributed to *CF*'s capability to address all types of inconsistencies in a single traversal, whereas *DR* necessitates separate resolutions for each inconsistency type. As depicted in Fig. 2.15, the CPU time for *CF* remains lower than that of other algorithms (insert, delete, and update) when the operations result in two distinct types of inconsistencies (types *II* and *III*), owing to its efficient resolution methods. This trend aligns with the findings presented in Fig. 2.14, where the efficiency of *CF* becomes markedly more apparent with an increase in temporal data. Figure 2.16 further compares the CPU time of the *CF* algorithm against that of *DR* when operations (insert, delete, and update) induce two types of inconsistencies (types *I* and *III*). Consistent with the results from Figs. 2.14, 2.15, and 2.16 confirms that the CPU time for *CF* consistently remains lower than that for *DR*. Additionally, it is observable from Fig. 2.16 that an increase in the volume of temporal XML data results in a substantial rise in CPU time consumption.

The resolution of multiple types of inconsistencies (types *I*, *II*, and *III*) results in a greater CPU time for the *DR* algorithm compared to the *CF* algorithm, as illustrated

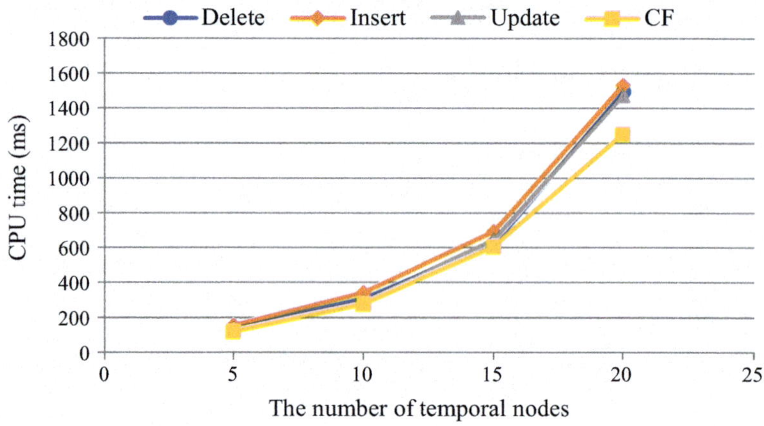

Fig. 2.14 CPU time of fixing inconsistencies of types *I* and *II*

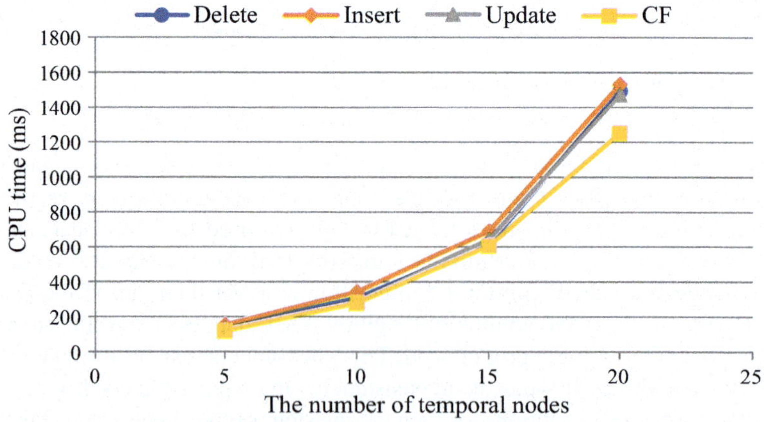

Fig. 2.15 CPU time of fixing inconsistencies of types *II* and *III*

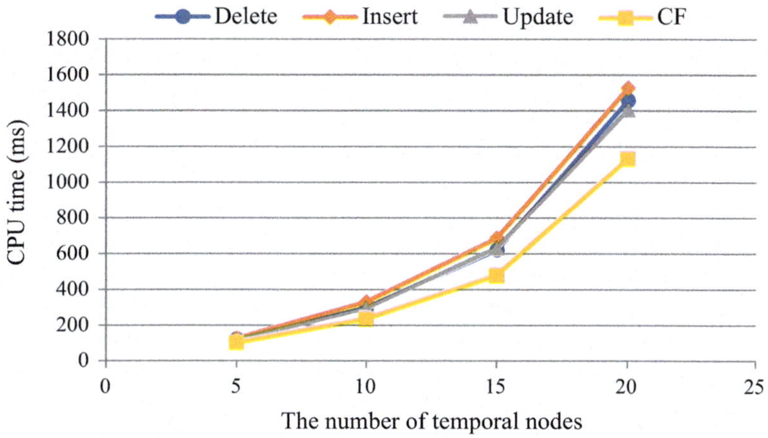

Fig. 2.16 CPU time of fixing inconsistencies of types *I* and *III*

in Fig. 2.17. Furthermore, the *CF* algorithm is capable of achieving a higher level of temporal consistency among data objects. While *CF* exhibits certain limitations when addressing a singular type of consistency, it is generally able to fix all inconsistencies in temporal XML documents with a single traversal. Consequently, we conclude that our *CF* algorithms demonstrate superior performance in addressing inconsistencies when multiple types are present in temporal XML documents.

In the third set of experiments, we conduct a comparative analysis of *TXCF* and our proposed algorithms (*CF*) in the context of temporal XML documents exhibiting various types of inconsistencies (specifically, inconsistencies of type *I*, *II* or *III*). Figures 2.18, 2.19, and 2.20 illustrate the CPU time required by *TXCF* and *DR* for fixing each type of inconsistency in temporal XML documents. As depicted in

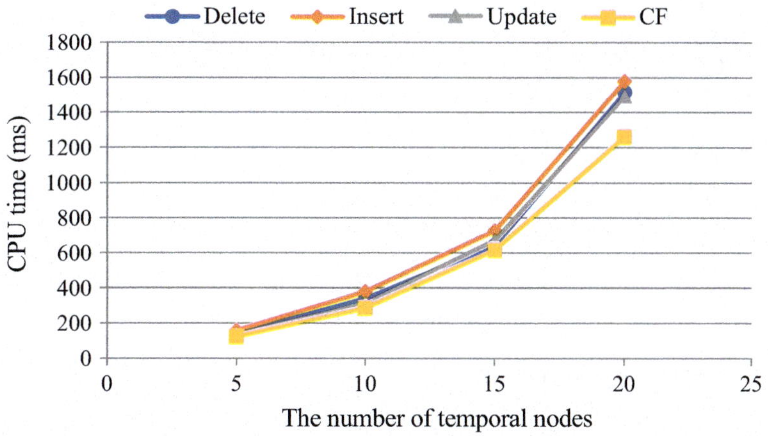

Fig. 2.17 CPU time of fixing inconsistencies of types *I*, *II* and *III*

Fig. 2.18, the CPU time for both our algorithm (*CF*) and *TXCF* increases with the number of temporal XML nodes. However, when the volume of temporal data is held constant, the CPU time for *CF* is comparatively lower. This discrepancy can be attributed to the capability of our algorithms to process all nodes simultaneously within the temporal XML documents, whereas *TXCF* is limited to addressing one node at a time in the presence of inconsistencies. Figure 2.19 presents a comparison of the two algorithms concerning the number of temporal XML nodes and the corresponding CPU time. Notably, *CF* demonstrates a clear advantage over *TXCF* in scenarios involving inconsistencies of type *II*. This superiority can be explained by *CF*'s efficient reduction of the candidate set for resolving inconsistencies, as it operates on two temporal XML nodes concurrently, thereby significantly enhancing performance in rectification. Figure 2.20 details the outcomes associated with inconsistencies of type *III*, where the total CPU time encompasses both the checking and fixing processes. The data in Fig. 2.20 indicates that *CF* utilizes less CPU time when the volume of temporal XML data is equivalent. This efficiency arises from the fact that *CF* generates additional inconsistencies subsequent to *TXCF*'s resolution of inconsistencies of type *III*.

TXCF lacks the capability to accommodate fuzzy and spatial information. When inconsistencies of this nature arise, fixing them requires a considerable number of steps. In contrast, our methodology is designed to support both fuzzy and spatial information, demonstrating a technical advantage in addressing and resolving inconsistencies.

Overall, experimental findings indicate that our algorithm (*CF*) demonstrates superior performances compared to *DR* and *TXCF* in fixing multiple inconsistencies in XML documents.

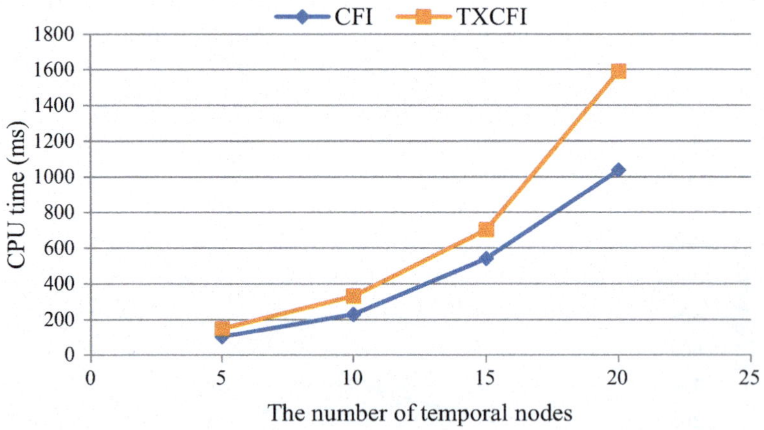

Fig. 2.18 Fixing inconsistency of type *I*

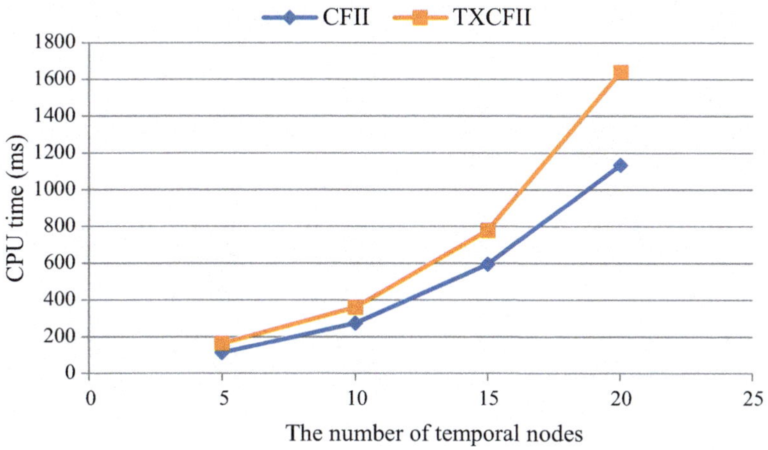

Fig. 2.19 Fixing inconsistency of type *II*

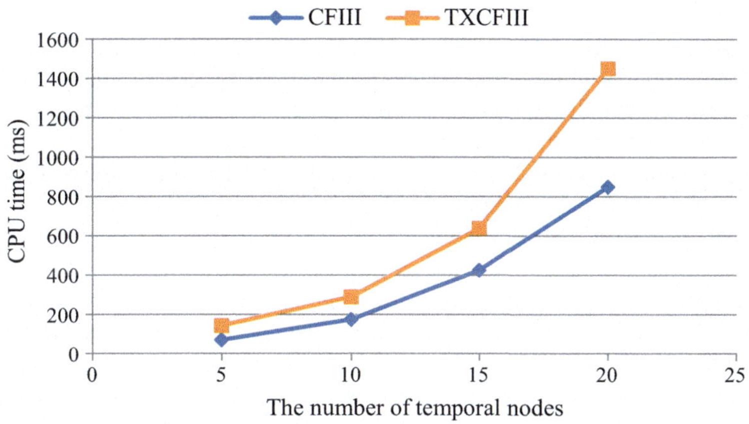

Fig. 2.20 Fixing inconsistency of type *III*

2.6.2.4 Comparison of Checking and Fixing Algorithms Between DOM and dom4j

In this subsection, we employ DOM and the dom4j to implement the algorithms for checking and fixing. Subsequently, we conduct a comparative analysis of the performance of DOM and dom4j, focusing on metrics such as CPU time and memory consumption.

In the initial set of experiments, we conducted a comparative analysis of the CPU time associated with the checking and fixing algorithms utilized in both DOM and dom4j. The results derived from the virtual dataset are presented in Fig. 2.21. Given that the outcomes from the storm dataset exhibit qualitatively similar trends,

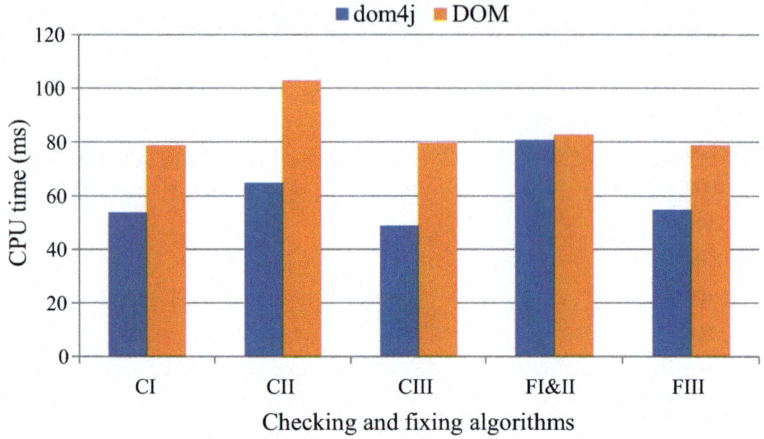

Fig. 2.21 CPU time (*ms*) of dom4j and DOM

the performance metrics from the storm dataset sufficiently illustrate the differences observed. As indicated in Fig. 2.21, the CPU time required for DOM is greater than that of dom4j in the context of the checking and fixing algorithms. This discrepancy can be attributed to the more complex procedures that DOM necessitates for the establishment of a fuzzy spatiotemporal XML document. In contrast, dom4j demonstrates greater flexibility in the creation of such documents, providing a more user-friendly approach to managing the elements of fuzziness, time, and space within fuzzy spatiotemporal XML documents.

In the subsequent series of experiments, we evaluate the memory costs associated with the algorithms for checking and fixing inconsistencies in both DOM and dom4j. Figure 2.22 illustrates the disparity in memory costs between the two frameworks. The findings indicate that the memory expenditure associated with dom4j exceeds that of DOM. This outcome further corroborates our earlier analysis, which posited that the utilization of dom4j entails a greater memory consumption.

In general, dom4j typically exhibits higher memory consumption and lower CPU utilization, which enhances its flexibility and efficiency in identifying and fixing inconsistencies in extensive fuzzy spatiotemporal XML documents. Consequently, we infer that our algorithm for checking and fixing inconsistencies, implemented using dom4j, demonstrates superior performance in addressing multiple inconsistencies in fuzzy spatiotemporal XML documents.

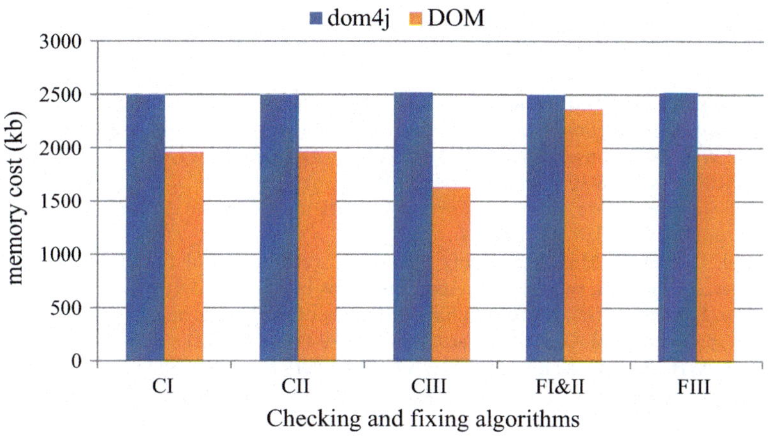

Fig. 2.22 Memory cost (*kb*) of dom4j and DOM

2.7 Summary

In this chapter, we propose a fuzzy spatiotemporal XML data model that accommodates dynamic updates. Building upon this model, we delineate the conditions for inconsistency and present algorithms designed to check and fix inconsistencies in fuzzy spatiotemporal XML documents. Subsequently, we conduct a performance comparison utilizing both a real fuzzy spatiotemporal XML dataset (specifically, a storm dataset) and a virtual fuzzy spatiotemporal XML dataset. The results of these comparisons facilitate a comprehensive discussion regarding inconsistencies. Furthermore, our findings indicate that the dom4j framework demonstrates greater flexibility and efficiency in fixing consistency issues in fuzzy spatiotemporal XML documents.

References

1. X. Chen, C. Zhang, X. Chen, N. Saunier, L. Sun, Discovering dynamic patterns from spatiotemporal data with time-varying low-rank autoregression. IEEE Trans. Knowl. Data Eng. **36**(2), 504–517 (2024)
2. B. Cardone, F.D. Martino, A novel spatiotemporal prediction method based on fuzzy transform: application to demographic balance data. Inf. Sci. **634**, 677–695 (2023)
3. Y. Yin, Y. Sheng, Y. He, J. Qin, Modeling vague spatiotemporal objects based on interval type-2 fuzzy sets. Int. J. Geogr. Inf. Sci. **36**(6), 1258–1273 (2022)
4. L. Bai, N. Li, L. Liu, X. Hao, Querying multi-source heterogeneous fuzzy spatiotemporal data. J. Intell. Fuzzy Syst. **40**(5), 9843–9854 (2021)
5. P. Veena, S.C. Bommisetty, R.U. Kiran, S. Agarwal, K. Zettsu, Discovering fuzzy frequent spatial patterns in large quantitative spatiotemporal databases, in *Proceedings of the 30th International Conference on Fuzzy Systems*, (2021), pp. 1–8

6. A. Sözer, A. Yazici, H. Oğuztüzün, F.E. Petry, "Querying fuzzy spatiotemporal databases: implementation issues, *Uncertainty Approach for Spatial Data Modeling and Processing*, (2010), pp. 97–116
7. E. Stefanakis, A unified framework for fuzzy spatio-temporal representation and reasoning, in *Proceedings of the 20th International Cartographic Conference*, (2001), pp. 2678–2687
8. World Wide Web Consortium (1998) Extensible markup language (XML) 1.0. http://www.w3.org/TR/REC-xml. W3C Recommendation 10 February
9. L. Bai, L. Yan, Z. Ma, Fuzzy spatiotemporal data modeling and operations in XML. Appl. Artif. Intell. **29**(3), 259–282 (2015)
10. L. Bai, L. Yan, Z. Ma, Determining topological relationship of fuzzy spatiotemporal data integrated with XML twig pattern. Appl. Intell. **39**(1), 75–100 (2013)
11. L. Zhu, J. Wang, L. Bai, A general characterization of integrating and querying heterogeneous fuzzy spatiotemporal XML data. Earth Sci. Inf. **16**(4), 3303–3321 (2023)
12. L. Bai, A. He, M. Liu, L. Zhu, Y. Xing, Adaptive query relaxation and result categorization of fuzzy spatiotemporal data based on XML. Expert Syst. Appl. **168**, 1–19 (2021)
13. L. Bai, X. Duan, B. Qin, Adaptive query relaxation and top-*k* result sorting of fuzzy spatiotemporal data based on XML. J. Intell. Fuzzy Syst. **37**(3), 2502–2520 (2022)
14. L. Bertossi, J. Chomicki, Query answering in inconsistent databases. Logics Emerg. Appl. Databases **3**(1), 393–424 (2010)
15. G. Greco, S. Greco, E. Zumpano, A logical framework for querying and repairing inconsistent databases. Knowl. Data Eng. **15**(6), 1389–1408 (2003)
16. W.N. Shuhadah, M.M. Deris, A. Noraziah, M.Y. Saman, M. Rabiei, Database consistency using update-ordering in distributed databases. J. Algorithms Comput. Technol. **1**(1), 17–43 (2007)
17. T.C. Wang, Y.H. Chen, Applying fuzzy linguistic preference relations to the improvement of consistency of fuzzy AHP. Inf. Sci. **178**(19), 3755–3765 (2008)
18. M. Falda, F. Rossi, K.B. Venable, Fuzzy conditional temporal problems: strong and weak consistency. Eng. Appl. Artif. Intell. **21**(5), 710–722 (2008)
19. S.J. Ho, T.W. Kuo, A.K. Mok, Similarity-based load adjustment for real-time data-intensive applications, in *Proceedings of the 18th IEEE Real-Time Systems Symposium*, (1997), pp. 144–153
20. M. Xiong, S. Han, D. Chen, Deferrable scheduling for temporal consistency: schedulability analysis and overhead reduction, in *Proceedings of 12th IEEE International Conference on Embedded and Real-Time Computing Systems and Applications*, (2006), pp. 117–124
21. M. Xiong, K. Ramamritham, Deriving deadlines and periods for real-time update transactions. Computers **53**(5), 567–583 (2004)
22. M. Xiong, Q. Wang, K. Ramamritham, On earliest deadline first scheduling for temporal consistency maintenance. Real-Time Syst. **40**(2), 208–237 (2008)
23. M.G. Del, M.A. Rodríguez, C. Claramunt, Modeling consistency of spatio-temporal graphs. Data Knowl. Eng. **84**(3), 59–80 (2013)
24. J.A. Huisman, L. Breuer, K. Eckhardt, Spatial consistency of automatically calibrated SWAT simulations in the Dill catchment and three of its sub-catchments, in *Proceedings of the 2003 International SWAT Conference*, (2003), pp. 168–173
25. J. Sun, Y.J. Xing, An effective image retrieval mechanism using family-based spatial consistency filtration with object region. Int. J. Autom. Comput. **7**(1), 23–30 (2010)
26. S. Servigne, T. Ubeda, A. Puricelli, A methodology for spatial consistency improvement of geographic databases. GeoInformatica **4**(1), 7–34 (2000)
27. G. Ferrari-Trecate, R. Rovatti, Consistent Sobolev regression via fuzzy systems with overlapping concepts. Fuzzy Sets Syst. **157**(8), 1075–1091 (2006)
28. O. Williams, M. Isard, J. MacCormick, Estimating disparity and occlusions in stereo video sequences, in *Proceedings of IEEE Computer Society Conference on Computer Vision and Pattern Recognition*, (2005), pp. 250–257
29. R. Khoshabeh, S.H. Chan, T.Q. Nguy, Spatio-temporal consistency in video disparity estimation, in *Proceedings of IEEE International Conference on Acoustics, Speech and Signal Processing*, (2011), pp. 885–888

30. L. Bejaoui, Y. Bédard, F. Pinet, Logical consistency for vague spatiotemporal objects and relations, in *Proceedings the 5th International Symposium on Spatial DataQuality*, (2007), pp. 1–8
31. Z. Ma, L. Bai, Y. Ishikawa, L. Yan, Consistencies of fuzzy spatiotemporal data in XML documents. Fuzzy Sets Syst. **343**, 97–125 (2018)
32. L. Bai, Z. Shao, Z. Lin, S. Cheng, Fixing inconsistencies of fuzzy spatiotemporal XML data. Appl. Intell. **47**(1), 257–275 (2017)
33. L.A. Zadeh, Fuzzy sets. Inf. Control. **8**(3), 338–353 (1965)
34. L.A. Zadeh, Fuzzy sets as a basis for a theory of possibility. Fuzzy Sets Syst. **100**(S1), 9–34 (1999)
35. Z. Ma, Fuzzy database modeling with XML, *Advances in Database Systems* vol. 29. (Kluwer, 2005)
36. J. Gong, R. Cheng, D.W. Cheung, Efficient management of uncertainty in XML schema matching. J. Very Large Data Bases **21**(3), 385–409 (2012)
37. C. Xu, Y. Wang, Y. Gu, Efficient fuzzy ranking queries in uncertain databases. Appl. Intell. **37**(1), 47–59 (2012)
38. Summary of Hurricane and Tropical Cyclones named Ingrid in North Atlantic in 2013
39. H. Hamrouni, Z. Brahmia, R. Bouaziz, An efficient approach for detecting and repairing data inconsistencies resulting from retroactive updates in multi-temporal and multi-version XML databases, in Bassiliades, N., et al. *New Trends in Database and Information Systems II. Advances in Intelligent Systems and Computing*, (Springer, 2015)
40. F. Rizzolo, A.A. Vaisman, Temporal XML: modeling, indexing, and query processing. Int. J. Very Large Data Bases **17**(5), 1179–1212 (2008)

Chapter 3
Transformation of Fuzzy Spatiotemporal Data Between XML and Object-Oriented Databases

3.1 Introduction

In the contemporary domain of computer science, the necessity for interoperability among autonomous databases has resulted in the emergence of multi-database systems, which may comprise either heterogeneous or homogeneous databases. Consequently, the ability to transform operations between databases that utilize distinct data models is essential in any heterogeneous multi-database system [1–5]. On the other hand, the swift advancement of the Internet has facilitated the emergence of XML (eXtensible Markup Language) as the prevailing standard for data representation and exchange on the Web, attributed to its simplicity, readability, and portability. Concurrently, object-oriented databases, which possess numerous advantageous features, have seen an increasing application across various domains. These databases are particularly well-suited for applications that necessitate highly structured data, extended transaction durations, and specialized data types for the storage of images and textual information. Consequently, numerous research endeavors are increasingly concentrating on data models that utilize XML [6–9] and object-oriented databases [10–13].

The utilization of XML and object-oriented databases is advantageous for effectively managing the fuzzy characteristics and spatiotemporal attributes of fuzzy spatiotemporal objects. This necessity arises from the demand for platform-independent exchange of content within object-oriented databases, as well as the requirement to store XML data in a structured database format. Consequently, significant efforts have been directed towards the transformation of fuzzy spatiotemporal data between XML and object-oriented databases to address this issue. On one hand, the transformation from XML to object-oriented databases is particularly appealing due to the shared characteristics between the object-oriented model and XML, which facilitates a more efficient mapping process. This mapping approach ensures the preservation of database-specific features. Naser et al. [14] present a methodology for reengineering XML into an object-oriented database framework. They assert

L. Bai and L. Zhu, *Fuzzy Spatiotemporal XML Data Management*, Studies in Computational Intelligence 1183, https://doi.org/10.1007/978-3-031-81033-6_3

that the object graph, which is derived from the schema's characteristics, can effectively encapsulate and summarize both complex and simple elements, as well as their interconnections. This approach allows for the direct representation of all components within a simulated object graph in an object-oriented database, thereby facilitating platform-independent content exchange among object-oriented databases. Liu et al. [15] address the challenge of migrating fuzzy XML to fuzzy object-oriented databases, emphasizing the necessity of processing imprecise and uncertain information prevalent in real-world applications. To identify the object-oriented schema that most accurately reflects the existing fuzzy XML schema (DTD), Naser et al. [16] propose a strategy based on a set of mapping rules. Such reengineering efforts not only enhance interoperability between fuzzy object-oriented databases and fuzzy XML modeling techniques but also contribute to the development of practical design methodologies for fuzzy object-oriented databases. The mapping process involves deriving the object graph from the XML schema's characteristics, which succinctly summarizes and incorporates all complex and simple elements along with their links, fundamental to the XML schema. Subsequently, these links are simulated through nesting to create a simulated object graph, ensuring that all elements within this graph can be directly represented in an object-oriented database. In [17], Bai et al. investigate the transformation of fuzzy spatiotemporal data from XML to object-oriented databases by proposing a set of transformation rules.

On the other hand, in the context of the transformation from object-oriented databases to XML, Terwilliger et al. [18] propose a framework that facilitates the conversion of object-based queries and updates into their XML counterparts through the application of flexible, declarative mappings between classes and XML schema types. These mappings are also employed to decompose XML fragments derived from query results into client-side objects. Naser et al. [19] present a methodology that utilizes an object graph based on the object-oriented model, which effectively summarizes and incorporates all nesting and inheritance relationships, thereby enabling the transformation of an existing object-oriented database into XML. This approach capitalizes on the shared characteristics of the object-oriented model and XML, particularly in managing complex data and relationships, to facilitate the transformation of common data and straightforward relationships. In [16], the object-oriented schema is represented through an object graph that encapsulates all nesting and inheritance links, fundamental components of the object-oriented model. Inheritance is then simulated through nesting to create a simulated object graph, ensuring that all elements within this graph can be directly represented in XML format. Liu and Ma [20] develop a formal transformation process from fuzzy object-oriented databases to fuzzy XML, utilizing mapping formalisms that encapsulate the semantics of both fuzzy XML Schema and fuzzy object-oriented schema. Additionally, the research conducted by Bai et al. [21] explores the transformation rules applicable to spatiotemporal data transitioning from fuzzy object-oriented databases to XML.

In this chapter, we investigate the methodologies for transforming fuzzy spatiotemporal data between XML and object-oriented databases. We initiate our discussion by delineating the representation of fuzzy spatiotemporal data within the frameworks of XML and object-oriented databases. Subsequently, we formulate a comprehensive set of rules designed to facilitate the transformation of fuzzy spatiotemporal data between XML and object-oriented databases.

3.2 Preliminaries

3.2.1 Characteristics of XML

XML is classified as a meta-markup language utilized for the purpose of data description. The term "markup" pertains to the arrangement of characters or symbols inserted at specific locations within a text or word processing document. XML data is characterized as self-describing or self-defining, indicating that the data's structural information is inherently included within the data itself. Consequently, there is no requirement to establish a predefined structure for data storage upon its arrival. XML serves as a straightforward and adaptable descriptive text markup format, enabling the definition of data structures, facilitating comprehension of their meaning and context, and offering methodologies for articulating the semantics and structure of information. Structured information encompasses both content and an indication of the role that content fulfills. A markup language functions as a mechanism for identifying structures within a document. The XML specification delineates a standardized approach for incorporating markup into documents.

The XML standard provides a versatile framework for the creation of information formats and the electronic exchange of structured data over the public Internet. XML is essentially a more straightforward and user-friendly subset of the Standard Generalized Markup Language (SGML), designed specifically for Internet applications. The fundamental component of an XML document is the element, which is delineated by tags. Each element comprises a beginning and an ending tag, and all elements within an XML document are encapsulated within a singular outermost element referred to as the root element. Furthermore, XML accommodates nested elements, allowing for the inclusion of elements within other elements, thereby facilitating the representation of hierarchical structures. The names of elements serve to convey the content they encapsulate, while the overall structure illustrates the interrelationships among the elements. Additionally, XML permits the definition of attributes for elements, enabling the specification of characteristics within the start tag of an element. The efficacy of XML lies in its inherent simplicity, which enables the aggregation of substantial amounts of information into a cohesive XML document.

3.2.2 Characteristics of Object-Oriented Database

Object-oriented databases integrate traditional database technologies with object-oriented methodologies. The foundational principle of the object-oriented paradigm is centered around the concepts of classes and objects, which are crucial for the design of database systems. In this model, entities from the real world are conceptualized as objects. Each object is assigned a unique object identifier, which serves to distinctly identify the object throughout its existence and remains constant over time. This object identifier is determined by the system and is independent of the object's value. Furthermore, each object possesses a collection of properties that represent its state, as well as a variety of methods that characterize its behavior. These properties and methods are encapsulated within the object itself. A class functions as a blueprint that defines the variables and methods associated with a specific type of object. It embodies the abstraction of a category of entities that share common characteristics in the real world; thus, a class signifies a collection of objects, while an object is a specific instance of that class.

An object-oriented database schema comprises a collection of classes, which are organized within an object-oriented data model that exhibits a hierarchical structure. Within this schema, a set of classes can establish a class hierarchy, and an object-oriented database may encompass multiple such hierarchies. In a class hierarchy, a class inherits all properties, methods, and messages from its superclass(es). The concept of object nesting is a significant aspect of object-oriented database systems, enabling various users to perceive objects at differing levels of granularity. The interplay between object-nested hierarchies and class hierarchical structures contributes to the formation of intricate horizontal and vertical frameworks.

3.2.3 Membership Function

Membership functions were first introduced by Zadeh [22, 23]. Within the framework of fuzzy set theory, Zadeh employs membership functions to analyze the entire range of potential values. The membership function associated with a fuzzy set serves as a generalization of the indicator function found in classical set theory. In the context of fuzzy logic, this function signifies the degree of truth, extending the concept of valuation. It is important to note that degrees of truth are frequently conflated with probabilities; however, they are fundamentally distinct concepts. Fuzzy truth pertains to the membership within imprecisely defined sets, rather than the likelihood of a specific event or condition occurring.

For any given set A, a membership function on A is any function from A to the real unit interval [0, 1]. These membership functions characterize fuzzy subsets of A. The membership function which represents a fuzzy set $\tilde{A}$ is usually denoted by μ_A. For an element x of A, the value $\mu_A(x)$ is referred to as the membership degree of x in the fuzzy set $\tilde{A}$. The membership degree $\mu_A(x)$ quantifies the grade of membership of the element x to the fuzzy set $\tilde{A}$. A membership degree of 0 indicates that x is not a member of the fuzzy set, while a degree of 1 signifies that x is fully included in the fuzzy set. Values that fall between 0 and 1 denote fuzzy members, which are only partially included in the fuzzy set. In the context of fuzzy set theory, traditional bivalent sets are commonly referred to as crisp sets. Fuzzy set theory conceptualizes set membership as a possibility distribution, which is a function that can be expressed through a mathematical formula. Specifically, the value of a possibility distribution is represented as a crisp real number within the interval [0, 1].

3.3 Fuzzy Spatiotemporal Data Model

3.3.1 Fuzzy Spatiotemporal Data Model Based on XML

In light of the attributes associated with fuzzy spatiotemporal data, we put forth a definition for the fuzzy spatiotemporal XML data model.

Definition 3.1 *FSTDM* (*Fuzzy spatiotemporal XML data model*). Fuzzy spatiotemporal XML data is a five-tuple $FSTDM = (KAT, DAT, FAT, FSP, FTM)$, where

- *KAT* is key attribute of spatiotemporal data.
- *DAT* is deterministic general attributes of spatiotemporal data.
- *FAT* is fuzzy general attributes of spatiotemporal data.
- *FSP* is fuzzy spatial attributes of spatiotemporal data.
- *FTM* is fuzzy temporal attributes of spatiotemporal data.

In Definition 3.1, *KAT* is characterized as a spatiotemporal entity that distinguishes itself from other types of objects. The transformations of *KAT* illustrate the process by which spatiotemporal entities evolve into different spatiotemporal forms. *DAT* is employed to articulate deterministic and static characteristics, whereas *FAT* is utilized to convey fuzzy and static attributes, as opposed to dynamic attributes such as *FSP* and *PST*. Spatiotemporal data may encompass one or more general attributes, which can be either deterministic or fuzzy in nature. Furthermore, *FSP* refers to fuzzy spatial attributes that encompass both fuzzy spatial position and fuzzy spatial motion. Depending on the specific application context, fuzzy spatial position can be categorized into fuzzy spatial points, fuzzy spatial lines, and fuzzy spatial regions. For the representation of fuzzy spatial points, a collection of points is utilized, each associated with varying degrees of possibility. In the case of fuzzy spatial lines, two points are initially employed to depict a straight line or an approximate straight line, with an offset applied to account for the fuzzy nature of the count. Fuzzy spatial regions are represented using the Minimum Bounding Rectangle (MBR), allowing for the depiction of fuzzy regions through two points. Regarding fuzzy motion direction, each potential direction is assigned a probability value, while fuzzy motion values are represented by an interval. Lastly, *FTM* encompasses fuzzy temporal attributes, which include fuzzy time points and fuzzy time intervals. The former is defined as a representation of a set of potential time points, each associated with a degree of possibility for the occurrence of an event, while the latter is characterized as a representation of a set of potential time intervals, also accompanied by a degree of possibility.

In summary, *FSTDM* encompasses two distinct forms of fuzziness. The first one is the fuzziness in attribute values of elements such as *FAT*, which is represented through possibility distributions. The second one is the fuzziness in elements such as *FSP* and *FTM*. To characterize these fuzzy elements within XML, we employ self-defined tags and data types. It is evident that the attribute values of fuzzy elements may also exhibit fuzziness.

In an XML document, it is essential to provide specific information. To achieve a more generalized structure, we employ XML Schema to define the element in *FSTDM*. Compliance with the stipulations of XML Schema, which encompasses fuzzy spatiotemporal information, is imperative for XML documents. Our analysis of the schema is conducted from the perspectives of *KAT*, *DAT*, *FAT*, *FSP*, and *FTM*, as illustrated in Fig. 3.1. Regarding the schema of *KAT*, the element designated as "*Elem*" is characterized by the *KAT* attribute, which is a requisite for the *Elem* element. In the context of the *DAT* schema, the term "*original-definition*" serves as a generic label that can be substituted with a specific data type in practical applications. Concerning the *FAT* component, the fuzzy value may either be a real number within the interval [0, 1] or a distribution function. We define "*Val*" to specify the type of fuzzy value, "*Poss*" to represent values within the range of [0, 1], and "*Func*" to denote the distribution function. With respect to the *FSP* component, several fuzzy elements exist, each of which must be represented by an appropriate XML Schema. The *FTM* component may manifest in the other four components. Notably, both fuzzy time points and fuzzy time intervals are classified as fuzzy elements, and their types are designated as "*xs:dataTime*," which is already established within XML Schema.

For each class *C* in *OOSDM* schema *S*, an element with the same name as *C* is created and then positioned under the root element. Fuzzy spatiotemporal XML documents can be categorized into two types: flat documents and nested documents. A flat fuzzy spatiotemporal XML document is characterized by its composition of simple elements, which serve as sub-elements (attributes) of the root element. In contrast, nested fuzzy spatiotemporal XML documents exhibit greater complexity, as they encapsulate relationships among various root elements. Furthermore, a fuzzy spatiotemporal XML document can be naturally represented as an ordered rooted tree. Consequently, we employ XML trees to illustrate both flat and nested fuzzy spatiotemporal XML documents.

Definition 3.2 For flat fuzzy spatiotemporal XML data tree, we have *FSTDT* = (*R*, *E*, *GA*, *SP*, *TM*, *VAL*, *DIS*, *TEX*), where

- *R* is root node in *FSTDT*. The root node serves as a virtual representation of a fuzzy spatiotemporal object, while all of its child nodes correspond to the intrinsic attributes associated with that object.

Schema of KAT

```
<xs:element name="Elem" minOccurs="0"
maxOccurs="unbounded">
 <xs:complexType>
  <xs:attribute name="KAT" type="xs:ID"
use="required"/>
 </xs:complexType>
</xs:element>
```

Schema of FAT (fuzzy value is a real number in interval [0, 1])

```
<xs:element name="FAT" minOccurs ="0"
maxOccurs ="unbounded">
 <xs:complexType>
  <xs:element name ="Val"/>
 </xs:complexType>
</xs:element>
<xs:element name="Val" minOccurs ="0"
maxOccurs ="unbounded">
 <xs:complexType>
  <xs:attribute name ="Poss" type ="xs:fuzzy"
default="1.0"/>
 </xs:complexType>
</xs:element>
```

Schema of FAT (fuzzy value is a distribution function)

```
<xs:element name="FAT" minOccurs ="0"
maxOccurs ="unbounded">
 <xs:complexType>
  <xs:element name ="Val"/>
 </xs:complexType>
</xs:element>
<xs:element name="Val" minOccurs ="0"
maxOccurs ="unbounded">
 <xs:complexType>
  <xs:attribute name ="Func" type
="xs:string"/>
 </xs:complexType>
</xs:element>
```

Schema of FSP (fuzzy motion direction)

```
<xs:element
name="MotionDirection">
 <xs:complexType>
  <xs:sequence>
   <xs:element name="degree"
type="xs:integer"/>
   <xs:element name ="Val"/>
  </xs:sequence>
 </xs:complexType>
</xs:element>
<xs:element name="Val" minOccurs ="0"
maxOccurs ="unbounded">
 <xs:complexType>
  <xs:attribute name ="Poss" type ="xs:fuzzy"
default="1.0"/>
 </xs:complexType>
</xs:element>
```

Schema of FSP(fuzzy spatial point)

```
<xs:element name="SpatialPoint">
 <xs:complexType>
  <xs:sequence>
   <xs:element name="No." type="xs:integer"/
>
   <xs:element ref="point"/>
  </xs:sequence>
 </xs:complexType>
</xs:element>
<xs:element name="point" minOccurs="1"
maxOccurs="unbounded">
 <xs:complexType>
  <xs:sequence>
   <xs:element name="Xaxis" type="
xs:integer"/>
   <xs:element name="Yaxis" type="
xs:integer"/>
  </xs:sequence>
 </xs:complexType>
</xs:element>
```

Schema of FSP(fuzzy spatial line)

```
<xs:element name="SpatialLine">
 <xs:complexType>
  <xs:sequence>
   <xs:element name="No." type="xs:integer"/
>
   <xs:element ref="point "/>
  </xs:sequence>
  <xs:attribute name="offset"
type="xs:integer"/>
 </xs:complexType>
</xs:element><xs:element name="point"
minOccurs="2" maxOccurs="2">
 <xs:complexType>
  <xs:sequence>
   <xs:element name="Xaxis"
type="xs:integer"/>
   <xs:element name="Yaxis"
type="xs:integer"/>
  </xs:sequence>
 </xs:complexType>
</xs:element>
```

Schema of FSP (fuzzy motion value)

```
<xs:element name="MotionValue">
 <xs:simpleType>
  <xs:restriction base="xs:integer">
   <xs:minInclusive value="valueStart"/>
   <xs:maxInclusive value="valueEnd"/>
  </xs:restriction>
 </xs:simpleType>
</xs:element>
```

Schema of DAT

```
<xs:element name="Elem" minOccurs ="0"
maxOccurs ="unbounded">
 <xs:complexType>
  <xs:element name ="DAT" type ="original-
definition"/>
 </xs:complexType>
</xs:element>
```

Schema of FSP (fuzzy spatial region)

```
<xs:element name="SpatialRegion">
 <xs:complexType>
  <xs:sequence>
  <xs:element name="No." type="xs:integer"/
>
   <xs:element ref="point"/>
  </xs:sequence>
 </xs:complexType>
</xs:element>
<xs:element name="point" minOccurs="2"
maxOccurs="2">
 <xs:complexType>
  <xs:sequence>
   <xs:element name="Xaxis"
type="xs:integer"/>
   <xs:element name="Yaxis"
type="xs:integer"/>
  </xs:sequence>
 </xs:complexType>
</xs:element>
```

Schema of FSP(fuzzy time point)

```
<xs:element name="TimePoint ">
 <xs:restriction base="dateTime">
  <xs:minInclusive value="valueStart"/>
  <xs:maxInclusive value="valueEnd"/>
 </xs:restriction>
</xs:element>
```

Schema of FSP (fuzzy time interval)

```
<xs:element name="TimeInterval">
 <xs:complexType>
  <xs:sequence>
    <xs:element name="timeStart"/>
    <xs:element name="timeEnd"/>
   </xs:sequence>
 </xs:complexType>
</xs:element>
<xs:element name="timeStart"
minOccurs="0" maxOccurs="unbounded">
 <xs:simpleType>
  <xs:restriction base="dateTime">
   <xs:minInclusive value="valueStart"/>
   <xs:maxInclusive value="valueEnd"/>
   </xs:restriction>
 <xs:simpleType>
</xs:element>
<xs:element name="timeEnd" minOccurs="0"
maxOccurs="unbounded">
 <xs:simpleType>
  <xs:restriction base="dateTime">
   <xs:minInclusive value="valueStart"/>
   <xs:maxInclusive value="valueEnd"/>
   </xs:restriction>
 <xs:simpleType>
</xs:element>
```

Fig. 3.1 Schema of *FSTDM*

- *E* is element node in *FSTDT*. An element node serves to represent an element within a flat fuzzy spatiotemporal XML document. Each element node is associated with several other nodes that function as its child nodes.
- *GA* is general attribute node in *FSTDT*. It may be crisp or fuzzy.
- *SP* is spatial attribute node in *FSTDT*. There exist five distinct categories of spatial attributes: fuzzy spatial point, fuzzy spatial line, fuzzy spatial region, fuzzy motion direction, and fuzzy motion value.
- *TM* is temporal attribute note in *FSTDT*. Fuzzy time points and fuzzy time intervals are classified as components of the temporal attribute node.
- *VAL* is fuzzy value node in *FSTDT. Func* and *Poss* are two child nodes of *VAL*.
- *TEX* is leaf node. It is value of its parent node. Data type of *TEX* may be *dataTime*, integer, and string, etc.

Definition 3.3 For nested fuzzy spatiotemporal XML data tree, we have *NSTDT* = (R_1, R_2, *RE*, *E*, *GA*, *SP*, *TM*, *VAL*, *DIS*, *TEX*), where

- R_1 and R_2 are two different root nodes in *NSTDT*.

- *RE* serves as a relationship node that can be utilized to depict the relationship between R_1 and R_2.
- The meaning of *RE*, *E*, *GA*, *SP*, *TM*, *VAL*, *DIS*, *TEX* are defined in Definition 3.2.

There exist three distinct relationships among various root nodes, namely fuzzy association, fuzzy inheritance, and fuzzy aggregation. Among these relationships, association is the most prevalent, while semantic association is characterized as the weakest. Fuzzy association solely denotes a dependency relationship in a semantic context and does not convey the directionality of this dependency (typically, associations are bidirectional). Furthermore, the relationships within associations can manifest as one-to-one, one-to-many, or many-to-many. In the context of XML trees, associations are represented through links, which serve as physical conceptual connections between root nodes. In practical applications, these relationships can be effectively identified through class involvement characterized by *n*:*m* relationships, where *n* and *m* are constrained by cardinality. Inheritance denotes a general/special relationship; specifically, if all attributes of root element *A* are present in root element *B*, while some attributes of root element *B* are not found in root element *A*, then element *A* is classified as a general element, and element *B* as a special class. Aggregation, on the other hand, signifies a whole/part relationship, with fuzzy aggregations indicating that one element constitutes a part of another element. The XML Schema illustrating fuzzy relationships is presented in Fig. 3.2.

In the schema of fuzzy association, the relationship between element *R* and *LR* is fuzzy association. To denote "*fuzzyAssociation*", the "*extension*" tag is employed to establish connection between *R* and *LR*. Element "*involvement*" denotes multiplicity of association, while attribute "*Poss*" indicates membership degree between *R* and *LR*. For the schema of fuzzy inheritance, "*GER*" and "*SPR*" are utilized to represent

Schema of fuzzy association	Schema of fuzzy inheritance	Schema of fuzzy aggregation
<xs:element name="R" minOccurs ="0" maxOccurs ="unbounded"> <xs:complexType> <xs:element name="Ri" type="original-definition"/> </xs:complexType> </xs:element> <xs:element name="LR" minOccurs ="0" maxOccurs ="unbounded"> <xs:complexType> <xs:element name ="fuzzyAssociation" > <xs:complexType> <xs:extension base= "R"/> <xs:element name="involvement" type="string"> <xs:element name="Val"> <xs:complexType> <xs:attribute name ="Poss" type ="xs:fuzzy" default="1.0"/> </xs:complexType> </xs:element> </xs:complexType> </xs:element> <xs:element name="LRi" type="original-definition"/> </xs:complexType> </xs:element>	<xs:element name="GER" minOccurs ="0" maxOccurs ="unbounded"> <xs:complexType> <xs:element name="GERi" type="original-definition"/> </xs:complexType> </xs:element> <xs:element name="SPR" minOccurs ="0" maxOccurs ="unbounded"> <xs:complexType> <xs:element name ="fuzzyInheritance" > <xs:complexType> <xs:extension base= "GER"/> <xs:element name="Val"> <xs:complexType> <xs:attribute name ="Poss" type ="xs:fuzzy" default="1.0"/> </xs:complexType> </xs:element> </xs:complexType> </xs:element> <xs:element name="SPRi" type="original-definition"/> </xs:complexType> </xs:element>	<xs:element name="PR" minOccurs ="0" maxOccurs="unbounded"> <xs:complexType> <xs:element name="PRi" type="original-definition"/> </xs:complexType> </xs:element> <xs:element name="WR" minOccurs ="0" maxOccurs="unbounded"> <xs:complexType> <xs:element name ="fuzzyAggregation" > <xs:complexType> <xs:element ref="PR"/> <xs:element name="Val"> <xs:complexType> <xs:attribute name="Poss" type="xs:fuzzy" default="1.0"/> </xs:complexType> </xs:element> </xs:complexType> </xs:element> <xs:element name="WRi" type="original-definition"/> </xs:complexType> </xs:element>

Fig. 3.2 Schema of fuzzy relationship

general element and special element, respectively. GER_i is sub-element of *GER* and SPR_i is sub-element of SPR_i. In schema of fuzzy aggregation, *PR* is a part element, and *WR* is a whole element. The "*ref*" tag serves to identify the relationship between *PR* and *WR*, with "*Poss*" similarly representing the membership degree.

3.3.2 Fuzzy Spatiotemporal Data Model Based on Object-Oriented Database

The fuzzy spatiotemporal data model, which is grounded in object-oriented database principles and referred to as *OOSDM*, is characterized by its representation of objects, classes, and relationships. In this context, an object signifies an element within the domain of interest, a class represents a collection of objects that share common attributes, and a relationship illustrates the connections among various classes. *OOSDM* serves as a class-based representation framework that facilitates the expression of diverse types of relationships, such as associations, aggregations, and generations, as well as constraints, including subclass and superclass constraints, that are pertinent to a range of applications.

Objects possess properties that can be categorized as either attributes (or methods) intrinsic to the object or as relationships between the object and one or more other objects. An object is considered fuzzy due to the presence of incomplete or uncertain information. Formally, an object qualifies as a fuzzy object if it contains at least one attribute whose value is represented as a fuzzy set. A class is deemed fuzzy if it includes fuzzy domains for certain attributes (or methods), or if it is a subclass derived from a fuzzy class through specialization. Similarly, a superclass that encompasses one or more fuzzy classes through generalization is also classified as fuzzy. Furthermore, the relationships among classes or objects may exhibit fuzziness. Consequently, we propose four distinct levels of fuzziness applicable to classes in *OOSDM*. These four levels of fuzziness are depicted as follows:

- At the initial level, classes and attribute sets of class may be fuzzy, indicating that they have a possibility to the model.
- The second level is related to the fuzzy occurrences of objects.
- The third level concerns the fuzzy values associated with the attributes (or methods) of special objects.
- The fourth level is about the relationships among various classes.
- Furthermore, *OOSDM* incorporates fuzzy spatial and temporal attributes. The fuzziness associated with spatiotemporal attributes is classified as the third level of fuzziness. A detailed representation of these fuzzy spatiotemporal attributes can be found in Table 3.1.

To effectively model the initial level of fuzziness, which pertains to an attribute or class characterized by a degree of possibility, it is essential to append the attribute or class name with the phrase "*WITH mem DEGREE*." Here, the variable *mem*,

Table 3.1 The representations of fuzzy spatiotemporal attributes

Fuzzy spatiotemporal attribute	The representation in *OOSDM*
Fuzzy time point	Fuzzy *TimePoint*
Fuzzy time interval	Fuzzy *TimeInterval*
Fuzzy spatial point	Fuzzy *SpatialPoint*
Fuzzy spatial line	Fuzzy *SpatialLine*
Fuzzy spatial region	Fuzzy *SpatialRegion*
Fuzzy motion direction_1	Fuzzy *Direction_1*
Fuzzy motion direction_2	Fuzzy *Direction_2*
Fuzzy motion value	Fuzzy *MotionValue*

constrained within the range of 0 to 1, serves as a scalar representation of the extent to which the attribute is associated with the class or the class is associated with the data model. Regarding the second level of fuzziness, it is necessary to specify the likelihood that an object belongs to a particular class. To facilitate this, an additional attribute is incorporated into the class to denote the degree of membership of the object to the class, with the attribute domain defined as [0, 1]. This specialized attribute is referred to as μ in this chapter. For the third level of fuzziness, the model introduces the keyword *FUZZY*. The fourth level addresses the relationships among various classes, which encompass fuzzy generations, fuzzy aggregations, and fuzzy associations. To represent these fuzzy relationships, we utilize a relationship membership degree, also confined to the domain of [0, 1].

3.4 Transformation of Fuzzy Spatiotemporal Data from XML to Object-Oriented Database

The fuzzy spatiotemporal XML schema is conceptualized as a structured tree that encompasses all necessary information. The fuzzy object-oriented spatiotemporal data model (*FOSDM*) is articulated through a class definition. In this subsection, we propose a methodology that leverages various types of nodes within the XML tree to facilitate the transformation of fuzzy spatiotemporal XML data into object-oriented databases.

The transformation architecture is illustrated in Fig. 3.3. Within the XML tree, there exist three categories of nodes: the root element, non-leaf elements, and leaf elements. In the context of this chapter, non-leaf elements are characterized by the inclusion of general attributes, spatial attributes, and temporal attributes. Conversely, leaf elements may exhibit either fuzzy or deterministic characteristics. Therefore, the objective is to develop transformation rules applicable to the various node elements.

A fuzzy spatiotemporal XML schema, referred to as *F*, should encompass a root element, a non-leaf element that incorporates a key attribute, general attributes, relationship attributes, and spatiotemporal attributes, as well as a leaf element

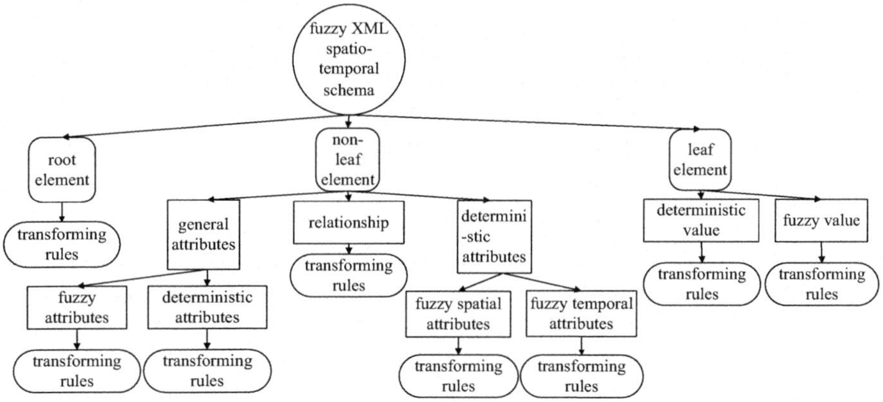

Fig. 3.3 Transforming architecture from XML to object-oriented database

whose values may be either deterministic or fuzzy. To facilitate the conversion of fuzzy spatiotemporal data from XML to an object-oriented database, the specific transformation rules are outlined in detail below.

Rule 3.1 In relation to the fuzzy spatiotemporal XML *F*, a class is established that shares the same designation as the root element of *F*.

Rule 3.2 Regarding the key attribute *KAT* in XML *F*, an attribute bearing the same designation as *KAT* is established as the object identifier (*OID*) for class F.

Rule 3.3 For each deterministic non-leaf element *DAT* in XML *F*, an attribute or method bearing the same designation as *DAT* is generated and incorporated into class *F*.

When all root elements and non-leaf elements exhibit deterministic characteristics, it is possible to apply Rules 3.1 to Rule 3.3 to convert data from XML Schema to *FOSDM* schema. Within XML, two types of fuzziness can be identified: fuzziness associated with elements and fuzziness related to the attribute values of those elements. For attributes that exhibit fuzziness, the subsequent rules may be employed.

Rule 3.4 Regarding to the fuzziness associated with leaf element, if the fuzzy value is a real number within the interval [0, 1], an attribute sharing the same name as *FVR* followed by a pair of words *WITH Poss*, where $0 \leq Poss \leq 1$ is a scalar and indicates the likelihood of the attribute is created.

Rule 3.5 Regarding to the fuzziness in leaf element, if the fuzzy value is characterized as a distribution function, a keyword *WITH Func* followed by an attribute with the same name as *FVF* is generated and incorporated within class *F*. *Func* represents a distribution function that can be expressed by mathematical formula.

Given that the focus of our research pertains to fuzzy spatiotemporal data, the aforementioned rules are employed for data generalization. As a result, we will discuss fuzzy spatial nodes and fuzzy temporal nodes. Despite all being classified as fuzzy non-leaf elements, their representations exhibit distinct characteristics.

With regard to fuzzy spatial nodes, the subsequent rules will be implemented for fuzzy spatial point nodes, fuzzy spatial line nodes, fuzzy spatial region nodes, fuzzy motion direction, and fuzzy motion value.

Rule 3.6 In our model, the non-leaf element of fuzzy spatial point node is characterized as fuzzy, in accordance with the specifications outlined in Definition 3.1. Then, a fuzzy spatial point attribute with the same name as *SpatialPoint* following a keyword *Fuzzy* is created and incorporated within class *F*.

Rule 3.7 In the context of the fuzzy spatial line within non-leaf elements, the XML Schema delineates that the representation of a fuzzy spatial line is achieved through a distribution function. Consequently, a keyword *Fuzzy* is introduced, accompanied by an attribute for the fuzzy spatial line that shares the same designation as *Spatialline*, and this is incorporated within class *F*.

Rule 3.8 Regarding the fuzzy spatial region node in XML, it is constructed based on the MBR model of fuzzy spatial regions, which allows for the representation of fuzzy spatial lines within an XML Schema. Consequently, a fuzzy spatial region attribute with the same name as *SpatialRegion* following a keyword *Fuzzy* is created and incorporated within class *F*.

Rule 3.9 Regarding the direction of fuzzy element motion, it is represented as a distribution function. Consequently, a fuzzy spatial region attribute with the same name as *MotionDirection* following a keyword *Fuzzy* is created and incorporated within class *F*.

Rule 3.10 Regarding to fuzzy spatial point node *MotionValue*, its distribution function owns both a maximum and a minimum. Then a keyword *Fuzzy* followed by a fuzzy spatial line attribute with the same name as *MotionValue* is created and incorporated within class *F*.

Rule 3.11 For fuzzy time point node in XML *F*, the value associated with a fuzzy time point node is inherently non-deterministic. To represent this value, we employ a distribution function. Consequently, a fuzzy time point attribute with the same name as *TimePoint* following a keyword *Fuzzy* is created and incorporated within class *F*.

Rule 3.12 The representation of the fuzzy time interval node in XML *F* is derived from Definition 3.1. According to the XML definition, the *TimeInterval* representation employs a distribution function. As a result, a keyword *Fuzzy* followed by a fuzzy time interval attribute with the same name as *TimeInterval* is created and incorporated within class *F*.

Rule 3.13 For all non-temporal and non-root nodes that possess temporal sub-nodes, it is necessary to combine the non-temporal attributes with their corresponding temporal attributes. In the *FOSDM* schema, these attributes are represented by a small dashed-outline rectangle.

In the following, we will present the transforming rules of the relationships, including fuzzy generalization, fuzzy association and fuzzy aggregations.

Rule 3.14 Regarding to a root node R_i, if it has a child node named *fuzzyAssociation*, the "*extension*" tag marks an element R_j which is a root element in other XML. Particularly, two classes with the same names as R_i and R_j are created firstly, and a fuzzy association named *fuzzyAssociation* on an *n*: *m* basis with element "*involvement*" is then created. Additionally, an element named "*Val*" is utilized to depict the degree of membership in fuzzy nested XML. It is advisable for us to contribute collectively alongside the *FuzzyAssociation.*

Rule 3.15 Concerning a root node R_i, if it has a child node named *fuzzyInheritance*, a sub-element of *fuzzyInheritance* is a tag "*extension*" marking "*base* = R_j". Specifically, a subclass with the same names as R_i is created firstly, followed by the creation of a super class named R_j. Furthermore, fuzzy inheritance relationship named *fuzzyInheritance* is created when transforming fuzzy XML in fuzzy object-oriented (*OO*) spatiotemporal database. Additionally, an element named "*Val*" is utilized to denote the degree of membership in fuzzy nested XML. It is recommended that this element be considered in conjunction with the *fuzzyInheritance.*

Rule 3.16 When considering a root node R_i in fuzzy XML, if it contains a child node named *fuzzyAggregation*, a sub-element of *fuzzyAggregation* is an element "*ref* = R_j" (potentially multiple references sub-element R_j, such as "*ref* = R_j", "*ref* = R_k"). Specifically, all of classes with the same name as R_i, R_j and R_k is created. Subsequently, a fuzzy aggregation among classes R_i, R_j and R_k is created. Finally, a sub-element of *fuzzyAggregation* is utilized to indicate the degree of membership in fuzzy nested XML. We should donate it together with the *fuzzyAggregation.*

In the fuzzy spatiotemporal XML document, all nodes are designed to convert the corresponding elements in *FOSDM*. Based on the aforementioned transformation rules, we propose an algorithm for the reengineering of fuzzy spatiotemporal XML into *FOSDM*. The pseudo-code for this algorithm is illustrated in Algorithm 3.1.

Algorithm 3.1 Transforming fuzzy XML to *FOSDM* schema

Input: fuzzy spatiotemporal XML Schema *F*
Output: corresponding *FOSDM* schema *S*
01 create a class in *S* by applying Rule 3.1
02 map the temporal element of the root in *F* to the temporal attribute within the generated class by applying Rule 3.12 and Rule 3.13
03 define the *OID* by applying Rule 3.2
04 transform the deterministic non-leaf attributes element *DAT* in *F* to the attributes with in the generated class by applying Rule 3.3
05 map non-spatiotemporal leaf element whose value is fuzzy *FVR* and *FVF* in *F* by applying Rule 3.4 and Rule 3.5 respectively
06 for fuzzy spatial position node, use Rule 3.6, Rule 3.7 and Rule 3.8 to map them
07 for fuzzy spatial motion node, apply Rule 3.9 and Rule 3.10
08 for non-root element owing temporal sub-element, use the Rule 3.13 to transform them
09 if root element owns sub-element *fuzzyAssociation*, transform fuzzy association applying Rule 3.14
10 if root element owns sub-element *fuzzyInheritance*, transform fuzzy inheritance applying Rule 3.15
11 if root element owns sub-element *fuzzyAggregation*, transform fuzzy aggregation applying Rule 3.16

To facilitate an intuitive comprehension of Algorithm 3.1, a flow diagram is presented in Fig. 3.4. The left column of the diagram delineates the various types of nodes, while the right column outlines the corresponding transformation rules. This flow diagram enhances the convenience and efficiency of the transformation process.

Utilizing Algorithm 3.1 and the corresponding flow diagram illustrated in Fig. 3.4, we will examine the following example, which is supported by the XML Schema depicted in Fig. 3.5, to facilitate the transformation process of the fuzzy XML Schema.

In this example, we apply the aforementioned rules systematically. Initially, we examine the XML Schema, which comprises seven root elements characterized by fuzzy relationships among them. Subsequently, we convert each root element into a class in accordance with Rule 3.1. Specifically, for the class city, *Fuzzy SpatialRegion* is transformed utilizing Rule 3.8, while the method *Own* is mapped through Rule 3.3. For the class *Cloud*, we employ Rule 3.2 to convert *KAT* to *OID*, and the attribute *Name* being a deterministic general attribute, is also mapped using Rule 3.3. The sub-element *SpatialPoint* of the root element *Cloud* is transformed according to Rule 3.6. Regarding the class *DarkCloud* which contains nodes for fuzzy time intervals, fuzzy motion values, and a fuzzy general attribute node *Rain* (where the fuzzy value is represented as a real number), we apply Rule 3.12, Rule 3.10 and Rule 3.4 for their mapping. The root element *WhiteCloud* includes a sub-element *Direction* which possesses a fuzzy time point attribute; we utilize Rule 3.9 for the transformation of *Direction* and Rule 3.13 for *TimePoint*. For the deterministic general method, Rule 3.3 is employed for mapping. The root element *LivingQuarter* contains three sub-elements: a fuzzy general element *Year* (with a fuzzy value represented as a distribution function), a deterministic general element *Market*, and a fuzzy spatial line element. Consequently, we apply Rule 3.5, Rule 3.3 and Rule 3.7 for their

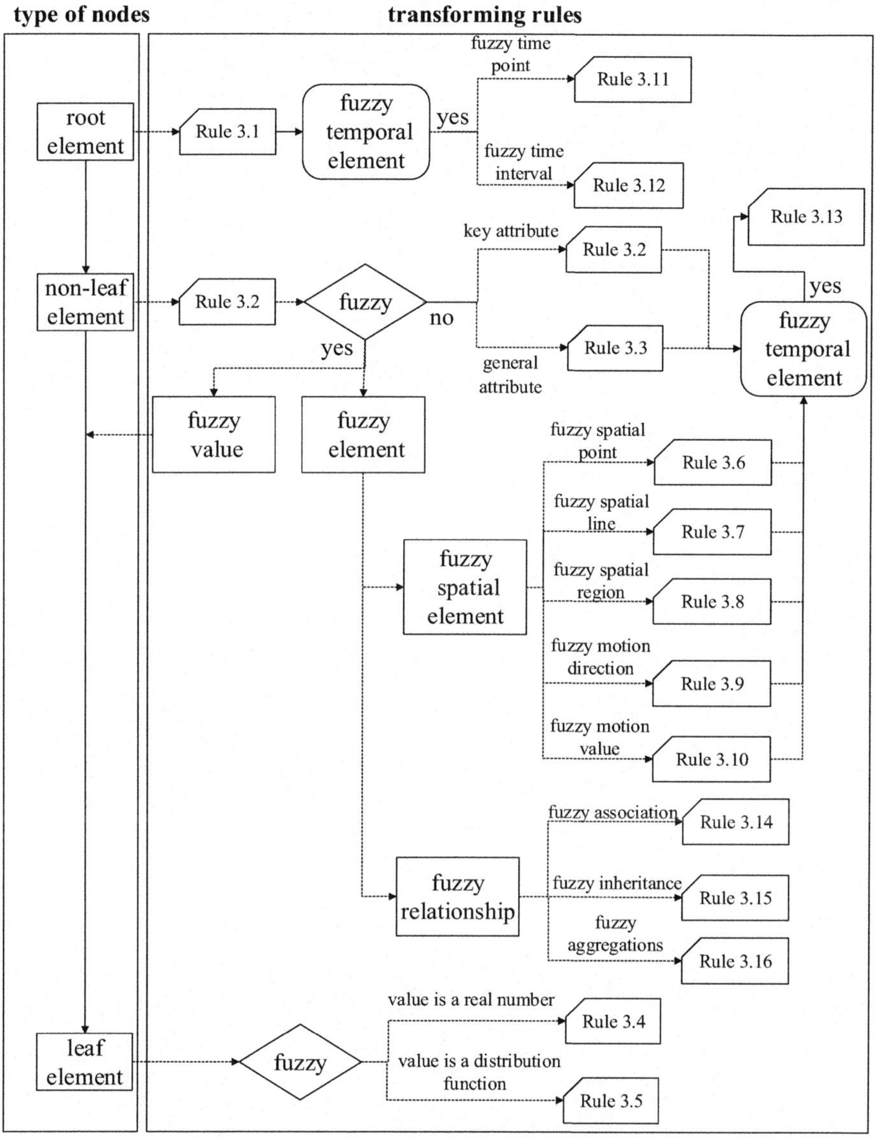

Fig. 3.4 Transforming flow diagram of Algorithm 3.1

respective transformations. In the case of the class *IndustrialDistrict*, the attributes *Product* and *Factory* can be reengineered based on Rule 3.2 within a fuzzy object-oriented spatiotemporal database. For the class *Land*, the attribute *Population* is transformed using Rule 3.2, while the Fuzzy *TimePoint* is mapped according to Rule 3.11. The results of the transformations for all classes are illustrated in Fig. 3.6.

1. Schema of root element City

```
<xs:element name="City" minOccurs="0" maxOccurs="unbounded">
 <xs:complexType>
  <xs:element name="Own" type ="original-definition"/>
  <xs:element name="SpatialRegion"/>
  </xs:complexType>
</xs:element>
<xs:element name="SpatialRegion" minOccurs="0" maxOccurs="unbounded">
 <xs:complexType>
  <xs:sequence>
   <xs:element ref="point "/>
  </xs:sequence>
 </xs:complexType>
</xs:element>
<xs:element name="point" minOccurs="2" maxOccurs="2">
 <xs:complexType>
  <xs:sequence>
   <xs:element name="Xaxis" type="xs:integer"/>
   <xs:element name="Yaxis" type="xs:integer"/>
  </xs:sequence>
 </xs:complexType>
</xs:element>
```

2. Schema of root element *Cloud*

```
<xs:element name="Cloud" minOccurs="0" maxOccurs="unbounded">
 <xs:complexType>
  <xs:element name="fuzzyAssociation">
   <xs:complexType>
    <xs:extension base="City"/>
    <xs:element name="Val">
      <xs:complexType>
       <xs:attribute name ="Poss" type ="xs:fuzzy" default="1.0"/>
      </xs:complexType>
    </xs:element>
   </xs:complexType>
  </xs:element>
  <xs:attribute name="OID" type ="xs:ID" use="required"/>
  <xs:element name="Name" type="original-definition"/>
  <xs:element name="SpatialPoint"/>
 </xs:complexType>
</xs:element>
<xs:element name="SpatialPoint" minOccurs="0" maxOccurs ="unbounded">
 <xs:complexType>
  <xs:sequence>
   <xs:element name="No." type="xs:integer"/>
   <xs:element ref="point"/>
  </xs:sequence>
 </xs:complexType>
</xs:element>
<xs:element name="point" minOccurs="1" maxOccurs="unbounded">
 <xs:complexType>
  <xs:sequence>
   <xs:element name="Xaxis" type="xs:integer"/>
   <xs:element name="Yaxis" type="xs:integer"/>
  </xs:sequence>
 </xs:complexType>
</xs:element>
```

3. Schema of root element *DarkCloud*

```
<xs:element name="DarkCloud" minOccurs="0" maxOccurs="unbounded">
 <xs:complexType>
  <xs:element name="fuzzyGeneration" >
   <xs:complexType>
    <xs:extension base="Cloud"/>
    <xs:element name="Val">
      <xs:complexType>
       <xs:attribute name ="Poss" type ="xsd:fuzzy" default="1.0"/>
      </xs:complexType>
    </xs:element>
   </xs:complexType>
 </xs:element>
 <xs:element name="Rain"/>
 <xs:element name="MotionValue"/>
 <xs:element name="TimeInterval"/>
 </xs:complexType>
</xs:element>
<xs:element name="Rain" minOccurs="0" maxOccurs="unbounded">
 <xs:complexType>
 <xs:element name="Val"/>
 </xs:complexType>
</xs:element>
<xs:element name="Val" minOccurs="0" maxOccurs="unbounded">
 <xs:complexType>
 <xs:attribute name="Poss" type="xsd:fuzzy" default="1.0"/>
 </xs:complexType>
</xs:element>
<xs:element name="MotionValue">
 <xs:simpleType>
  <xs:restriction base="xs:integer">
   <xs:minInclusive value="valueStart"/>
   <xs:maxInclusive value="valueEnd"/>
  </xs:restriction>
  </xs:simpleType>
 </xs:element>
<xs:element name="TimeInterval">
 <xs:complexType>
  <xs:sequence>
   <xs:element name="timeStart"/>
   <xs:element name="timeEnd"/>
  </xs:sequence>
 </xs:complexType>
</xs:element>
<xs:element name="timeStart" minOccurs="0" maxOccurs="unbounded">
 <xs:simpleType>
  <xs:restriction base="xs:dateTime">
  <xs:minInclusive value="valueStart"/>
  <xs:maxInclusive value="valueEnd"/>
   </xs:restriction>
 <xs:simpleType>
</xs:element>
<xs:element name="timeEnd" minOccurs="0" maxOccurs="unbounded">
 <xs:simpleType>
  <xs:restriction base="xs:dateTime">
  <xs:minInclusive value="valueStart"/>
  <xs:maxInclusive value="valueEnd"/>
   </xs:restriction>
 <xs:simpleType>
</xs:element>
<xs:element name="Direction_1" minOccurs="1" maxOccurs="unbounded">
 <xs:complexType>
  <xs:element name="TimePoint"/>
  <xs:sequence>
   <xs:element ref="degree"/>
  </xs:sequence>
 </xs:complexType>
</xs:element>
<xs:element name="degree" minOccurs="2" maxOccurs="2">
 <xs:simpleType>
  <xs:restriction base="xs:integer">
   <xs:minInclusive value="degreeStart"/>
   <xs:maxInclusive value="degreeEnd"/>
  </xs:restriction>
 </xs:simpleType >
</xs:element>
<xs:element name="TimePoint" minOccurs="0" maxOccurs="unbounded">
 <xs:restriction base="xs:dateTime">
  <xs:minInclusive value="valueStart"/>
  <xs:maxInclusive value="valueEnd"/>
 </xs:restriction>
</xs:element>
```

4. The subclass *WhiteCloud* is mapped as follow:

```
<xs:element name="WhiteCloud" minOccurs="0" maxOccurs ="unbounded">
 <xs:complexType>
  <xs:element name="fuzzyGeneration" >
   <xs:complexType>
    <xs:extension base="Cloud"/>
    <xs:element name="Val">
      <xs:complexType>
       <xs:attribute name ="Poss" type ="xs:fuzzy" default="1.0"/>
      </xs:complexType>
    </xs:element>
   </xs:complexType>
  </xs:element>
  <xs:element name="Sunshade" type="original-definition"/>
  <xs:element name="Direction_2"/>
 </xs:complexType>
</xs:element>
 <xs:element name="Direction_2" minOccurs="0" maxOccurs="unbounded">
 <xs:complexType>
  <xs:element name="MotionValue">
        <xs:simpleType>
         <xs:restriction base="xs:integer">
          <xs:minInclusive value="valueStart"/>
          <xs:maxInclusive value="valueEnd"/>
         </xs:restriction>
        </xs:simpleType>
  </xs:element>
  <xs:sequence>
   <xs:element name="degree" type="xs:integer"/>
   <xs:element name ="val"/>
  </xs:sequence>
 </xs:complexType>
</xs:element>
<xs:element name="Val" minOccurs="0" maxOccurs="unbounded">
 <xs:complexType>
  <xs:attribute name="Poss" type="xs:fuzzy" default="1.0"/>
 </xs:complexType>
</xs:element>
```

5. The linked and part-class *LivingQuarter* is mapped as follow:

```
<xs:element name="LivingQuarter" minOccurs="0" maxOccurs ="n">
 <xs:complexType>
  <xs:element name="fuzzyAssociation">
   <xs:complexType>
    <xs:extension base="City"/>
    <xs:element name="Val">
      <xs:complexType>
       <xs:attribute name ="Poss" type ="xs:fuzzy" default="1.0"/>
      </xs:complexType>
    </xs:element>
   </xs:complexType>
  </xs:element>
  <xs:element name="Year"/>
  <xs:element name="Market" type="original-definition"/>
  <xs:element name="SpatialLine"/>
 </xs:complexType>
</xs:element>
<xs:element name="Year" minOccurs="0" maxOccurs ="unbounded">
 <xs:complexType>
  <xs:element name ="Dist"/>
 </xs:complexType>
</xs:element>
<xs:complexType name="DistType" minOccurs="0" maxOccurs="unbounded">
 <xs:sequence>
  <xs:element name ="Val"/>
 </xs:sequence>
</xs:complexType>
<xs:element name="Val" minOccurs="0" maxOccurs="unbounded">
 <xs:complexType>
  <xs:element name="Vi" type="original-definition"/>
 </xs:complexType>
</xs:element>
<xs:element name="Spatialline">
 <xs:complexType>
  <xs:sequence>
   <xs:element name="No." type="xs:integer"/>
   <xs:element ref="point "/>
  </xs:sequence>
  <xs:attribute name="offset" type="xs:integer"/>
 </xs:complexType>
</xs:element>
<xs:element name="point" minOccurs="2" maxOccurs="2">
 <xs:complexType>
  <xs:sequence>
   <xs:element name="Xaxis" type="xs:integer"/>
   <xs:element name="Yaxis" type="xs:integer"/>
  </xs:sequence>
 </xs:complexType>
</xs:element>
```

6. The part-class *IndustrialDistrict* is mapped as follow:

```
<xs:element name="IndustrialDistrict" minOccurs="0" maxOccurs ="unbounded">
 <xs:complexType>
  <xs:element name="Produce" type="original-definition"/>
  <xs:element name="Factory" type="original-definition"/>
  <xs:element name="SpatialLine"/>
 </xs:complexType>
</xs:element>
<xs:element name="Spatialline">
 <xs:complexType>
  <xs:sequence>
   <xs:element name="No." type="xs:integer"/>
   <xs:element ref="point "/>
  </xs:sequence>
  <xs:attribute name="offset" type="xs:integer"/>
 </xs:complexType>
</xs:element>
<xs:element name="point" minOccurs="2" maxOccurs="2">
 <xs:complexType>
  <xs:sequence>
   <xs:element name="Xaxis" type="xs:integer"/>
   <xs:element name="Yaxis" type="xs:integer"/>
  </xs:sequence>
 </xs:complexType>
</xs:element>
```

7. The whole-class *Land* is mapped as follow:

```
<xs:element name="Land" minOccurs="0" maxOccurs ="unbounded">
 <xs:complexType>
  <xs:element name ="fuzzyAggregation" >
   <xs:complexType>
    <xs:element ref = "LivingQuarter"/>
    <xs:element ref = "IndustrialDistrict"/>
    <xs:element name="Val" minOccurs ="0" maxOccurs ="unbounded">
     <xs:complexType>
      <xs:attribute name ="Poss" type ="xs:fuzzy" default="1.0"/>
     </xs:complexType>
    </xs:element>
   </xs:complexType>
  </xs:element>
 </xs:complexType>
</xs:element>
 <xs:element name="Population" type="original-definition"/>
 <xs:element name="TimePoint"/>
 </xs:complexType>
</xs:element>
<xs:element name="TimePoint" minOccurs="0" maxOccurs="unbounded">
 <xs:restriction base="xs:dateTime">
  <xs:minInclusive value="valueStart"/>
  <xs:maxInclusive value="valueEnd"/>
 </xs:restriction>
</xs:element>
```

Fig. 3.5 Schema of the example

At last, when all of the classes are transformed, it is imperative to examine the interrelationships among these classes. The XML Schema, with *Cloud* as its root element, indicates that the relationship between the *Cloud* class and the *City* class is characterized as a fuzzy association. Similarly, the XML Schema with *LivingQuarter* as its root element also reveals a fuzzy association between the *LivingQuarter* class

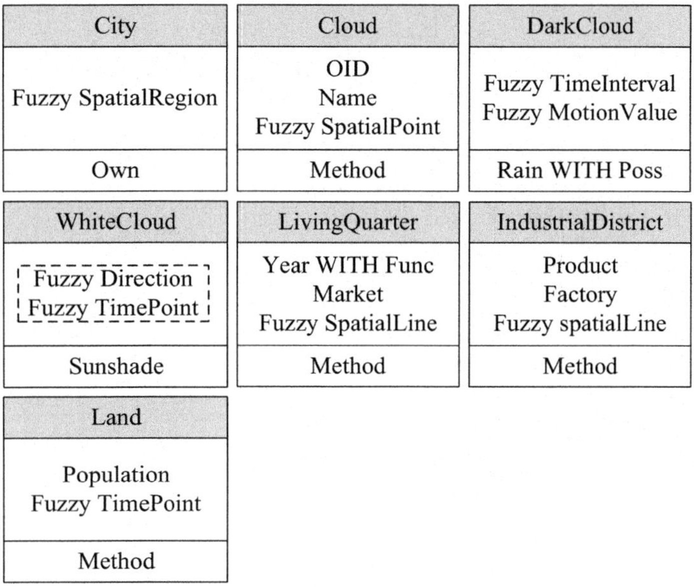

Fig. 3.6 All of transforming classes

and the *City* class, which is mapped using Rule 3.14. Furthermore, the XML Schemas with *WhiteCloud* and *DarkCloud* as their root elements demonstrate that the *Cloud* class functions as a superclass, while *WhiteCloud* and *DarkCloud* are identified as subclasses. This fuzzy inheritance is mapped according to Rule 3.15. Additionally, the XML Schema with *Land* as its root element confirms that the *Land* class represents a whole-class, with *LivingQuarter* and *IndustrialDistrict* serving as two part-classes. The transformation of this fuzzy aggregation is executed using Rule 3.16. The resultant transformations are illustrated in Fig. 3.7.

3.5 Transformation of Fuzzy Spatiotemporal Data from Object-Oriented Database to XML

OOSDM schema is typically characterized as a class definition, while the XML schema is depicted as a hierarchical tree structure. The process of translating the *OOSDM* schema into XML involves converting a class-based definition into a tree-based schema representation. This transformation encompasses two primary components. First, the conversion of fuzzy spatiotemporal data from an object-oriented database into a flat XML format, which is a simplified XML structure consisting solely of basic leaf elements and attributes as sub-elements of non-leaf elements. Second, the transformation of fuzzy spatiotemporal data from an object-oriented database into a nested XML format, which is a more complex XML structure that

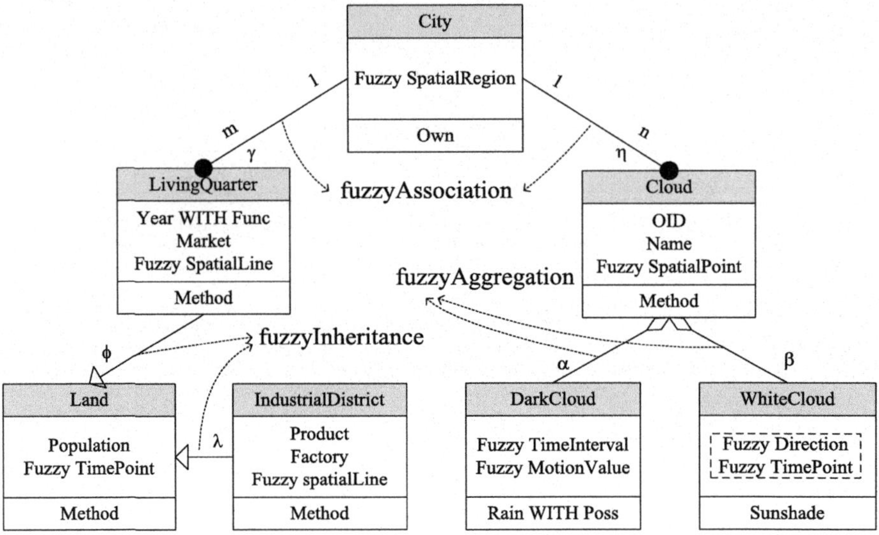

Fig. 3.7 Transforming result from fuzzy XML to fuzzy OO spatiotemporal database

captures the relationships between multiple root elements. The architecture of this transformation is illustrated in Fig. 3.8. As indicated in Fig. 3.8, the objective is to establish transformation rules that delineate the relationships among various classes, as well as between classes and fuzzy attributes (or methods), deterministic attributes (or methods), fuzzy spatial attributes, and fuzzy temporal attributes within the class. The subsequent subsections will elaborate on all of the transformation rules.

3.5.1 Transforming Fuzzy Spatiotemporal Data from Object-Oriented Database to Flat XML

We consider an *OOSDM* schema denoted as *S*, which encompasses various components including classes, object identifiers (OIDs), fuzzy temporal attributes, fuzzy spatial attributes, general attributes (which are neither OID nor spatiotemporal), and associated methods. It is important to note that the general attributes and methods may exhibit fuzziness. As illustrated in Fig. 3.8, numerous rules are to be established, with the fundamental mapping rules outlined as follows.

Rule 3.17 For an *OOSDM* schema *S*, a root element named *XMLRoot* is created in the corresponding XML Schema.

Rule 3.18 For each class *C* in *OOSDM* schema *S*, an element with the same name as *C* is created and then put under the root element.

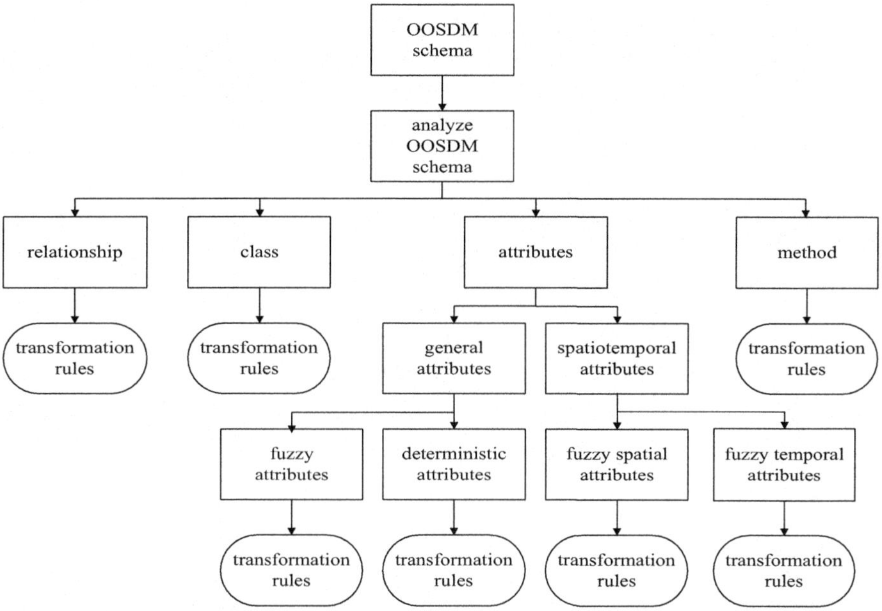

Fig. 3.8 Transforming architecture

Rule 3.19 For each object identifier (*OID*) with the name O_i in class *C* of *OOSDM* schema *S*, the xs:ID type is used for creating an attribute of the element for *C*.

Rule 3.20 For each general attribute (or method) *Ai* in class *C* of *OOSDM* schema *S*, an element with the corresponding name and data type is created. This element is subsequently positioned as a child element of *C*.

Rule 17 to Rule 20 can be directly applied to facilitate the transformation from *OOSDM* schema to XML Schema, provided that all the classes, attributes (or methods) are deterministic. In instances where there is fuzziness in *OOSDM*, the following rules (Rule 3.21 to Rule 3.23) should be empolyed.

Rule 3.21 If there exists the first level of fuzziness in *OOSDM* (recall the definition of fuzziness in *OOSDM*), for each general attribute (or method) A_i in class *C* of *OOSDM* schema *S*, the following strategy is utilized. Specifically, an element *Val* is created firstly as a child element of *C*, and then an element is placed as a child element of *Val* with the corresponding name and data type is created.

Rule 3.22 If there exists the second level of fuzziness in *OOSDM* (recall the definition of fuzziness in *OOSDM*), for each general attribute (or method) A_i in class *C* of *OOSDM* schema *S*, the following strategy is utilized. Specifically, an element *Dist* is created firstly as a child element of *C*, an element *Val* is then created as a child element of *Dist* and finally an element placed as a child element of *Val* with the corresponding name and data type is created.

Rule 3.23 If there exists the third level of fuzziness in *OOSDM* (recall the definition of fuzziness in *OOSDM*), for each general attribute (or method) *Ai* in class *C* of *OOSDM* schema *S*, the transforming strategy is utilized. Specifically, an element *Dist* is created firstly as a child element of A_i, then element *Val* is created as a child element of *Dist*, and last element V_i which is the fuzzy value is created as a child element of *Val*.

The aforementioned rules address the transformation of general attributes from the *OOSDM* schema to XML Schema. However, it is important to note that the *OOSDM* schema also encompasses certain spatial and temporal attributes. Therefore, an examination of the rules governing fuzzy spatiotemporal attributes will be conducted. As a result, Rule 3.24 to Rule 3.26 will be implemented in relation to fuzzy spatial points, fuzzy spatial lines, and fuzzy spatial regions.

Rule 3.24 For an *OOSDM* schema *S*, *SpatialPoint* is the fuzzy spatial point attributes, and an element with the same name *SpatialPoint* is created as a child element of *C*.

Rule 3.25 For an *OOSDM* schema *S*, *Spatialline* is the fuzzy spatial line attributes, and the element *Spatialline* in XML is created as a child element of *C*.

Rule 3.26 For an *OOSDM* schema *S*, *SpatialRegion* is the fuzzy spatial line attributes, and the corresponding XML Schema of an element *SpatialRegion* is created.

For fuzzy spatial attributes, in addition to fuzzy spatial position, there is the concept of fuzzy spatial motion, which encompasses both fuzzy motion direction and fuzzy motion value. Fuzzy motion direction is characterized by two conditions: one condition specifies that the angle of the fuzzy motion direction falls within a certain range, while the other condition allows for each motion direction to possess a potential value. The mapping of these conditions is facilitated by Rule 3.27 and Rule 3.28.

Rule 3.27 For an *OOSDM* schema *S*, *Direction_1* is the first-type fuzzy motion direction attributes, and an element with the same name *Direction_1* is created as a child element of *C*.

Rule 3.28 For an *OOSDM* schema S, *Direction_2* is the second-type fuzzy motion direction attributes, and the element *Direction_2* in XML is created as a child element of *C*.

Rule 3.29 For an *OOSDM* schema *S*, *Motionvalue* is the fuzzy motion value attributes, and the corresponding XML Schema of an element *Motionvalue* is created.

In the subsequent discussion, we will examine fuzzy temporal attributes, specifically focusing on fuzzy time points and fuzzy time intervals.

Rule 3.30 For an *OOSDM* schema *S*, *TimePoint* is the fuzzy time point attributes, and the corresponding XML Schema of an element *TimePoint* is created.

Rule 3.31 For an *OOSDM* schema *S*, *TimeInterval* is the fuzzy time interval attributes, and an element with the same name *TimeInterval* is created as a child element of *C*.

Utilizing the established transformation rules, one can readily derive the transformation flow outlined in Algorithm 3.2. To facilitate an intuitive comprehension of Algorithm 3.2, a corresponding flow diagram is provided in Fig. 3.9. The left column of this diagram delineates the sequence of transformations, while the right column enumerates the associated transformation rules. The principal transformation is highlighted within the dotted box. Furthermore, the comprehensive XML Schema pertaining to Rule 3.17 to Rule 3.31 is illustrated in Fig. 3.10.

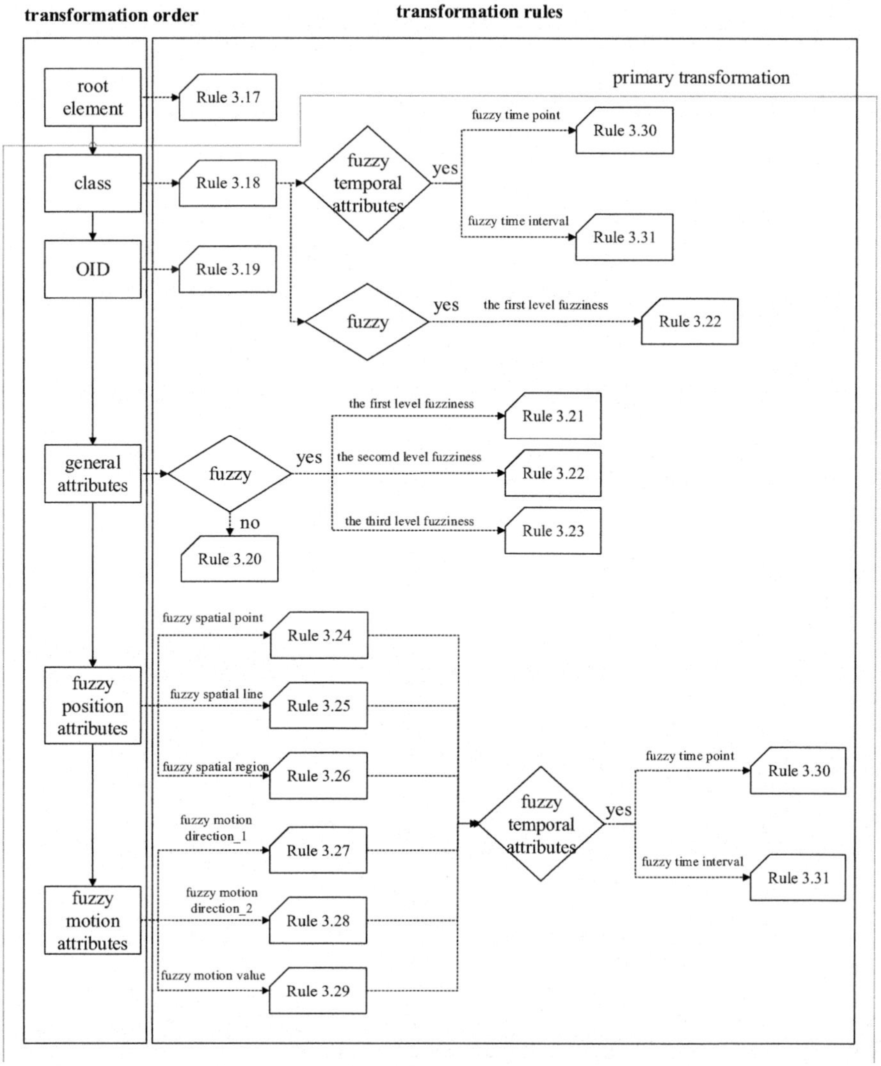

Fig. 3.9 Transforming flow diagram of Algorithm 3.2

Rule 3.17

```
<?xml version="1.0"
encoding="UTF-8"?>
<xs:schema xmlns:xs="http://
www.w3.org/2001/XMLSchema"

elementFormDefault="qualified"

attributeFormDefault="unqualified"
>
 <xs:element name="C">
  <xs:complexType>
   <xs:sequence>
    <!--sub-elements of C -->
   </xs:sequence>
  </xs:complexType>
 </xs:element >
</xs:schema>
```

Rule 3.18

```
<xs:element name="C"
minOccurs="0"
maxOccurs="unbounded">
 <xs:complexType>
  <xs:attribute name="Oi"
type="xs:ID" use="required"/>
 </xs:complexType>
</xs:element>
```

Rule 3.19

```
<xs:element name="C"
minOccurs ="0"
maxOccurs ="unbounded">
 <xs:complexType>
  <xs:element name ="Oi"
type ="xs:ID" use ="required"/
>
 </xs:complexType>
</xs:element>
```

Rule 3.20

```
<xs:element name="C"
minOccurs ="0"
maxOccurs ="unbounded">
 <xs:complexType>
  <xs:element name ="Ai"
type ="original-definition"/>
 </xs:complexType>
</xs:element>
```

Rule 3.21

```
<xs:element name="Ai"
minOccurs ="0" maxOccurs
="unbounded">
 <xs:complexType>
  <xs:element name ="Val"/>
 </xs:complexType>
</xs:element>
<xs:element name="Val"
minOccurs ="0" maxOccurs
="unbounded">
 <xs:complexType>
  <xs:attribute name ="Poss"
type ="xs:fuzzy" default="1.0"/
>
 </xs:complexType>
</xs:element>
```

Rule 3.22

```
<xs:element name="C"
minOccurs ="0" maxOccurs
="unbounded">
 <xs:complexType>
  <xs:element name ="Dist"/>
 </xs:complexType>
</xs:element>
<xs:complexType
name="DistType" minOccurs
="0" maxOccurs
="unbounded">
 <xs:sequence>
 <xs:element name="Val" />
 </xs:sequence>
</xs:complexType>
<xs:element name="Val"
minOccurs ="0" maxOccurs
="unbounded">
 <xs:complexType>
  <xs:element name="Ai" type
="Original-definition" />
  <xs:attribute name ="Poss"
type ="xs:fuzzy" default="1.0"/
>
 </xs:complexType>
</xs:element>
```

Rule 3.23

```
xs:element name="Ai" minOccurs
="0" maxOccurs ="unbounded">
 <xs:complexType>
  <xs:element name ="Dist"/>
 </xs:complexType>
</xs:element>
<xs:complexType name="Dist"
minOccurs ="0" maxOccurs
="unbounded">
 <xs:sequence>
  <xs:element name="Val" />
 </xs:sequence>
</xs:complexType>
<xs:element name="Val"
minOccurs ="0" maxOccurs
="unbounded">
 <xs:complexType>
  <xs:element name="Vi" type
="Original-definition" />
 </xs:complexType>
</xs:element>
```

Rule 3.24

```
<xs:element name="SpatialPoint">
 <xs:complexType>
  <xs:sequence>
   <xs:element name="No."
type="xs:integer"/>
   <xs:element ref="Point"/>
  </xs:sequence>
 </xs:complexType>
</xs:element>
<xs:element name="Point"
minOccurs="1"
maxOccurs="unbounded">
 <xs:complexType>
  <xs:sequence>
   <xs:element name="Xaxis"
type=" xs:integer"/>
   <xs:element name="Yaxis"
type=" xs:integer"/>
  </xs:complexType>
</xs:element>
```

Rule 3.25

```
<xs:element name="Spatialline">
 <xs:complexType>
  <xs:sequence>
   <xs:element name="No."
type="xs:integer"/>
   <xs:element ref="Point "/>
  </xs:sequence>
  <xs:attribute name="Offset"
type="xs:integer"/>
 </xs:complexType>
</xs:element>
<xs:element name="Point"
minOccurs="2" maxOccurs="2">
 <xs:complexType>
  <xs:sequence>
   <xs:element name="Xaxis"
type="xs:integer"/>
   <xs:element name="Yaxis"
type="xs:integer"/>
  </xs:sequence>
 </xs:complexType>
</xs:element>
```

Rule 3.26

```
<xs:element
name="SpatialRegion">
 <xs:complexType>
  <xs:sequence>
   <xs:element name="No."
type="xs:integer"/>
   <xs:element ref="Point"/>
  </xs:sequence>
 </xs:complexType>
</xs:element>
<xs:element name="Point"
minOccurs="2" maxOccurs="2">
 <xs:complexType>
  <xs:sequence>
   <xs:element name="Xaxis"
type="xs:integer"/>
   <xs:element name="Yaxis"
type="xs:integer"/>
  </xs:sequence>
 </xs:complexType>
</xs:element>
```

Rule 3.27

```
<xs:element name="Direction_1">
 <xs:complexType>
  <xs:sequence>
   <xs:element ref="Degree"/>
  </xs:sequence>
 </xs:complexType>
</xs:element>
<xs:element name="Degree"
minOccurs="2" maxOccurs="2">
 <xs:simpleType>
  <xs:restriction base="xs:integer">
   <xs:minInclusive
value="degreeStart"/>
   <xs:maxInclusive
value="degreeEnd"/>
  </xs:restriction>
 </xs:simpleType >
</xs:element>
```

Rule 3.28

```
<xs:element name="Direction_2">
 <xs:complexType>
  <xs:sequence>
   <xs:element name="Degree"
type="xs:integer"/>
   <xs:element name ="Val"/>
  </xs:sequence>
 </xs:complexType>
</xs:element>
<xs:element name="Val" minOccurs
="0" maxOccurs ="unbounded">
 <xs:complexType>
  <xs:attribute name ="Poss" type
="xs:fuzzy" default="1.0"/>
 </xs:complexType>
</xs:element>
```

Rule 3.29

```
<xs:element name="MotionValue">
 <xs:simpleType>
  <xs:restriction base="xs:integer">
   <xs:minInclusive
value="valueStart"/>
   <xs:maxInclusive
value="valueEnd"/>
  </xs:restriction>
 </xs:simpleType>
</xs:element>
```

Rule 3.30

```
<xs:element name="TimePoint ">
<xs:restriction base="dateTime">
  <xs:minInclusive
value="valueStart"/>
  <xs:maxInclusive
value="valueEnd"/>
 </xs:restriction>
</xs:element>
```

Rule 3.31

```
<xs:element name="TimeInterval">
 <xs:complexType>
  <xs:sequence>
<xs:element name="timeStart"/>
<xs:element name="timeEnd"/>
</xs:sequence>
</xs:complexType>
</xs:element>
<xs:element name="timeStart"
minOccurs="0"
maxOccurs="unbounded">
 <xs:simpleType>
  <xs:restriction base="dateTime">
   <xs:minInclusive
value="valueStart"/>
   <xs:maxInclusive
value="valueEnd"/>
</xs:restriction>
 <xs:simpleType>
</xs:element>
<xs:element name="timeEnd"
minOccurs="0"
maxOccurs="unbounded">
 <xs:simpleType>
 <xs:restriction base="dateTime">
   <xs:minInclusive
value="valueStart"/>
   <xs:maxInclusive
value="valueEnd"/>
</xs:restriction>
 <xs:simpleType>
</xs:element>
```

Rule 3.32

```
<xs:element name="C" minOccurs ="0"
maxOccurs ="unbounded">
 <xs:complexType>
  <!--sub-elements of C-->
 </xs:complexType>
</xs:element>

<xs:element name="LC" minOccurs
="0" maxOccurs ="unbounded">
 <xs:complexType>
 <xs:element name ="fuzzyGeneration"
>
    <xs:complexType>
     <xs:extension base="C"/>
     <xs:element name="Val">
        <xs:complexType>
         <xs:attribute name ="Poss" type
="xs:fuzzy" default="1.0"/>
        </xs:complexType>
     </xs:element>
    </xs:complexType>
  </xs:element>
  <xs:element name="Si" type="original-
definition"/>
 </xs:complexType>
</xs:element>
```

Rule 3.33

```
<xs:element name="C" minOccurs ="0"
maxOccurs ="unbounded">
 <xs:complexType>
  <!--sub-elements of C -->
 </xs:complexType>
</xs:element>

<xs:element name="LC" minOccurs
="0" maxOccurs ="unbounded">
 <xs:complexType>
 <xs:element name ="fuzzyAssociation"
>
    <xs:complexType>
    <xs:extension base= "C"/>
    <xs:element name="Val">
        <xs:complexType>
         <xs:attribute name ="Poss" type
="xs:fuzzy" default="1.0"/>
        </xs:complexType>
    </xs:element>
    </xs:complexType>
  </xs:element>
  <xs:element name="Si" type="original-
definition"/>
 </xs:complexType>
</xs:element>
```

Rule 3.34

```
<xs:element name="PCi" minOccurs
="0" maxOccurs="unbounded">
 <xs:complexType>
  <!--sub-elements of PCi -->
 </xs:complexType>
</xs:element>

<xs:element name="WC" minOccurs
="0" maxOccurs="unbounded">
 <xs:complexType>
  <xs:element name
="fuzzyAggregation" >
    <xs:complexType>
     <xs:element ref="PCi"/>
     <xs:element name="Val">
        <xs:complexType>
         <xs:attribute name="Poss"
type="xs:fuzzy" default="1.0"/>
        </xs:complexType>
     </xs:element>
    </xs:complexType>
  </xs:element>
  <xs:element name="Si" type="original-
definition"/>
 </xs:complexType>
</xs:element>
```

Fig. 3.10 The XML schema of Rule 3.17 to Rule 3.34

Algorithm 3.2 Transforming *OOSDM* Schema to flat XML

Input: OOSDM schema *S*
Output: transformed fuzzy XML Schema *X*
01 create root element *XMLRoot* for S by applying Rule 3.17
02 get next fuzzy *OOSDM* class *C*, return generated fuzzy XML element by applying Rule 3.18
03 transform the *OID* by applying Rule 3.19
04 transform the deterministic general attributes (or method) by applying Rule 3.20
05 if there exists fuzziness of the in *C*, then map the object with the three levels of fuzziness by applying by Rule 3.21, Rule 3.22, and Rule 3.23
06 if the class *C* and general attributes own fuzzy temporal attributes, use Rule 3.30 and Rule 3.31 to map them
07 for the fuzzy spatial position attributes, use Rule 3.24, Rule 3.25, and Rule 3.26 to map them, if they simultaneously own fuzzy temporal attributes, apply the Rule 3.30 and Rule 3.31
08 for the fuzzy spatial motion attributes, by applying Rule 3.27, Rule 3.28, and Rule 3.29, if they simultaneously own fuzzy temporal attributes, use the Rule 3.30 and Rule 3.31 to map them

3.5.2 Transforming Fuzzy Spatiotemporal Data from Object-Oriented Database to Nested XML

In this subsection, we will delineate the transformation rules governing relationships such as fuzzy generalization, fuzzy association, and fuzzy aggregation among fuzzy *OOSDM* classes. The concept of generalization within the *OOSDM* framework establishes a subclass/superclass relationship between classes, wherein one class, referred to as the superclass, provides a broader characterization of a collection of other classes, known as subclasses. Specifically, we will outline the following rule.

Rule 3.32 For each subclass *SC* in *OOSDM*, let *C* denote its corresponding superclass. An element similar to Rule 3.18 and with the same name as *SC* is created. In order to represent the fuzzy generation, an element named *fuzzyGeneration* is defined and this element contains two sub-elements: one utilizes the "extension" tag to inherit the properties of *C*, while the other represents the degree of membership. S_i is the attribute (or method) in class *SC*.

Associations among *OOSDM* database schema are depicted through the use of links. In practical implementations, these relationships can be effectively identified by class involvement with *n*:*m*. Here, *n* and *m* are subject to specific constraints. In particular, we have the following rule.

Rule 3.33 For every class *C* in *OOSDM*, let *LC* denote its linked class. An element similar to Rule 3.18 and with the same name as *LC* is created. In order to represent the fuzzy association, an element named *fuzzyAssociation* is defined and this element contains two sub0elements: one utilizes the "extension" tag to identity relationship between *C* and *LC*, while the other represents the degree of membership. For *LC*, *maxOccurs* = "*m*" indicates the involvement 1:*m* and *maxOccurs* = "unbounded" represent involvement unknown. S_i is the attribute (or method) in class *LC*.

Aggregation serves as an abstraction that allows for the treatment of objects within composition relationships as higher-level entities. In the context of XML Schema, the transformation of aggregation involves the organization of sub-elements (corresponding to the part-class in fuzzy *OOSDM*) under a singular element, which aligns with the whole-class in fuzzy *OOSDM*. Specifically, the following rule is established.

Rule 3.34 For each part-class PC_i in *OOSDM*, let *WC* denote its whole-class. An element similar to Rule 2 and with the same name as *WC* is created. In order to represent the fuzzy association, an element named *fuzzyAggregation* is defined and this element contains two sub-elements: one utilizes the "ref" tag to refer to PC_i, while the other represents the degree of membership. S_i is the attribute (or method) in class *WC*.

In accordance with the aforementioned transformation rules, we present the algorithm designed for transforming the *OOSDM* schema into a nested XML Schema. The pseudo-code for this algorithm is illustrated in Algorithm 3.3. Furthermore, the comprehensive XML Schema of Rule 3.32 to Rule 3.34 is presented in Fig. 3.10.

Algorithm 3.3 Transforming *OOSDM* Schema to Nested XML

Input: *OOSDM* schema *S*
Output: transformed nested XML Schema *X*

```
create root element XMLRoot for S by applying Rule 3.17
transform each class and attributes in class by applying Algorithm 3.2
for fuzzy generalization, use Rule 3.32 to map it
for fuzzy association, use Rule 3.33 to map it
for fuzzy aggregations, use Rule 3.34 to map it
```

As delineated in Algorithm 3.3, the flow diagram illustrating the transformation process is depicted in Fig. 3.11. This flow diagram makes reference to the principal transformation outlined in Fig. 3.9.

3.5.3 Example

In order to understand the procedures outlined in Algorithm 3.2 and Algorithm 3.3, we provide a comprehensive example as illustrated in Fig. 3.12. This example encompasses various types of relationships among objects, including fuzzy relationships, deterministic general attributes, fuzzy general attributes, fuzzy spatial attributes, and fuzzy temporal attributes. By utilizing this example, we aim to apply all the rules that have been previously introduced.

This analysis focuses on the fuzzy *OOSDM* database schema as illustrated in Fig. 3.12. Initially, we employ Rule 3.17 to establish the root element. Subsequently, we utilize Rule 3.18 to Rule 3.31 to map the entire class presented in the example.

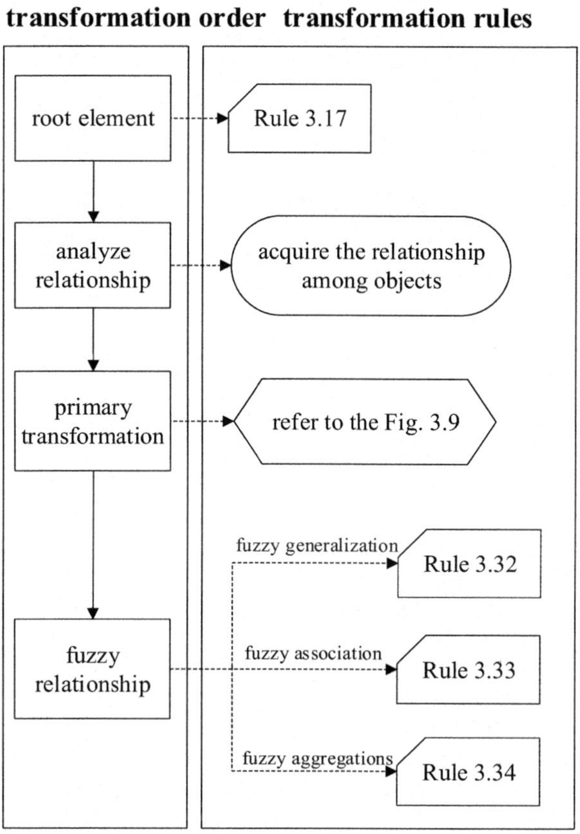

Fig. 3.11 Transforming flow diagram of Algorithm 3.3

Regarding the class *Cloud*, we first apply Rule 3.18 for class mapping, followed by Rule 3.19 for the mapping of *OID*. The attribute *name* is classified as a general deterministic attribute, necessitating the application of Rule 3.20 for its mapping. The fuzzy spatial point is mapped using Rule 3.24. For the class *DarkCloud*, we implement Rule 3.31 to convert the fuzzy time interval. The method *Rain* being a fuzzy method, is transformed using Rule 3.21. Finally, we apply Rules 3.30 and 3.28 to convert the fuzzy time point and fuzzy motion *direction_2*, respectively.

The class *City* is recognized as a fuzzy class. Consequently, we initially apply Rule 3.18 and Rule 3.22. The fuzzy spatial region associated with the attribute is subsequently transformed utilizing Rule 3.26. For the fuzzy attribute *Year* within the class *LivingQuarter*, we employ Rule 3.23 for mapping purposes. The transformation of the fuzzy *SpatialLine* is conducted through Rule 3.25. Our methodology effectively facilitates the transformation of the classes *Land* and *IndustrialDistrict*. Furthermore, we will proceed to map the fuzzy relationships. The subclasses *DarkCloud* and *WhiteCloud* fall under the superclass *Cloud*. The degree of membership

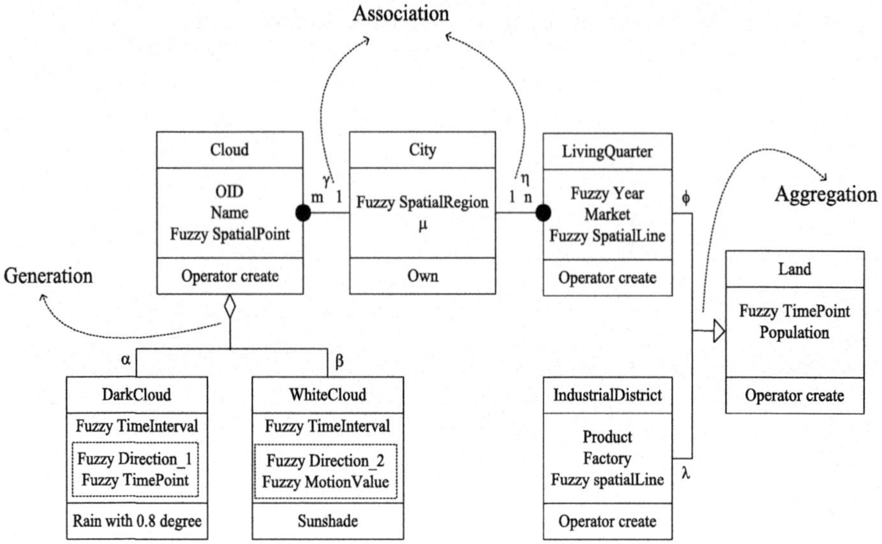

Fig. 3.12 Fuzzy OOSDM database schema

of *DarkCloud* to *Cloud* is denoted as α, while the degree of membership of *WhiteCloud* to *Cloud* is represented as β. The classes *LivingQuarter* and *IndustrialDistrict* are identified as part-classes, with *Land* serving as their whole-class. The degrees of membership between *LivingQuarter* and *Land*, as well as between *IndustrialDistrict* and *Land*, are indicated as ϕ and λ, respectively. The relationship between the class *Cloud* and the class *City* is characterized as a fuzzy association, with a degree of membership denoted as γ and an involvement ratio of $1{:}m$. Similarly, the relationship between the class *LivingQuarter* and the class *City* is also classified as a fuzzy association, with a degree of membership represented as η and an involvement ratio of $1{:}n$. Rule 3.32 to Rule 3.34 are employed to map the relationships among the classes. The final transformation results are illustrated in Fig. 3.13.

To demonstrate the efficacy of our methodology, we apply it to meteorological phenomena, illustrating the application of our model in recording and transforming fuzzy spatiotemporal data. Figure 3.14 depicts the trajectory of Post-Tropical Cyclone Sandy [24], which occurred from October 22, 2010, to October 30, 2010. In the following, we will conduct an analysis and modeling of Cyclone Sandy, culminating in the transformation from *OOSDM* to XML.

To accurately model the post-tropical cyclone Sandy, we will conduct an analysis of the statistical data pertaining to Sandy utilized in our experiments. In Fig. 3.14, several black points labeled *P1*, *P2*, and *P3* are identified, each representing distinct phases of Sandy's development. Specifically, *P1* corresponds to the Tropical Storm phase, *P2* denotes a Category 1 Hurricane, and *P3* signifies a Category 2 Hurricane. It is important to note that the wind speed and pressure values vary across these different stages, with each stage exhibiting a specific range of values. Table 3.2 presents the

1. The class *City* is mapped as follow:

```
<xs:element name="City" minOccurs="0" maxOccurs="unbounded">
 <xs:complexType>
  <xs:element name="Dist"/>
 </xs:complexType>
</xs:element>
<xs:complexType name="Dist" minOccurs="0" maxOccurs="unbounded">
 <xs:sequence>
  <xs:element name ="Val"/>
 </xs:sequence>
</xs:complexType>
<xs:element name="Val" minOccurs="0" maxOccurs="unbounded">
 <xs:complexType>
  <xs:element name="Own" type ="original-definition"/>
  <xs:element name="SpatialRegion"/>
  <xs:attribute name ="Poss" type ="xs:fuzzy" default="1.0"/>
 </xs:complexType>
</xs:element>
<xs:element name="SpatialRegion" minOccurs="0" maxOccurs="unbounded">
 <xs:complexType>
  <xs:sequence>
   <xs:element ref="point "/>
  </xs:sequence>
 </xs:complexType>
</xs:element>
<xs:element name="point" minOccurs="2" maxOccurs="2">
 <xs:complexType>
  <xs:sequence>
   <xs:element name="Xaxis" type="xs:integer"/>
   <xs:element name="Yaxis" type="xs:integer"/>
  </xs:sequence>
 </xs:complexType>
</xs:element>
```

2. The linked and superclass *Cloud* is mapped as follow:

```
<xs:element name="Cloud" minOccurs="0" maxOccurs="unbounded">
 <xs:complexType>
  <xs:element name="fuzzyAssociation">
    <xs:complexType>
     <xs:extension base="City"/>
     <xs:element name="Val">
       <xs:complexType>
        <xs:attribute name ="Poss" type ="xs:fuzzy" default="1.0"/>
       </xs:complexType>
     </xs:element>
    </xs:complexType>
  </xs:element>
  <xs:attribute name="OID" type ="xs:ID" use="required"/>
  <xs:element name="Name" type="original-definition"/>
  <xs:element name="SpatialPoint"/>
 </xs:complexType>
</xs:element>
<xs:element name="SpatialPoint" minOccurs="0" maxOccurs ="unbounded">
 <xs:complexType>
  <xs:sequence>
   <xs:element name="No." type="xs:integer"/>
   <xs:element ref="point"/>
  </xs:sequence>
 </xs:complexType>
</xs:element>
<xs:element name="point" minOccurs="1" maxOccurs="unbounded">
 <xs:complexType>
  <xs:sequence>
   <xs:element name="Xaxis" type="xs:integer"/>
   <xs:element name="Yaxis" type="xs:integer"/>
  </xs:sequence>
 </xs:complexType>
</xs:element>
```

3. The subclass *DarkCloud* is mapped as follow:

```
<xs:element name="DarkCloud" minOccurs="0" maxOccurs="unbounded">
 <xs:complexType>
  <xs:element name="fuzzyGeneration" >
    <xs:complexType>
     <xs:extension base="Cloud"/>
     <xs:element name="Val">
       <xs:complexType>
        <xs:attribute name ="Poss" type ="xsd:fuzzy" default="1.0"/>
       </xs:complexType>
     </xs:element>
    </xs:complexType>
  </xs:element>
  <xs:element name="Rain"/>
  <xs:element name="TimeInterval"/>
  <xs:element name="Direction_1"/>
 </xs:complexType>
</xs:element>
<xs:element name="Rain" minOccurs="0" maxOccurs="unbounded">
 <xs:complexType>
  <xs:element name="Val"/>
 </xs:complexType>
</xs:element>
<xs:element name="Val" minOccurs="0" maxOccurs="unbounded">
 <xs:complexType>
  <xs:attribute name="Poss" type="xsd:fuzzy" default="1.0"/>
 </xs:complexType>
</xs:element>
<xs:element name="TimeInterval">
 <xs:complexType>
  <xs:sequence>
    <xs:element name="timeStart"/>
    <xs:element name="timeEnd"/>
   </xs:sequence>
 </xs:complexType>
</xs:element>
<xs:element name="timeStart" minOccurs="0" maxOccurs="unbounded">
 <xs:simpleType>
  <xs:restriction base="xs:dateTime">
  <xs:minInclusive value="valueStart"/>
  <xs:maxInclusive value="valueEnd"/>
   </xs:restriction>
 <xs:simpleType>
</xs:element>
<xs:element name="timeEnd" minOccurs="0" maxOccurs="unbounded">
 <xs:simpleType>
  <xs:restriction base="xs:dateTime">
  <xs:minInclusive value="valueStart"/>
  <xs:maxInclusive value="valueEnd"/>
   </xs:restriction>
 <xs:simpleType>
</xs:element>
<xs:element name="Direction_1" minOccurs="1" maxOccurs="unbounded">
 <xs:complexType>
  <xs:element name="TimePoint"/>
  <xs:sequence>
  <xs:element ref="degree"/>
  </xs:sequence>
 </xs:complexType>
</xs:element>
<xs:element name="degree" minOccurs="2" maxOccurs="2">
 <xs:simpleType>
  <xs:restriction base="xs:integer">
  <xs:minInclusive value="degreeStart"/>
  <xs:maxInclusive value="degreeEnd"/>
  </xs:restriction>
 </xs:simpleType >
</xs:element>
<xs:element name="TimePoint" minOccurs="0" maxOccurs="unbounded">
 <xs:restriction base="xs:dateTime">
  <xs:minInclusive value="valueStart"/>
  <xs:maxInclusive value="valueEnd"/>
 </xs:restriction>
</xs:element>
```

4. The subclass *WhiteCloud* is mapped as follow:

```
<xs:element name="WhiteCloud" minOccurs="0" maxOccurs ="unbounded">
 <xs:complexType>
  <xs:element name="fuzzyGeneration" >
    <xs:complexType>
     <xs:extension base="Cloud"/>
     <xs:element name="Val">
       <xs:complexType>
        <xs:attribute name ="Poss" type ="xs:fuzzy" default="1.0"/>
       </xs:complexType>
     </xs:element>
    </xs:complexType>
  </xs:element>
  <xs:element name="Sunshade" type="original-definition"/>
  <xs:element name="Direction_2"/>
 </xs:complexType>
</xs:element>
 <xs:element name="Direction_2" minOccurs="0" maxOccurs="unbounded">
 <xs:complexType>
  <xs:element name="MotionValue">
       <xs:simpleType>
        <xs:restriction base="xs:integer">
         <xs:minInclusive value="valueStart"/>
         <xs:maxInclusive value="valueEnd"/>
        </xs:restriction>
       </xs:simpleType>
  </xs:element>
  <xs:sequence>
   <xs:element name="degree" type="xs:integer"/>
   <xs:element name ="val"/>
  </xs:sequence>
 </xs:complexType>
</xs:element>
<xs:element name="Val" minOccurs="0" maxOccurs="unbounded">
 <xs:complexType>
  <xs:attribute name="Poss" type="xs:fuzzy" default="1.0"/>
 </xs:complexType>
</xs:element>
```

5. The linked and part-class *LivingQuarter* is mapped as follow:

```
<xs:element name="LivingQuarter" minOccurs="0" maxOccurs ="n">
 <xs:complexType>
  <xs:element name="fuzzyAssociation">
    <xs:complexType>
     <xs:extension base="City"/>
     <xs:element name="Val">
       <xs:complexType>
        <xs:attribute name ="Poss" type ="xs:fuzzy" default="1.0"/>
       </xs:complexType>
     </xs:element>
    </xs:complexType>
  </xs:element>
  <xs:element name="Year"/>
  <xs:element name="Market" type="original-definition"/>
  <xs:element name="SpatialLine"/>
 </xs:complexType>
</xs:element>
<xs:element name="Year" minOccurs="0" maxOccurs ="unbounded">
 <xs:complexType>
  <xs:element name ="Dist"/>
 </xs:complexType>
</xs:element>
<xs:complexType name="DistType" minOccurs="0" maxOccurs="unbounded">
 <xs:sequence>
  <xs:element name ="Val"/>
 </xs:sequence>
</xs:complexType>
<xs:element name="Val" minOccurs="0" maxOccurs="unbounded">
 <xs:complexType>
  <xs:element name="Vi" type="original-definition"/>
 </xs:complexType>
</xs:element>
<xs:element name="Spatialline">
 <xs:complexType>
  <xs:sequence>
   <xs:element name="No." type="xs:integer"/>
   <xs:element ref="point "/>
  </xs:sequence>
  <xs:attribute name="offset" type="xs:integer"/>
 </xs:complexType>
</xs:element>
<xs:element name="point" minOccurs="2" maxOccurs="2">
 <xs:complexType>
  <xs:sequence>
   <xs:element name="Xaxis" type="xs:integer"/>
   <xs:element name="Yaxis" type="xs:integer"/>
  </xs:sequence>
 </xs:complexType>
</xs:element>
```

6. The part-class *IndustrialDistrict* is mapped as follow:

```
<xs:element name="IndustrialDistrict" minOccurs="0" maxOccurs ="unbounded">
 <xs:complexType>
  <xs:element name="Produce" type="original-definition"/>
  <xs:element name="Factory" type="original-definition"/>
  <xs:element name="SpatialLine"/>
 </xs:complexType>
</xs:element>
<xs:element name="Spatialline">
 <xs:complexType>
  <xs:sequence>
   <xs:element name="No." type="xs:integer"/>
   <xs:element ref="point "/>
  </xs:sequence>
  <xs:attribute name="offset" type="xs:integer"/>
 </xs:complexType>
</xs:element>
<xs:element name="point" minOccurs="2" maxOccurs="2">
 <xs:complexType>
  <xs:sequence>
   <xs:element name="Xaxis" type="xs:integer"/>
   <xs:element name="Yaxis" type="xs:integer"/>
  </xs:sequence>
 </xs:complexType>
</xs:element>
```

7. The whole-class *Land* is mapped as follow:

```
<xs:element name="Land" minOccurs="0" maxOccurs ="unbounded">
 <xs:complexType>
  <xs:element name ="fuzzyAggregation" >
   <xs:complexType>
    <xs:element ref = "LivingQuarter"/>
    <xs:element ref = "IndustrialDistrict"/>
    <xs:element name="Val" minOccurs ="0" maxOccurs ="unbounded">
     <xs:complexType>
      <xs:attribute name ="Poss" type ="xs:fuzzy" default="1.0"/>
     </xs:complexType>
    </xs:element>
   </xs:complexType>
  </xs:element>
 </xs:complexType>
</xs:element>
 <xs:element name="Population" type="original-definition"/>
 <xs:element name="TimePoint"/>
 </xs:complexType>
</xs:element>
<xs:element name="TimePoint" minOccurs="0" maxOccurs="unbounded">
 <xs:restriction base="xs:dateTime">
 <xs:minInclusive value="valueStart"/>
 <xs:maxInclusive value="valueEnd"/>
 </xs:restriction>
</xs:element>
```

Fig. 3.13 The final transforming results

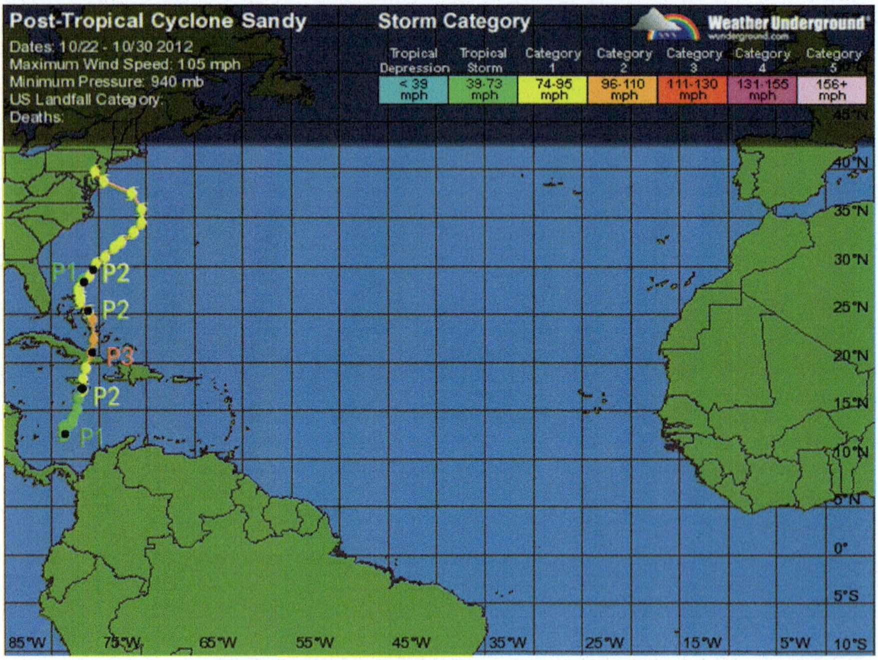

Fig. 3.14 Track of post-tropical cyclone Sandy

range of wind speeds for *P1*, *P2*, and *P3*, while Table 3.3 outlines the corresponding pressure ranges for these stages. Our analysis indicates that wind speed values exhibit a consistent pattern, whereas pressure values demonstrate irregularity. Consequently, wind speed emerges as a critical factor in our modeling approach. Given that each stage possesses distinct attributes, *P1*, *P2*, and *P3* can be classified as three separate categories; however, they maintain interrelatedness among the classes.

Regarding *P1*, it is classified as the initial stage. In this initial stage, *P1* is regarded as *OID* due to its designation as a hurricane, which is inherently unique. The attributes of the spatiotemporal data may encompass multiple factors, such as wind speed and atmospheric pressure. Additionally, the wind speed values typically fall within a standard range. Concerning *FAT*, the extent of wind damage remains uncertain, and the likelihood of occurrence is low, indicating a primary level of fuzziness. In terms of *FST*, the effects of wind speed and the forces of aggregation are relatively weak,

Table 3.2 The range of wind speed in *P1*, *P2* and *P3*

Class	Wind speed (mph)	
	minValue	maxValue
P1	39	73
P2	74	95
P3	96	110

Table 3.3 The range of pressure in *P1*, *P2* and *P3*

Class	Pressure (mb)	
	minValue	maxValue
P1	986	999
P2	940	973
P3	960	963

resulting in a non-deterministic focus of the wind. To represent this state, we can employ fuzzy spatial points and a secondary fuzzy motion direction. Latitude and longitude coordinates can be utilized to denote the fuzzy spatial point. For *FTM*, the onset of the hurricane is not precisely defined, necessitating the use of fuzzy time points for modeling purposes.

Regarding *P2*, it is currently in the developmental phase. During this phase, *P2* is classified as *OID* and is characterized by its uniqueness. In the context of *AT*, the primary attributes of a hurricane include wind speed and pressure. For *FAT*, it is imperative to closely examine the damage caused by wind. Consequently, wind damage is incorporated into the model, thereby enhancing the likelihood of accurate predictions. This aspect is categorized as the first level of fuzziness. In the context of *FST*, as wind speed and aggregation forces intensify, the impact area of the hurricane expands, and its trajectory is observed at an angle. To represent this state, we can utilize fuzzy spatial lines and the initial fuzzy motion direction. Latitude and longitude coordinates can effectively represent these fuzzy spatial lines. For *FTM*, while the precise development time of a hurricane is indeterminate, it is understood to occur after the initiation phase and prior to the formation phase; thus, a fuzzy time interval is employed for modeling purposes.

Regarding *P3*, it represents the forming stage. During this stage, *P2* is regarded as *OID*, which is characterized by its uniqueness. In the context of *AT*, we select wind speed and hurricane pressure as key variables. For *FAT*, we incorporate wind damage into the model; however, the destructive force is characterized by a degree of fuzziness, indicating a third level of uncertainty. In *FST*, both the speed and the aggregative forces of the wind intensify, resulting in an expanded influence area of the hurricane. To represent this state, we can utilize fuzzy spatial regions along with the initial fuzzy motion direction. Concurrently, it is essential to consider the motion value. For *FTM*, the application of a fuzzy time interval is deemed appropriate.

The relationship among *P1*, *P2*, and *P3* can be characterized as a process of incremental development. It is recommended that a method termed *Develop* be incorporated into the model. The association among *P1*, *P2*, and *P3* is identified as a fuzzy relationship, exhibiting a high degree of membership. Based on the analysis conducted, the *OOSDM* Schema is illustrated in Fig. 3.15.

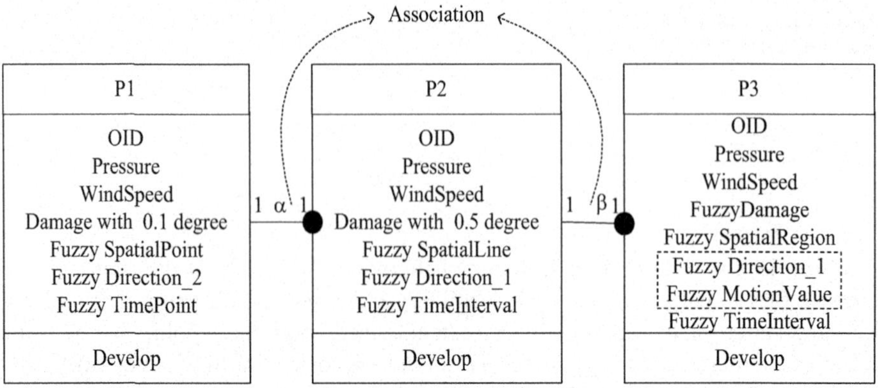

Fig. 3.15 The *OOSDM* schema of post-tropical cyclone Sandy

Next, we will undertake the conversion of post-tropical cyclone Sandy from the *OOSDM* to XML. The model comprises three distinct classes, each of which encompasses *OID*, general attributes (or methods), fuzzy general attributes, fuzzy spatial position, fuzzy spatial motion, and fuzzy temporal attributes. Interrelationships exist among these classes. Specifically, the class *P1* possesses numerous attributes (or methods), including *OID* and general attributes such as pressure and wind speed, which range from 39 to 73 mph, representing the first level of fuzziness. Additionally, *P1* includes fuzzy spatial point attribute, fuzzy motion direction_2 attribute, fuzzy time point attribute and develop method. The representation of class *P1* is illustrated in Fig. 3.16.

As illustrated in Fig. 3.16, the XML Schema for *P1* is deemed valid. Rule 3.17 to Rule 3.34 are articulated in formal language; however, adjustments to the XML Schema are necessary to align with real-world data applications. For instance, the general attribute "pressure" has undergone a modification in its data type from its original definition to an integer, while the general method "develop" has transitioned from its original definition to a string. Additionally, the type for *OID* is designated as a string. In the context of fuzzy spatial points, *Xaxis* corresponds to *Lat* and *Yaxis* corresponds to *Lon*, both of which are classified as float types. The post-tropical cyclone Sandy occurred from October 22, 2010, to October 30, 2010, thereby establishing the fuzzy time point as spanning from 2012–10–22T00:00:00 to 2012–10–31T23:59:59. Finally, the generated XML document for *P1* is presented in Fig. 3.17.

The circumstances of *P1*, *P2*, and *P3* are similar, so we only present the results of *P2* and *P3*. The class *P2* is mapped in Fig. 3.18, while the generating XML document of *P2* is illustrated in Fig. 3.19. The class *P3* is mapped in Fig. 3.20, and the generating XML document of *P3* is represented in Fig. 3.21.

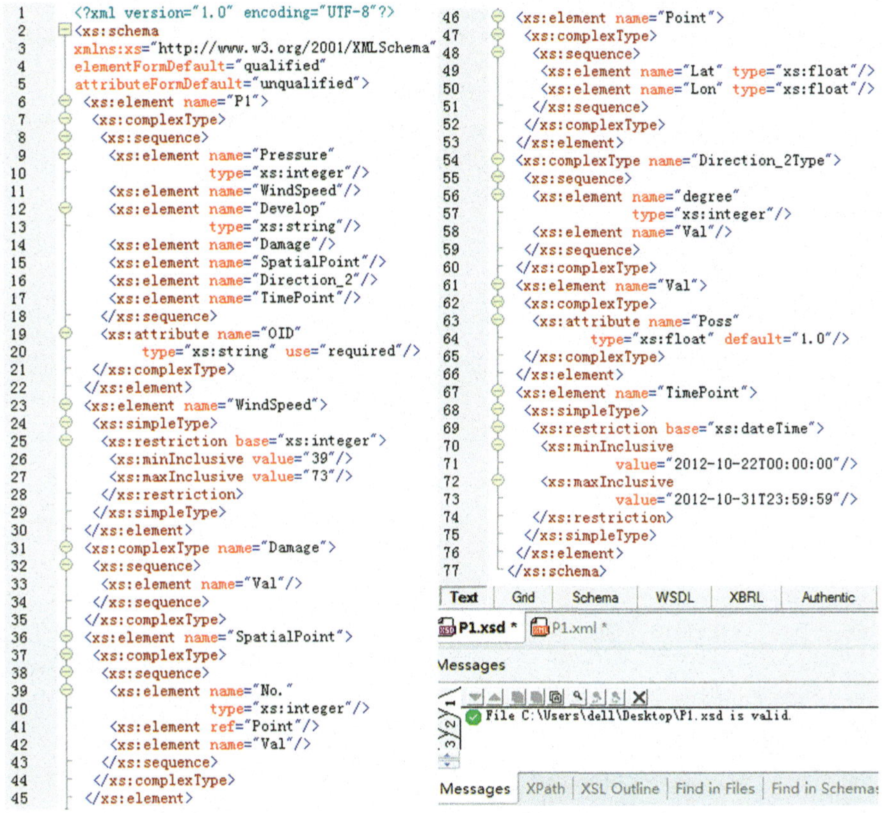

Fig. 3.16 The XML schema of *P1*

3.6 Summary

This chapter presents a framework for representing fuzzy spatiotemporal data utilizing XML and object-oriented databases. Subsequently, we explore the methodologies for transforming fuzzy spatiotemporal data from XML to object-oriented databases and vice versa. We propose a comprehensive set of rules to facilitate the transformation of fuzzy spatiotemporal data between XML and object-oriented databases. Furthermore, we outline the procedural steps involved in the transformation process.

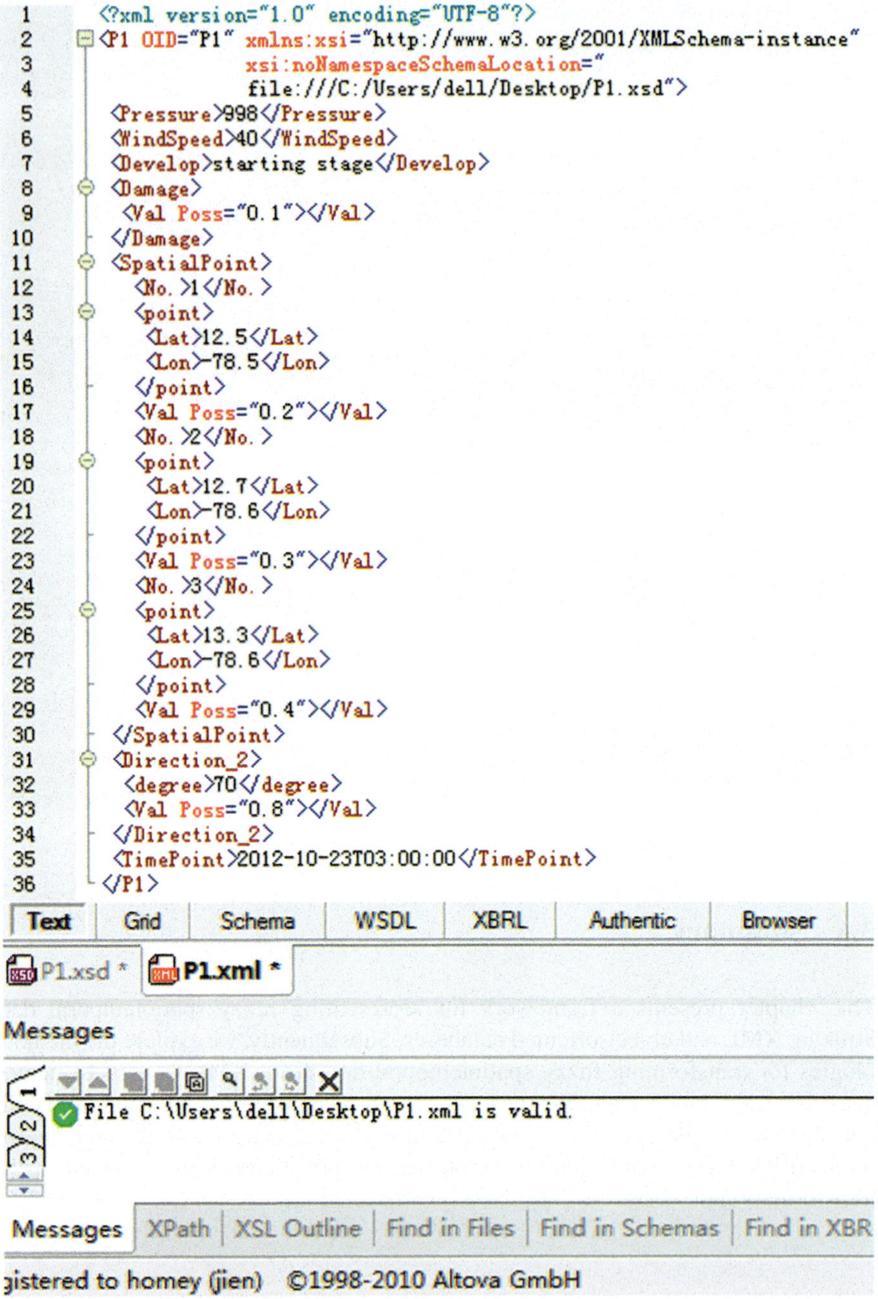

Fig. 3.17 The XML document of *P1*

```xml
<?xml version="1.0" encoding="UTF-8"?>
<xs:schema
xmlns:xs="http://www.w3.org/2001/XMLSchema"
elementFormDefault="qualified"
attributeFormDefault="unqualified">
 <xs:element name="P2">
  <xs:complexType>
   <xs:sequence>
    <xs:element name="fuzzyAssociation">
     <xs:complexType>
      <xs:sequence>
       <xs:extension base="P1"/>
       <xs:element name="Val">
        <xs:complexType>
         <xs:attribute name="Poss"
          type="xs:float" default="1.0"/>
        </xs:complexType>
       </xs:element>
      </xs:sequence>
     </xs:complexType>
    </xs:element>
    <xs:element name="Pressure"
                type="xs:integer"/>
    <xs:element name="WindSpeed"/>
    <xs:element name="Develop"
                type="xs:string"/>
    <xs:element name="Damage"/>
    <xs:element name="SpatialLine"/>
    <xs:element name="Direction_1"/>
    <xs:element name="TimeInterval"/>
   </xs:sequence>
   <xs:attribute name="OID"
        type="xs:string" use="required"/>
  </xs:complexType>
 </xs:element>
 <xs:element name="WindSpeed">
  <xs:simpleType>
   <xs:restriction base="xs:integer">
    <xs:minInclusive value="74"/>
    <xs:maxInclusive value="95"/>
   </xs:restriction>
  </xs:simpleType>
 </xs:element>
 <xs:complexType name="DamageType">
  <xs:sequence>
   <xs:element name="Val"/>
  </xs:sequence>
 </xs:complexType>
 <xs:element name="Val">
  <xs:complexType>
   <xs:attribute name="Poss"
           type="xs:float" default="1.0"/>
  </xs:complexType>
 </xs:element>
 <xs:complexType name="SpatiallineType">
  <xs:sequence>
   <xs:element name="No."
              type="xs:integer"/>
   <xs:element ref="point"/>
  </xs:sequence>
  <xs:attribute name="offset"
                type="xs:integer"/>
 </xs:complexType>
 <xs:element name="point">
  <xs:complexType>
   <xs:sequence>
    <xs:element name="Lat" type="xs:float"/>
    <xs:element name="Lon" type="xs:float"/>
   </xs:sequence>
  </xs:complexType>
 </xs:element>
 <xs:complexType name="Direction_1Type">
  <xs:sequence>
   <xs:element ref="degree"/>
  </xs:sequence>
 </xs:complexType>
 <xs:element name="degree">
  <xs:simpleType>
   <xs:restriction base="xs:integer">
    <xs:minInclusive value="0"/>
    <xs:maxInclusive value="360"/>
   </xs:restriction>
  </xs:simpleType>
 </xs:element>
 <xs:complexType name="TimeIntervalType">
  <xs:sequence>
   <xs:element name="timeStart"/>
   <xs:element name="timeEnd"/>
  </xs:sequence>
 </xs:complexType>
 <xs:element name="timeStart">
  <xs:simpleType>
   <xs:restriction base="xs:dateTime">
    <xs:minInclusive
           value="2012-10-22T00:00:00"/>
    <xs:maxInclusive
           value="2012-10-31T23:59:59"/>
   </xs:restriction>
  </xs:simpleType>
 </xs:element>
 <xs:element name="timeEnd">
  <xs:simpleType>
   <xs:restriction base="xs:dateTime">
    <xs:minInclusive
           value="2012-10-22T00:00:00"/>
    <xs:maxInclusive
           value="2012-10-31T23:59:59"/>
   </xs:restriction>
  </xs:simpleType>
 </xs:element>
</xs:schema>
```

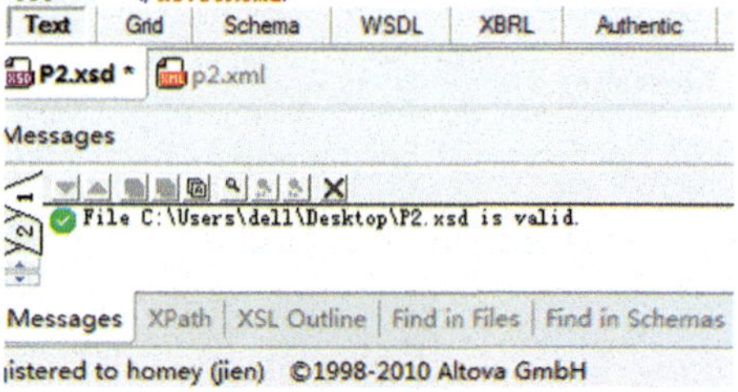

Fig. 3.18 The XML Schema of *P2*

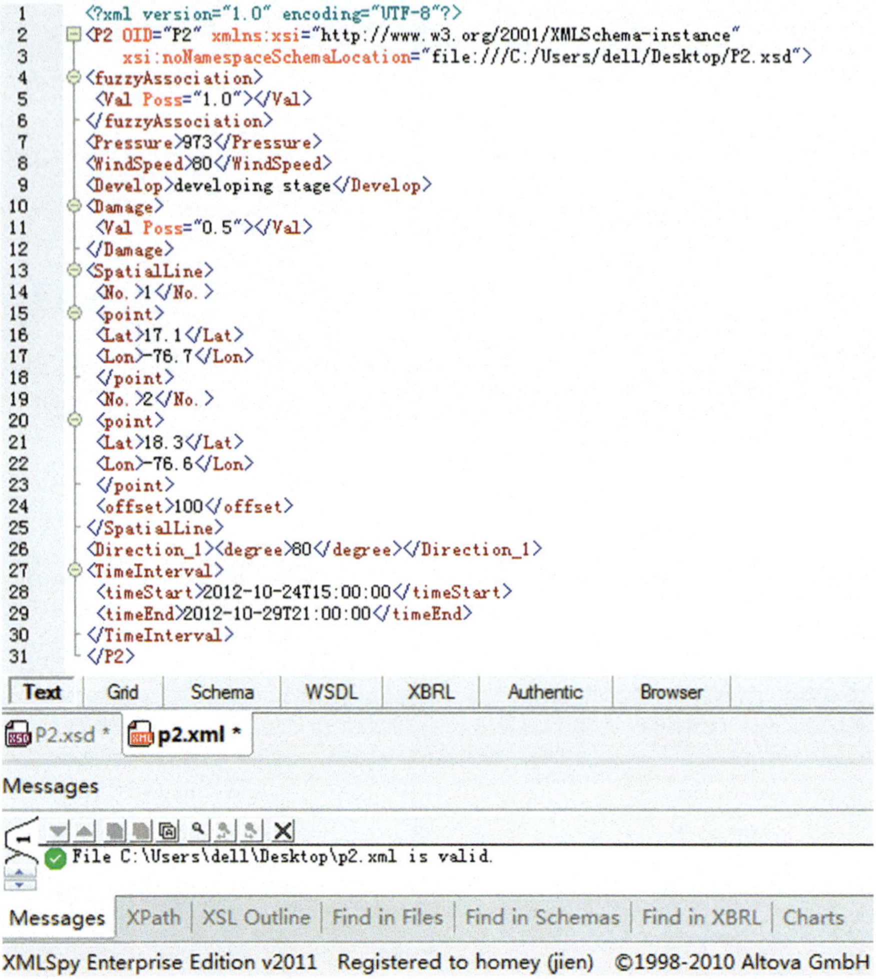

Fig. 3.19 The XML document of *P2*

```xml
<?xml version="1.0" encoding="UTF-8"?>
<xs:schema
 xmlns:xs="http://www.w3.org/2001/XMLSchema"
 elementFormDefault="qualified"
 attributeFormDefault="unqualified">
 <xs:element name="P3">
  <xs:complexType>
   <xs:sequence>
    <xs:element name="fuzzyAssociation">
     <xs:complexType>
      <xs:sequence>
       <xs:extension base="P2"/>
       <xs:element name="Val">
        <xs:complexType>
         <xs:attribute name="Poss"
          type="xs:float" default="1.0"/>
        </xs:complexType>
       </xs:element>
      </xs:sequence>
     </xs:complexType>
    </xs:element>
    <xs:element name="Pressure"
             type="xs:integer"/>
    <xs:element name="WindSpeed"/>
    <xs:element name="Develop"
             type="xs:string"/>
    <xs:element name="Damage"/>
    <xs:element name="SpatialRegion"/>
    <xs:element name="Direction_1"/>
    <xs:element name="TimeInterval"/>
   </xs:sequence>
   <xs:attribute name="OID"
      type="xs:string" use="required"/>
  </xs:complexType>
 </xs:element>
 <xs:element name="WindSpeed">
  <xs:simpleType>
   <xs:restriction base="xs:integer">
    <xs:minInclusive value="96"/>
    <xs:maxInclusive value="110"/>
   </xs:restriction>
  </xs:simpleType>
 </xs:element>
 <xs:complexType name="DamageType">
  <xs:sequence>
   <xs:element name="Dist"/>
  </xs:sequence>
 </xs:complexType>
 <xs:complexType name="DistType">
  <xs:sequence>
   <xs:element name="Val"/>
  </xs:sequence>
 </xs:complexType>
 <xs:element name="Val">
  <xs:complexType>
   <xs:sequence>
    <xs:element name="Vi"
             type="xs:string"/>
   </xs:sequence>
  </xs:complexType>
 </xs:element>
 <xs:element name="SpatialRegion">
  <xs:complexType>
   <xs:sequence>
    <xs:element name="No." type="xs:integer"
    <xs:element ref="point"/>
   </xs:sequence>
  </xs:complexType>
 </xs:element>
 <xs:element name="point">
  <xs:complexType>
   <xs:sequence>
    <xs:element name="Lat" type="xs:float"/>
    <xs:element name="Lon" type="xs:float"/>
   </xs:sequence>
  </xs:complexType>
 </xs:element>
 <xs:complexType name="Direction_1Type">
  <xs:sequence>
   <xs:element name="MotionValue">
    <xs:simpleType>
     <xs:restriction base="xs:integer">
      <xs:minInclusive value="0"/>
      <xs:maxInclusive value="1000"/>
     </xs:restriction>
    </xs:simpleType>
   </xs:element>
   <xs:sequence>
    <xs:element ref="degree"/>
   </xs:sequence>
  </xs:sequence>
 </xs:complexType>
 <xs:element name="degree">
  <xs:simpleType>
   <xs:restriction base="xs:integer">
    <xs:minInclusive value="0"/>
    <xs:maxInclusive value="360"/>
   </xs:restriction>
  </xs:simpleType>
 </xs:element>
 <xs:complexType name="TimeIntervalType">
  <xs:sequence>
   <xs:element name="timeStart"/>
   <xs:element name="timeEnd"/>
  </xs:sequence>
 </xs:complexType>
 <xs:element name="timeStart">
  <xs:simpleType>
   <xs:restriction base="xs:dateTime">
    <xs:minInclusive
             value="2012-10-22T00:00:00"/>
    <xs:maxInclusive
             value="2012-10-31T23:59:59"/>
   </xs:restriction>
  </xs:simpleType>
 </xs:element>
 <xs:element name="timeEnd">
  <xs:simpleType>
   <xs:restriction base="xs:dateTime">
    <xs:minInclusive
             value="2012-10-22T00:00:00"/>
    <xs:maxInclusive
             value="2012-10-31T23:59:59"/>
   </xs:restriction>
  </xs:simpleType>
 </xs:element>
</xs:schema>
```

Text | Grid | Schema | WSDL | XBRL | Authentic

P3.xsd * | p3.xml

Messages

File C:\Users\dell\Desktop\P3.xsd is valid.

Messages | XPath | XSL Outline | Find in Files | Find in Schemas

jistered to homey (jien) ©1998-2010 Altova GmbH

Fig. 3.20 The XML Schema of *P3*

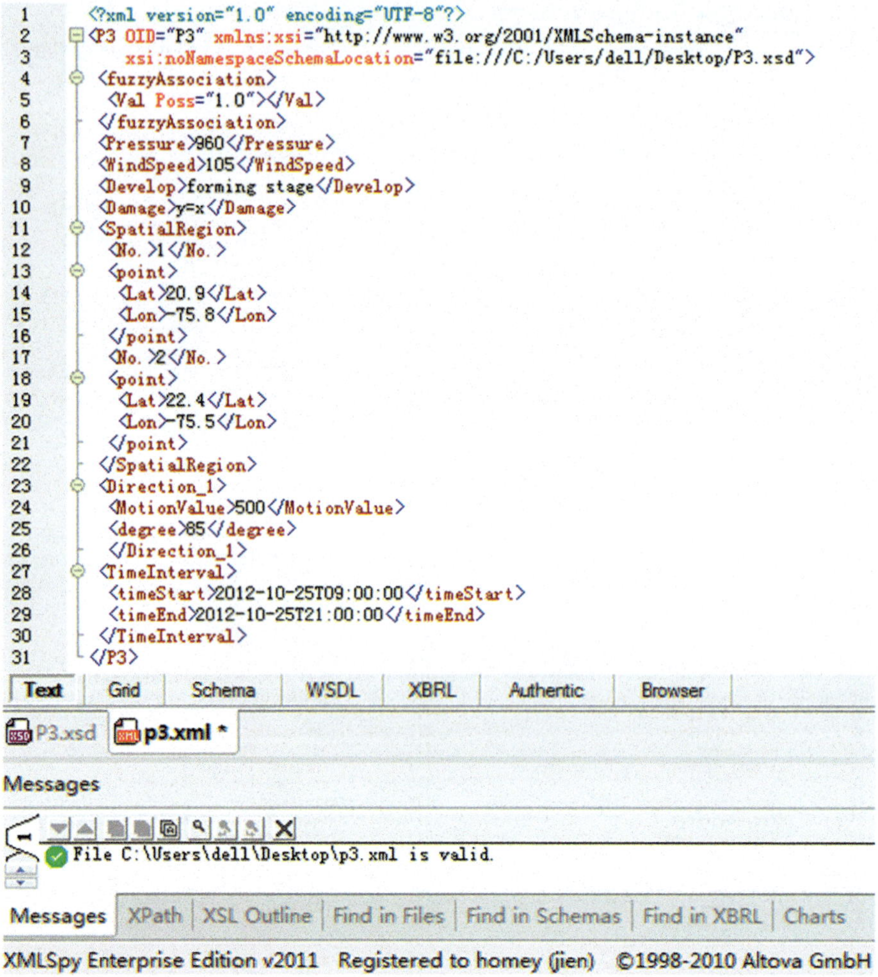

Fig. 3.21 The XML document of *P3*

References

1. H.L. Bhandari, R. Chitrakar, Enhancement of a transformation algorithm to migrate SQL database into NoSQL graph database. Data Sci. J. **23**, 1–22 (2024)
2. N. Bansal, S. Sachdeva, L.K. Awasthi, Query-based denormalization using hypergraph (QBDNH): a schema transformation model for migrating relational to NoSQL databases. Knowl. Inf. Syst. **66**(1), 681–722 (2024)
3. N. Zaidi, I. Ishak, F. Sidi, L.S. Affendey, An efficient schema transformation technique for data migration from relational to column-oriented databases. Comput. Syst. Sci. Eng. **43**(3), 1175–1188 (2022)
4. O. Alotaibi, E. Pardede, Transformation of schema from relational database (RDB) to NoSQL databases. Data **4**(4), 148 (2019)

5. S. Candrlic, M.A. Katic, M. Pavlic, A system for transformation of sentences from the enriched formalized Node of Knowledge record into relational database. Expert Syst. Appl. **115**, 442–464 (2019)
6. R. Chen, G. Cai, J. Chen, Y. Hong, Integrated method for distributed processing of large XML data. Clust. Comput. **27**(2), 1375–1399 (2024)
7. J. Vidakovic, S. Ristic, S. Kordic, I. Lukovic, Extended tuple constraint type as a complex integrity constraint type in XML data model—definition and enforcement. Comput. Sci. Inf. Syst. **15**(3), 821–843 (2018)
8. B. Oliboni, G. Pozzani, Representing fuzzy information by using XML schema, in *Proceedings of DEXA* (2008), pp. 683–687
9. L. Bai, L. Yan, Z. Ma, Fuzzy spatiotemporal data modeling and operations in XML. Appl. Artif. Intell. **29**(3), 259–282 (2015)
10. J. Shan, J. Lu, X. Chen, L. Yan, Z. Ma, A semantics-preserving approach for extracting RDF knowledge from object-oriented databases. J. Web Eng. **22**(2), 197–220 (2023)
11. C. Ma, J. Yang, W. Xia, J. Liu, Y. Zhang, X. Sui, A model for expressing industrial information based on object-oriented industrial heat sources detected using multi-source thermal anomaly data in China. Remote Sens. **14**(4), 835 (2022)
12. M.L. Glukharev, M.F. Solomatova, Access differentiation in object-oriented databases based on the extended object-oriented Harrison-Ruzzo-Ullman model. Autom. Control. Comput. Sci. **54**(8), 1007–1012 (2020)
13. A. Frihida, D.J. Marceau, M. Theriault, Spatio-temporal object-oriented data model for disaggregate travel behavior. Trans. GIS **6**(3), 277–294 (2002)
14. T. Naser, R. Alhajj, M.J. Ridley, Reengineering XML into object-oriented database, in *Proceedings of the 2008 IEEE International Conference on Information Reuse and Integration*, (2008), pp. 1–6
15. J. Liu, Z. Ma, X. Feng, Formal approach for reengineering fuzzy XML in fuzzy object-oriented databases. Appl. Intell. **38**(4), 541–552 (2013)
16. T. Naser, R. Alhajj, M.J. Ridley, Two-way mapping between object-oriented databases and XML. Informatica **33**(3), 297–308 (2009)
17. L. Bai, Z. Jia, J. Liu, Transformation of fuzzy spatiotemporal data from XML to object-oriented database. Earth Sci. Inf. **11**(3), 449–461 (2018)
18. J.F. Terwilliger, P.A. Bernstein, S. Melnik, Full-fidelity flexible object-oriented XML access. Proc. VLDB Endowment **2**(1), 1030–1041 (2009)
19. T. Naser, K. Kianmehr, R. Alhajj, M.J. Ridley, Transforming object-oriented databases into XML, in *Proceedings of the 2007 IEEE International Conference on Information Reuse and Integration*, (2007), pp. 600–605
20. J. Liu, Z. Ma, Formal transformation from fuzzy object-oriented databases to fuzzy XML. Appl. Intell. **39**(3), 630–641 (2013)
21. L. Bai, Z. Jia, J. Liu, Reengineering object-oriented fuzzy spatiotemporal data into XML. IEEE Access **6**(1), 12686–12699 (2018)
22. L.A. Zadeh, Fuzzy sets. Inf. Control. **8**(3), 338–353 (1965)
23. L.A. Zadeh, Fuzzy sets as a basis for a theory of possibility. Fuzzy Sets Syst. **1**(1), 3–28 (1978)
24. Weather Underground. https://www.wunderground.com

Chapter 4
Topological Relations of Fuzzy Spatiotemporal XML Data Over Time

4.1 Introduction

Among the various operations performed on fuzzy spatiotemporal data, determining topological relations is particularly significant. To illustrate this, consider the challenge of monitoring weather conditions in Hong Kong. The inquiry regarding the presence of rainfall in Hong Kong possesses a spatiotemporal dimension, necessitating an evaluation of cloud cover over the region across a sequence of events and time intervals. This involves analyzing various scenarios, such as periods when the cloud and Hong Kong are disjoint, instances when they intersect at specific time points, durations of overlap, situations where the cloud is contained within Hong Kong, and eventual returns to disjoint states. However, relying solely on topological relations between the cloud and Hong Kong may lead to erroneous conclusions regarding rainfall. For instance, the presence of clouds does not necessarily indicate rain, particularly if the cloud cover is minimal. A cloud cover of 2% may suggest clear conditions, while a 30% cloud cover could indicate overcast skies without precipitation. Therefore, to enhance the precision of determining topological relations within fuzzy spatiotemporal data, the integration of fuzzy logic into the assessment of cloud cover is proposed. Consequently, a variety of methodologies have been proposed for the determination of topological relations, which primarily encompass point-set topology [1], dimensional-extended methods [2], minimum bounding rectangles [3], region connection calculus [4], the CBM approach [5], the 2D-STRING model [6], the Voronoi method [7], and the generalized intersection model [8]. Notably, the well-known 4-Intersection approach [9, 10], the 9-Intersection approach [11], and the Intersection and Difference (ID) model [12, 13] have been developed to formalize topological relations between simple and precise regions.

Fuzzy temporal relations are essential in various real-world applications, prompting inquiries into the appropriate formalization of these relations to ascertain their fuzzy characteristics. To address this issue, significant efforts have been made in recent years to develop fuzzy temporal relations, as evidenced by numerous

L. Bai and L. Zhu, *Fuzzy Spatiotemporal XML Data Management*, Studies in Computational Intelligence 1183, https://doi.org/10.1007/978-3-031-81033-6_4

studies [14–18]. In their work, Badaloni and Giacomin [15] incorporate the concepts of flexibility and fuzziness into Allen's interval-based temporal framework, resulting in a novel formalism termed IA^{fuz}, which serves to extend classical Interval Algebra (IA). Additionally, Badaloni and Falda [14] explore the representation of diverse forms of fuzzy temporal relations, aiming to establish a more comprehensive representation methodology. Ribarić et al. [16] introduce a geometric approach to validate the possibility and necessity measures associated with temporal relations between two fuzzy time points, as well as between a fuzzy time point and a fuzzy time interval. However, a significant limitation of these advancements is the computational expense associated with evaluating fuzzy temporal relations, which poses challenges for practical applications. To mitigate this issue, Schockaert et al. [18] examine how the evaluation of fuzzy temporal interval relations for piecewise linear fuzzy intervals can be simplified to that of linear fuzzy intervals, offering a characterization that is both efficient for evaluation and straightforward to implement. Furthermore, Schockaert and De Cock [17] propose a framework for reasoning about qualitative and metric temporal relations among fuzzy time intervals, leveraging established results to address disjunctive temporal reasoning challenges.

Spatial relations are essential in a variety of application domains, including Geographic Information Systems (GIS) [19] and image interpretation [20]. These relations can be categorized into three types: topological (e.g., "overlap," "meet", etc.), directional (e.g., "north of," "south of", etc.), and metric (e.g., "3 km away from", etc.) [21]. Given the critical importance of topological relations in spatial reasoning and analysis, there has been a growing focus on their quantitative assessment. Notably, both the Region Connection Calculus (RCC) [22] and the 9-intersection model [23, 24] offer formal frameworks for characterizing qualitative spatial relations.

Spatial data in real-world applications often exhibit a degree of fuzziness due to the subjective nature of their values [25, 26]. Consequently, a significant challenge arises in modeling fuzzy spatial data to ascertain their fuzzy topological relationships. To address this issue, several definitions of spatial relations between fuzzy spatial data have been proposed, which either extend the RCC or the 9-intersection model. These definitions conceptualize a fuzzy region as comprising two distinct components: one that includes points definitively within the fuzzy region, and another that encompasses points definitively outside of it [27–29].

Building upon these foundational models, research has expanded to explore topological relations involving indeterminate spatial objects [30–32] and fuzzy spatial objects [33–37]. In [38], a basic fuzzification of the definitions pertaining to RCC-8 relations is introduced. They presents a fuzzy adaptation of the crisp spatial logic originally formulated by Randell et al. [22], which utilizes the single relation "connected-with" as its foundational element. However, this approach results in the loss of numerous properties inherent to the original RCC-8 relations. Furthermore, the methodology for applying the definitions outlined in [38] to compute the values of fuzzy spatial relations between two specified fuzzy regions remains ambiguous. To address these issues, Schockaert [39] et al. expand upon the RCC framework by offering generalized definitions of spatial relations as fuzzy relations, thereby enabling the articulation of the extent to which a specific spatial relation between two regions is valid.

Specifically, the extended 9-Intersection model substitutes the crisp boundary with a broader boundary, and the 3×3 Intersection matrix has been augmented to 4×4 and 5×5 Intersection matrices, which account for the interior, boundary, and exterior of simple fuzzy spatial regions. Furthermore, while Tang [37] and Bejaoui et al. [36] have proposed 152 and 242 relations, respectively, between two fuzzy regions, their studies do not consider the continuous determination of these regions. In addition to the static determination of topological relations, there has been an extension of the two-dimensional Egg/Yolk model [28] into three dimensions, which aims to describe the approximate topological relations of indeterminate evolving regions over time [40]. However, the existing researches predominantly rely on traditional databases, which limits their capacity to effectively manage the complexities associated with fuzzy attribute extension and fuzzy temporal extension. In addition, there are less reports addressing the connections between fuzzy spatial relations and spatial relations, particularly from a mathematical perspective. While fuzzy spatial relations have been formalized within the frameworks of both the RCC [38, 39] and the 9-intersection model [33–35], there remains a limited number of studies focusing on a comprehensive algorithm for identifying all fuzzy relations. It is noteworthy that there are a total of 23 fuzzy relations, which will be discussed in a subsequent section of this chapter, despite the existence of specific methodologies that have been proposed in prior researches [34, 39].

Topological relations exhibit invariance under continuous transformations, thereby reflecting the global characteristics of fuzzy spatiotemporal data. Consequently, these relations are crucial in the process of change detection for fuzzy spatiotemporal data represented in GIS databases, such as those pertaining to coastal and forest environments, as well as in satellite imagery databases, including TM and SPOT images. Given their significance in various practical applications, several efficient methodologies for describing topological relations have been proposed. Furthermore, it is important to note that spatiotemporal data in real-world applications often exhibit fuzziness and may undergo changes over time. Therefore, the modeling of topological relations among fuzzy spatiotemporal data across temporal dimensions presents a critical area of inquiry. Topological relationships among fuzzy regions characterize the spatial intersections or relationships that exist between these regions in a given space, a topic that has been the subject of extensive research. The Egg/Yolk model, which employs an adaptation of the RCC theory, provides fundamental relations for representing qualitative spatial regions, as articulated by J.E.P.E. Furthermore, the 9-Intersection model has been introduced for the analysis of topological relations, focusing on the intersections among the interior, boundary, and exterior components of the regions in question. Building upon this framework, the model has been extended to accommodate fuzzy regions. However, the aforementioned researchers focus exclusively on topological relations within a two-dimensional framework, neglecting the dimension of time in their analyses. Consequently, several studies [5, 40] have sought to adapt these findings to a three-dimensional context to incorporate temporal evolution. However, this transition to three-dimensional space has resulted in the omission of certain cases. For instance, Bao et al. [40] extend the two-dimensional indeterminate topological model introduced in [28] to derive 46 three-dimensional volume pairs from 46 basic two-dimensional relations. This approach fails to account for specific cases, such as those involving "sometimes" and "always" topological relations. Furthermore, those studies do not differentiate between various spatiotemporal relations over time when represented by the same matrix.

The rapid advancement of the Internet has facilitated the emergence of XML (Extensible Markup Language) as a prominent medium for the integration and exchange of data from diverse sources. Furthermore, XML significantly impacts data management due to its ability to maintain a tree structure [41], allowing for the consideration of nodes or subtrees as metadata. XML data is structured in a tree format, and queries are typically formulated as twig patterns, which may include additional predicates related to the contents or attribute values of the tree nodes [42]. This structure simplifies the determination of topological relationships, as desired twigs can be identified through twig pattern matching, and relationships can be computed through operations on the nodes within the XML framework. Consequently, it is essential to explore the application of XML in the determination of fuzzy topological relations, a topic that has been addressed in our previous research on fuzzy spatiotemporal XML efforts [43–45]. Notably, despite XML's advantages in representing and manipulating fuzzy spatiotemporal data, and the potential for extending new data types [46] and fuzzy concepts [47, 48] to support such data, research focused on the determination of fuzzy topological relations among fuzzy spatiotemporal objects in XML has only recently commenced and warrants further investigation.

In this chapter, after presenting fuzzy temporal semantics and fuzzy spatial semantics, we build upon the 9-Intersection model to develop a framework for the identification of continuous topological relationships over time in XML. Based on this framework, we delineate all potential fuzzy topological relations that may exist between two fuzzy spatiotemporal regions over time, along with their associated fuzzy topological relation *R*. Subsequently, we propose a quantitative algorithm designed to identify fuzzy topological relations between two fuzzy spatiotemporal regions over time, utilizing the concept of neighborhood relations.

4.2 Related Work

4.2.1 Topological Relations Between Crisp Spatiotemporal Regions

In modeling crisp spatiotemporal regions, both discrete and continuous approaches have been employed to determine topological relations.

The discrete methodologies outlined in Refs. [49] and [50] delineate discrete topological relationships within spatiotemporal data, conceptualizing time as an additional dimension. Chirigati et al. [49] introduce the concept of Data Polygamy, a scalable framework grounded in topology that enables users to query statistically significant relationships among spatiotemporal datasets. This research addresses the challenge of identifying potential relationships between such datasets and facilitates user engagement in the data analysis and exploration process by permitting the formulation of relationship queries. In [50], a data model predicated on linear constraints is developed for the representation and manipulation of multidimensional data. To maintain low complexity in query evaluation, the orthographic dimension of an object $O = O_1 \times O_2 \times \ldots \times O_n$ is constrained. This approach allows for the independent processing of queries across each component, thereby achieving an effective balance between design simplicity, the expressive capacity of the query language, and the efficiency of query evaluation.

On the other hand, continuous approaches elucidate the continuous topological relationships of spatiotemporal data over time, treating time as an integral component of the coordinate system. To develop a continuous spatiotemporal data model, the research presented in [51] proposes a framework that characterizes the geometry, topology, and properties of geographic objects as functions of time along a valid real-time axis. These functions are determined by their values at a discrete set of temporal instances, along with user-defined interpolation methods. Within this framework, two distinct types of evolutions can be identified: those that maintain all topological relationships among the objects, such as homeomorphisms, and those that modify the topological relationships of the objects. Furthermore, Erwig and Schneider [52] introduce the concept of a temporal function as a foundational element of an algebraic model for spatiotemporal data types. An element of type $\tau(\alpha) = time \rightarrow \alpha$ represents all total functions mapping time to α, which is associated with one of the spatial data types: *point*, *line*, or *region*. Consequently, the outcomes include *moving points* as values of τ (*point*), *evolving lines* as values of type τ (*line*), and *evolving regions* as values of type τ (*region*).

4.2.2 Topological Relations Between Fuzzy Spatiotemporal Regions

Spatiotemporal regions often exhibit characteristics of ambiguity rather than being distinctly defined. Numerous attempts have been undertaken to conceptualize the topological relationships among fuzzy spatiotemporal regions, employing various frameworks such as point-set topology, broad boundary topology, rough-set topology, three-valued topology, and fuzzy-set topology, among others.

Point-set topology, also referred to as set-theoretic topology or general topology, encompasses the examination of the abstract characteristics of continuity and proximity within various spaces. Fundamental concepts within point-set topology include continuity, dimension, compactness, and connectedness. Schneider [53] introduces a Spatial Plateau Algebra, which offers spatial plateau data types as a means of implementing fuzzy spatial data types. Each spatial plateau object is composed of a finite number of crisp counterparts that are either adjacent or disjoint from one another, each associated with distinct membership values, thereby creating different plateaus. This algebra functions as a type system that encompasses spatial plateau data types such as *fpoint*, *fline*, and *fregion*, along with an extensive array of spatial plateau operations.

Clementini and Felice [32] introduce an expansive boundary model that builds upon the existing framework for regions through a geometric lens. This model incorporates both inner and outer boundaries to delineate the extent of a region [54]. Points situated within the inner boundary are definitively classified as belonging to the region, whereas those located beyond the outer boundary are excluded from it. The area between these two boundaries constitutes the broad boundary, which encompasses points that may or may not be part of the region. In [55], the authors investigate decentralized algorithms aimed at computing topological information. This methodology produces comprehensive global topological data regarding regions by utilizing three fundamental boundary structures: boundary nodes, boundary cycles, and boundary orientation. Furthermore, a suite of decentralized algorithms is proposed, designed to efficiently address snapshot queries related to the topology of spatial regions, including inquiries concerning containment and adjacency.

A rough set can be considered as a formal approximation of a crisp set, characterized by a pair of sets that represent the lower and upper approximations of the original set. The research conducted by [56] integrates rough set theory with concepts from topology. Specifically, the topology induced by binary relations is employed to generalize fundamental rough set concepts. Within the approximation space, this approach examines two operators: the upper and lower approximations of subsets. Furthermore, rough membership functions facilitate the expression of fuzzy theory within topological spaces. Building upon this foundation, Qin and Pei [57] explore the relationship between fuzzy rough set models and fuzzy topologies within a finite universe. They propose an axiom for fuzzy topology and demonstrate that the collection of all lower approximation sets of fuzzy sets, based on a reflexive and transitive fuzzy relation, constitutes a fuzzy topology that adheres to the proposed axiom. Conversely, it is established that a fuzzy topology satisfying this axiom corresponds precisely to the set of all lower approximation sets under a reflexive and transitive fuzzy relation.

Three-valued logic refers to various many-valued logic systems characterized by the presence of three distinct truth values: true, false, and an indeterminate third value. In their work, Pauly and Schneider [58] utilize a three-valued logic system that incorporates an additional value, "maybe," to address the nuances of vagueness. Consequently, a vague topological predicate can yield one of these three truth values: it returns true when the relationship is definitively established, false when the relationship is unequivocally absent, and maybe when insufficient information precludes a definitive conclusion. This methodology facilitates a seamless transition from precise to vague spatial concepts, allowing for their analysis within a unified framework.

Fuzzy set theory represents an advancement over conventional set theory by assigning a membership grade, ranging from 0 to 1, to each element within a set [54]. A membership grade of 0 signifies that the element is not a member of the set. Dilo et al. [59] propose a model grounded in fuzzy theory that accounts for gradual transitions. This model employs a standard set of operators applicable to crisp objects and extends their definitions to accommodate vague objects. The operators that yield spatial types are categorized as regularized fuzzy set operators, which include union, intersection, and difference, alongside two topological operators: boundary and frontier, as well as two operators pertinent to vague partitions: overlay and fusion. Furthermore, a method is introduced [60] to generate verbal descriptors concerning topological relations among fuzzy regions, utilizing the fuzzy 4-intersection as a basis. The derivation of verbal terms related to the topological relationship between two fuzzy regions is contingent upon the computation of similarity within the fuzzy 4-intersection. Additionally, a generalization of the RCC is presented [61], which is informed by fuzzy set theory, and it explores how reasoning tasks such as satisfiability and entailment checking can be reformulated as linear programming problems. In this framework, C is defined as a fuzzy relation, whereby for each pair of regions (u, v), $C(u, v)$ denotes a degree within the interval [0, 1], indicating the extent of connectivity between regions u and v.

Table 4.1 presents a comparative analysis of various features. The findings indicate that existing research predominantly emphasizes qualitative assessments of topological relationships among fuzzy regions, while insufficient attention is given to the development of a robust and efficient model for quantitatively identifying these relationships. Furthermore, current studies primarily address discrete topological relations between fuzzy regions, neglecting the formulation of a model for identifying continuous topological relations among fuzzy regions over time.

4.3 Fuzzy Temporal Data Semantics

In this subsection, we delineate the fundamental principles of representing fuzzy temporal relations from two perspectives: fuzzy time points and fuzzy time intervals. Within each perspective, we define fuzzy temporal data and their corresponding relationships.

An instantaneous event may have its occurrence time imprecisely defined, referred to as a fuzzy time point (*FTP*). This concept of a fuzzy time point is understood as a representation of a collection of potential time points, accompanied by the associated degrees of likelihood for the occurrence of the event.

Table 4.1 Comparative study in terms of different features

References	Qualitative	Quantitative	Discrete	Continuous	Crisp	Fuzzy
Clementini and Felice [32]	✓		✓			✓
Bejaoui et al. [36]	✓		✓			✓
Tang [37]	✓		✓			✓
Grumbach et al. [50]		✓	✓		✓	
Le et al. [51]	✓			✓	✓	
Erwig and Schneider Schneider[52]	✓			✓	✓	
Schneider [53]	✓		✓			✓
Verstraete [54]	✓		✓			✓
Duckham et al. [55]		✓	✓			✓
Lashin et al. [56]	✓		✓			✓
Qin and Pei [57]	✓		✓			✓
Pauly and Schneider [58]	✓		✓			✓
Dilo et al. [59]	✓		✓			✓
Bjorke [60]	✓		✓			✓
Schockaert [61]	✓		✓			✓
Bai et al. [45]		✓		✓		✓

Definition 4.1 (*Fuzzy time point*) For a fuzzy time point, we have $FTP = (T, \delta)$, where:

- T represents a fuzzy time point
- represents the membership degree of the fuzzy time point corresponding to the time T, where $0 \leq \delta \leq 1$.

The fuzzy relations between two fuzzy time points can be categorized into three distinct cases: fuzzy equal, fuzzy disjoint, and fuzzy overlap, which are represented as *FTPequal*, *FTPdisjoint*, *FTPoverlap*, respectively. Here, a mathematic symbol *supp* is introduced, where $supp\, A = \{u \mid u \in U, A\,(u) > 0\}$.

Definition 4.2 (*Fuzzy relations of fuzzy time points*) For two fuzzy time points $FTP_1 = (T_1, \delta_1)$ and $FTP_2 = (T_2, \delta_2)$, $(t_{1i}, \delta_{1i}) \in FTP_1$ and $(t_{2i}, \delta_{2i}) \in FTP_2$, we have:

- $FTPequal(FTP_1, FTP_2) : \min(supp(t_{li}, \delta_{li})) = \min(supp(t_{2i}, \delta_{2i}))$

  $\vee \max(supp(t_{2i}\delta_{2i})) = \max(supp\,(t_{li}, \delta_{li}))$.
- $FTPdisjoint(FTP_1, FTP_2) : \max(supp(t_{li}, \delta_{li})) < \min(supp(t_{2i}, \delta_{2i}))$

  $\vee \max(supp(t_{2i}, \delta_{2i})) < \min(supp\,(t_{li} \cdot \delta_{li}))$.

- $FTPoverlap(FTP_1, FTP_2) : (\min(supp(t_{li}, \delta_{li})) < \min(supp(t_{2i}, \delta_{2i})))$

$\wedge(\min(supp(t_{2i}, \delta_{2i})) < \max(supp(t_{li}, \delta_{li})) < \max supp(t_{2i}, \delta_{2i})$

$\vee(\min(supp(t_{2i}, \delta_{2i})) < \min(supp(t_{1i}, \delta_{li}))) \wedge (\min(supp(t_{1i}, \delta_{li}))$

$< \max(supp(t_{2i}, \delta_{2i})) < \max(supp(t_{li}, \delta_{li})) \vee \max(supp(t_{li}, \delta_{li}))$

$= \min(supp(t_{2i}, \delta_{2i})) \vee \min(supp(t_{li}, \delta_{li})) = \max(supp(t_{2i}, \delta_{2i}))$

$\vee \min(supp(t_{li}, \delta_{li})) < \min(supp(t_{2i}, \delta_{2i})) < \max(supp(t_{2i}, \delta_{2i}))$

$\min(supp(t_{li}, \delta_{li})) \vee \min(supp(t_{2i}, \delta_{2i})) < \min(supp(t_{li}, \delta_{li}))$

$< \max(supp(t_{li}, \delta_{li})) < \min(supp(t_{2i}, \delta_{2i}))$.

Next, several significant characteristics of fuzzy time points are presented, grounded in the provided definitions.

Theorem 4.1 *For two fuzzy points $FTP_1 = (T_1, \delta_1)$ and $FTP_2 = (T_2, \delta_2)$. Here, $\propto$ represents precedence.*

- $FTP_1 = FTP_2 \Leftrightarrow T_1 = T_2 \wedge \delta_l = \delta_2$
- $FTP_1 \propto FTP_2 \Leftrightarrow supp(T_1, \delta_1) < supp(T_2, \delta_2)$

Proof Fuzzy time points encompass all potential temporal instances, thereby making it evident that two fuzzy time points $FTP_1 = FTP_2$ if and only if each $t_i \in T_1$ and $t_j \in T_2$ that $t_i = t_j$ and $\delta_i = \delta_j$. In the case of precedence, similarly, if and only if each $t_i \in T_1$ and $t_j \in T_2$ with positive possibility that $t_i < t_j$ (in other words, there is no $t_i \in T_1$ and $t_j \in T_2$ that $t_j \propto t_i$), $FTP_1 \propto FTP_2$. □

Theorem 4.2 *Let $FTP_1 = (T_1, \delta_1)$ and $FTP_2 = (T_2, \delta_2)$ be two fuzzy time points and poss (E) represents the possibility distribution of event E being true.*

- $\max(poss(FTP_1 = FTP_2), poss(FTP_1 \propto FTP_2), poss(FTP_2 \propto FTP_1)) = 1$.
- $\min(poss(FTP_1 = FTP_2), poss(FTP_1 \propto FTP_2), poss(FTP_2 \propto FTP_1)) = 0$.

Proof It is entirely possible that either $FTP_1 = FTP_2$, or ($FTP_1 \propto FTP_2$), or ($FTP_2 \propto FTP_1$), and is not possible that $FTP_1 = FTP_2$, and ($FTP_1 \propto FTP_2$), and ($FTP_2 \propto FTP_1$). This interpretation is both intuitive and comprehensible.

A fuzzy time interval (*FTI*) is represented as a temporal interval characterized by fuzzy boundaries, which serves to represent a collection of potential time intervals accompanied by varying degrees of possibility. □

Definition 4.3 (*Fuzzy time interval*) For a fuzzy time interval, we have $FTI = (T_s, \delta, T_e, \delta')$, where:

- T_s and T_e represent fuzzy starting and ending time point respectively, where $T_s \propto T_e$.
- and δ' represent the membership degrees associated with the fuzzy starting time point and ending time point, referred to as T_s and T_e, respectively.

A fuzzy time interval is determined by two fuzzy endpoints, and consequently, the degree of membership of the fuzzy interval is fundamentally influenced by the membership degrees of these two endpoints.

Theorem 4.3 *Let $FTI = (T_s, \delta, T_e, \delta')$ be a fuzzy time interval, the membership degree of the fuzzy time interval δ_{FTI} is:*

- if $\delta = 1, \delta_{FTI} = \delta'$.
- if $\delta' = 1, \delta_{FTI} = \delta$.
- if $0 < \delta < 1$ and $0 < \delta' < 1, \delta_{FTI} = \delta \times \delta'$

Proof The assessment of membership degrees in fuzzy time interval calculations can be categorized into three distinct cases:

- Case 1: In the case where the starting time point of the fuzzy time interval is crisp, we denote this as $\delta_{FTI} = 1 \times \delta' = \delta'$.
- Case 2: In the case where the ending time point of the fuzzy time interval is crisp, we denote this as $\delta_{FTI} = \delta \times 1 = \delta$.
- Case 3: In the case where both the two ending time points of the fuzzy time interval are all fuzzy, we denote this as $\delta_{FTI} = \delta \times \delta'$ because the two membership degrees associated with the two ending time points are considered to be independent of one another.

The fuzzy relations between two fuzzy time intervals encompass seven distinct cases: fuzzy before, fuzzy equal, fuzzy meet, fuzzy overlap, fuzzy during, fuzzy start, and fuzzy finish. These cases are denoted as *FTIbefore*, *FTIequal*, *FTImeet*, *FTIoverlap*, *FTIduring*, *FTIstart*, *FTIfinish*, respectively. □

Definition 4.4 (*Fuzzy relations of fuzzy time intervals*) For two fuzzy time intervals $FTI_1 = (T_{1s}, \delta_1, T_{1e}, \delta_1')$ and $FTI_2 = (T_{2s}, \delta_2, T_{2e}, \delta_2')$, we have:

- $FTIbefore(FTI_1, FTI_2) : supp(T_{le}, \delta_1') < supp(T_{2s}, \delta_2)$.
- $FTIequal(FTI_1, FTI_2) : supp(T_{ls}, \delta_l) = supp(T_{2s}, \delta_2) \wedge supp(T_{le}, \delta_l') = supp(T_{2e}, \delta_2')$.
- $FTImeet(FTI_1, FTI_2) : supp(T_{1e}, \delta_1') = supp(T_{2e}, \delta_2')$.
- $FTIoverlap(FTI_1, FTI_2) : supp(T_{2s}, \delta_2) < supp(T_{le}, \delta_l') < supp(T_{2e}, \delta_2')$
  $\wedge supp(T_{ls}, \delta_l) < supp(T_{2s}, \delta_2) < supp(T_{le}, \delta_l')$.
- $FTIduring(FTI_1, FTI_2) : supp(T_{2s}, \delta_2) < supp(T_{1s}, \delta_l) < supp(T_{2e}, \delta_2')$
  $\wedge supp(T_{2s}, \delta_2) < supp(T_{le}, \delta_l') < supp(T_{2e}, \delta_2')$.
- $FTIstart(FTI_1, FTI_2) : supp(T_{1s}, \delta_1) = supp(T_{2s}, \delta_2)$.
- $FTIfinish(FTI_1, FTI_2) : supp(T_{le}, \delta_l') = supp(T_{2e}, \delta_2')$.

Next, we delineate the correspondences between fuzzy and crisp temporal relations, drawing upon the research discussed previously. These correspondences are derived from two scenarios: the correspondences between fuzzy time points and the correspondences between fuzzy time intervals.

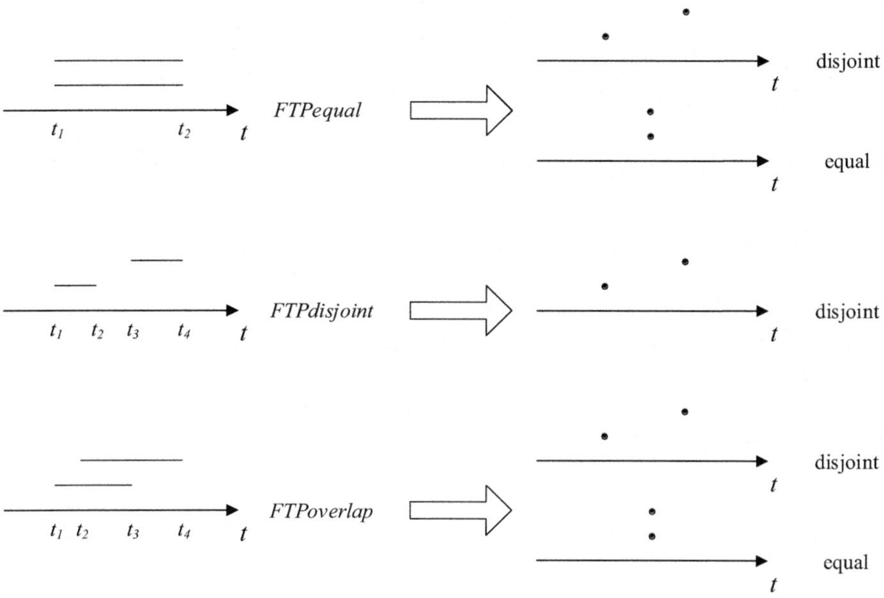

Fig. 4.1 Fuzzy relations of fuzzy time points

Figure 4.1 illustrates three fuzzy relations pertaining to two fuzzy time points, along with their respective crisp relations. For instance, in the case of *FTPequal*, the potential fuzzy time intervals for FTP_1 and FTP_2 are identical, spanning from t_1 to t_2. However, the associated crisp relations may either be disjoint or equal, as the fuzzy nature of the time points allows for their existence at any moment within the defined fuzzy time interval (from t_1 to t_2). For *FTPdisjoint*, the fuzzy time intervals of FTP_1 and FTP_2 are disjoint. In this scenario, the corresponding crisp relation must also be disjoint, as the fuzzy time intervals are disjoint $((FTP_1 = [t_1, t_2]) \cap (FTP_2 = [t_3, t_4]) = \emptyset)$. Lastly, When there exists a common area within the potential time intervals, it is *FTPoverlap*. The corresponding crisp relations for *FTPoverlap* may similarly be either disjoint or equal, akin to *FTPequal*. It is important to note that *FTPoverlap* is distinct from *FTPequal*, despite the potential for both to yield disjoint or equal crisp relations. This distinction can be analyzed from two perspectives: one perspective considers similarity as *FTPdisjoint* if $t \in (t_1, t_2) \subset FTP_1 \wedge t' \in (t_3, t_4) \subset FTP_2$, while the other perspective considers similarity as *FTPequal* if $t \in [t_2, t_3] \subset FTP_1 = t' \in [t_2, t_3] \subset FTP_2$. It should be emphasized that the illustration provided pertains to a specific instance of overlap between T_1 and T_2; other relational cases, such as meeting and containing, can be similarly extended.

Figure 4.2 illustrates the fuzzy relations associated with fuzzy time intervals alongside their corresponding crisp relations. The dashed region delineates the fuzzy time area for each fuzzy time interval, which may encompass any sub-interval within the fuzzy area. The relation denoted as *FTIbefore* indicates that any potential time interval of one fuzzy time interval precedes the other fuzzy time interval. Conversely,

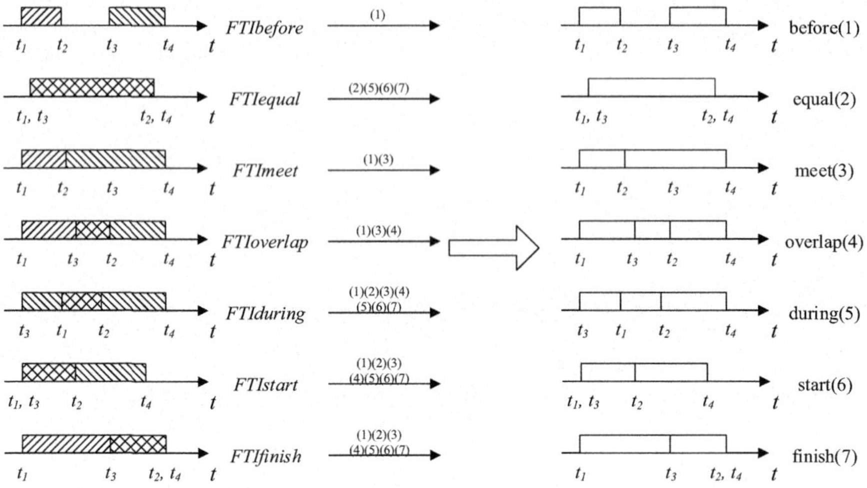

Fig. 4.2 Fuzzy relations of fuzzy time intervals

FTIequal signifies that the minimum and maximum possible time intervals of the two fuzzy time intervals are identical. The *FTImeet* relation occurs when the minimum possible time point of one fuzzy time interval coincides with the maximum possible time point of the other fuzzy time interval. The *FTIoverlap* relation is characterized by the minimum possible time point of one fuzzy time interval being situated between the possible time intervals of the other fuzzy time interval, while the maximum possible time point of the latter also lies within the former's possible time interval. The *FTIduring* relation is defined by the containment of any possible time intervals of one fuzzy time interval within the possible time intervals of the other fuzzy time interval. The *FTIstart* relation arises when the minimum possible time points of the two fuzzy time intervals are equal, whereas the *FTIfinish* relation is established when the maximum possible time points of the two fuzzy time intervals are equal.

The crisp temporal relations corresponding to fuzzy relations among fuzzy time points and fuzzy time intervals are illustrated in parentheses in Figs. 4.1 and 4.2. To facilitate a more intuitive understanding of the relations between fuzzy and crisp temporal relations, Fig. 4.3 illustrates these correspondences. In Fig. 4.3, a cross indicates the absence of correspondences, a tick signifies the presence of correspondences where the fuzzy relation has a corresponding crisp relation, and a blank space denotes correspondences that exist, yet the fuzzy relation lacks a corresponding crisp relation.

In the followings, we delineate the methodology for determining fuzzy temporal relations. Initially, a comprehensive formalized algorithm for the determination of fuzzy temporal relations is presented, followed by an illustrative example to elucidate its application. There exist ten distinct fuzzy temporal relations; due to the common characteristics among their respective determination algorithms, we have formalized a generalized algorithm. Subsequently, should a specific algorithm for

1 2 3 4 5 6 7 8	*FTPdisjoint*				*FTPequal*				*FTPoverlap*				*FTIbefore*				*FTIequal*			
disjoint (1) equal (2) overlap (3) before (4) meet (5) during (6) start (7) finish (8)	√		×	×	√	√	×	×	√	√	×	×	×			√	×	√		
	×	×	×	×	×	×	×	×	×	×	×	×						√	√	√
	FTImeet				*FTIoverlap*				*FTIduring*				*FTIstart*				*FTIfinish*			
	×	√		√	×		√	√	×	√	√	√	×	√	√	√	×	√	√	√
					√				√	√	√	√	√	√	√	√	√	√	√	√

Fig. 4.3 Correspondences between fuzzy and crisp temporal relations

a particular fuzzy temporal relation be necessary, the generalized algorithm can be adapted accordingly.

Algorithm 4.1 *Frelation Y, Z*

```
for (k = 1; k ≤ X; k ++)
let δ_k = 0
end for
for (m = 1; m ≤ i; i + +)
    for (n = 1; n ≤ m; n ++)
        for (r = 1; r ≤ X; r++)
        if F_r relation (Y, Z)
        δ_r = δ_r + δ_X
        end for
    end for
end for
if true (F_T relation)
        return δ_T Σ_{i=1}^{X} δ_i
```

The *Frelation* algorithm serves as a comprehensive method for identifying fuzzy temporal relations. It incorporates two iterative loops to facilitate the comparison of all potential fuzzy temporal relations. The computation of the membership degrees associated with these relations is conducted in a cumulative manner. Ultimately, the resultant value must be converted into a relative form, as it is derived from two membership degrees corresponding to two fuzzy time points, which can be considered a specific instance of fuzzy time intervals.

Algorithm 4.1 requires clarification on several key components: Y and Z represent two sets of fuzzy temporal data, which can be further elucidated by their corresponding time points. X denotes the quantity of fuzzy relations that exist between Y and Z. The function $F_r relation$ (Y, Z) evaluates these fuzzy relations X times, in accordance with Definition 4.2 or Definition 4.4. The variable T signifies the count of instances that fulfill the fuzzy relation, with values ranging from 1 to X. The resultant value is then normalized by dividing it by the total possible membership degrees.

Based on the proposed general algorithm, it is possible to extend any algorithm designed for the determination of fuzzy temporal relations. Due to constraints related to space, we are unable to present a comprehensive overview of all algorithms pertaining to fuzzy temporal relations. Instead, we focus on the specific case of fuzzy temporal relations between two fuzzy time intervals as a means to illustrate the extension of the general algorithm to a specific instance, as well as to elucidate the methodology for determining fuzzy temporal relations.

Let FTI_1 = {[2, 0.6, 5, 0.7], [3, 0.4, 5, 0.7], [3, 0.4, 6, 0.3]} and FTI_2 = {[3, 0.8, 4, 0.6], [4, 0.2, 5, 0.2], [4, 0.2, 6, 0.2]}. Since there are two distinct fuzzy time intervals, Y and Z are FTI_1 and FTI_2, and their representing time points are (T_{1si}, δ_{1i}, T_{1ei}, δ'_{1i}), and (T_{2sj}, δ_{2j}, T_{2ej}, δ'_{2j}). There exist seven fuzzy relations between two fuzzy time intervals (*FTIbefore*, *FTIequal*, *FTImeet*, *FTIoverlap*, *FTIduring*, *FTIstart*, and *FTIfinish*), resulting in a total of X equal to 7. Ultimately, the result values are normalized by dividing them by the seven membership degrees.

In accordance with Definition 4.4, we can obtain that the fuzzy relation between FTI_1 and FTI_2 is *FTIfinish*. It is important to note that the two fuzzy time intervals considered in this example are discrete. In the case of continuous intervals, we would need to account for all potential time intervals, evaluating them point by point along the time axis, as they conceptualize time as isomorphic to natural numbers. Consequently, we identify the following relation pairs: for the "*during*" relation, the pairs are {[2, 0.6, 5, 0.7], [3, 0.8, 4, 0.6]}, {[3, 0.4, 6, 0.3], [4, 0.2, 5, 0.2]}; for the "*finish*" relation, the pairs are {[2, 0.6, 5, 0.7], [4, 0.2, 5, 0.2]}, {[3, 0.4, 5, 0.7], [4, 0.2, 5, 0.2]}, {[3, 0.4, 6, 0.3], [4, 0.2, 6, 0.2]}; for the "*overlap*" relation, the pairs are {[2, 0.6, 5, 0.7], [4, 0.2, 6, 0.2]}, {[3, 0.4, 5, 0.7], [4, 0.2, 6, 0.2]}; and for the "*start*" relation, the pairs are {[3, 0.4, 5, 0.7], [3, 0.8, 4, 0.6]}, {[3, 0.4, 6, 0.3], [3, 0.8, 4, 0.6]}. Subsequently, we compute the membership degrees for each relation as follows: $\delta_5 = 0 + 0.6 \times 0.7 \times 0.8 \times 0.4 + 0.4 \times 0.3 \times 0.2 \times 0.2 = 0.1392$; $\delta_7 = 0 + 0.6 \times 0.7 \times 0.2 \times 0.2 + 0.4 \times 0.7 \times 0.2 \times 0.2 + 0.4 \times 0.3 \times 0.2 \times 0.2 = 0.0328$; $\delta_4 = 0 + 0.6 \times 0.7 \times 0.2 \times 0.2 + 0.4 \times 0.7 \times 0.2 \times 0.2 = 0.028$; $\delta_6 = 0 + 0.4 \times 0.7 \times 0.8 \times 0.6 + 0.4 \times 0.3 \times 0.8 \times 0.6 = 0.192$. Finally, we derive the probabilities for each relation: $\delta_5/(\delta_1 + \delta_2 + \delta_3 + \delta_4 + \delta_5 + \delta_6 + \delta_7) \approx 0.355$; $\delta_7/(\delta_1 + \delta_2 + \delta_3 + \delta_4 + \delta_5 + \delta_6 + \delta_7) \approx 0.084$; $\delta_4/(\delta_1 + \delta_2 + \delta_3 + \delta_4 + \delta_5 + \delta_6 + \delta_7) \approx 0.071$; $\delta_6/(\delta_1 + \delta_2 + \delta_3 + \delta_4 + \delta_5 + \delta_6 + \delta_7) \approx 0.490$. Consequently, the likelihood that the relation between the two fuzzy time intervals is classified as "*during*" is approximately 0.355, "*finish*" is approximately 0.084, "*overlap*" is approximately 0.071, and "*start*" is approximately 0.490.

The concept of a fuzzy time point can be considered a specific instance of a fuzzy time interval. Consequently, many characteristics associated with fuzzy time intervals are also relevant to fuzzy time points.

4.4 Fuzzy Spatial Data Semantics

In this subsection, we study the fundamental principles of representing fuzzy spatial topological relations through three dimensions: fuzzy points, fuzzy lines, and fuzzy regions. Fuzzy relations is employed to denote fuzzy spatial topological relations for the sake of clarity.

A fuzzy point is denoted as a point for which the precise location is indeterminate; however, the potential locations are identifiable within a specified region. In this context, a fuzzy point can be conceptualized as a point within two-dimensional Euclidean space, utilizing a membership degree that reflects the value of its membership function to denote the possible positions of the fuzzy point.

Definition 4.5 (*Fuzzy point*) For a fuzzy point (represented as *FP*), we have $FP = (x, y, \delta)$, where:

- x represents the projection value of the position to *x-axis*.
- y represents the projection value of the position to *y-axis*.
- δ represents the membership degree of the point being the position (x, y) (represented as $\delta_{(x, y)}$), where $0 \leq \delta \leq 1$.

A fuzzy point, denoted as (x_0, y_0, δ_0), signifies that the likelihood of the point being situated at the coordinates (x_0, y_0) is represented by the value δ_0. For instance, a fuzzy point expressed as (3, 9, 0.75) indicates that the probability of the point being located at (3, 9) is 0.75. It is important to note that if $\delta_{(x, y)} = 1$ equals 1 and is 0 in all other cases, then the point (x, y, δ) is classified as a crisp point.

The fuzzy relations between two fuzzy points can be categorized into two distinct cases: fuzzy equal, referred to as *FPPequal*, and fuzzy disjoint, denoted as *FPPdisjoint*. In the subsequent definition, a formal characterization of each case is provided. Additionally, the mathematical notation *supp* is introduced, where $supp\ A = \{u|u \in U, A\ (u) > 0\}$.

Definition 4.6 (*Fuzzy relations of fuzzy points*) For two fuzzy points $FP_1 = (x_1, y_1, \delta_1)$ and $FP_2 = (x_2, y_2, \delta_2)$, then we have:

- $FPPequal\ (FP_1, FP_2) : \exists(supp(x_1, \delta_l) = supp(x_2, \delta_2) \wedge supp(y_1, \delta_1)$
  $= supp(y_2, \delta_2))$.
- $FPPdisjoint\ (FP_1, FP_2) : \forall(\neg(supp(x_1, \delta_1) = supp(x_2, \delta_2) \wedge supp(y_1, \delta_1)$
  $= supp(y_2, \delta_2)))$.

A fuzzy line is a line for which the precise position or length is indeterminate, although the region it occupies is identifiable. In this context, the semantics of a line can be understood as a set of points situated between two terminal points. Consequently, a fuzzy line can be conceptualized as a line within a two-dimensional Euclidean space, characterized by two membership degrees that correspond to the values of the membership functions of the two terminal points. These membership degrees serve to delineate the potential locations of the fuzzy line.

Definition 4.7 (*Fuzzy line*) For a fuzzy line (represented as *FL*), we have $FL = (x_l, y_l, \delta, x_r, y_r, \delta')$, where:

- x_l and y_l represent the minimum projection values of the fuzzy line onto the *x-axis* and *y-axis*, respectively (left ending point).

- x_r and y_r represent the maximum projection values of the fuzzy line onto *x-axis* and *y-axis* respectively (right ending point).
- and δ' represent the membership degrees associated with the two ending points, where $0 \le \delta \le 1$ and $0 \le \delta' \le 1$.

The membership degree of a fuzzy line is fundamentally influenced by the membership degrees of its two defining endpoints, as the fuzzy line itself is delineated by these endpoints.

Theorem 4.4 *For a fuzzy line* $FL = (x_l, y_l, \delta, x_r, y_r, \delta')$, *the membership degree of the fuzzy line (represented as* δ_l*) is:*

- if $\delta = 1, \delta_l = \delta'$.
- if $\delta' = 1, \delta_l = \delta$.
- if $0 < \delta < 1$ and $0 < \delta' < 1, \delta_l = \delta \times \delta'$

Proof The membership degree of a fuzzy line is contingent upon the membership degrees of its two terminal points, allowing for the categorization of this relationship into three distinct cases:

- Case 1: If the left ending point of the fuzzy line is crisp, we have $\delta_l = 1 \times \delta' = \delta'$.
- Case 2: If the right ending point of the fuzzy line is crisp, we have $\delta_l = \delta \times 1 = \delta$.
- Case 3: If the two ending points of the fuzzy line are all fuzzy, we have $\delta_l = \delta \times \delta'$ due to the membership degrees of the two ending points are independent of one another. □

It is observed that the two terminal fuzzy points are considered to be independent. Consequently, the variables δ and δ' are also regarded as independent. For example, the membership degree of a fuzzy line denoted as $FL = (4, 6\ 0.7, 7, 9, 0.9)$ can be calculated as $0.7 \times 0.9 = 0.63$.

The fuzzy relations between a fuzzy point and a fuzzy line can be categorized into three distinct cases: fuzzy meet (represented as *FLPmeet*), fuzzy contain (represented as *FLPcontain*), and fuzzy disjoint (represented as *FLPdisjoint*). □

Definition 4.8 (*Fuzzy relations of fuzzy point and fuzzy line*) For a fuzzy point $FP = (x_0, y_0, \delta_0)$ and a fuzzy line $FL = (x_l, y_l, \delta, x_r, y_r, \delta')$, then we have:

- $FLPmeet(FL, FP) : \neg supp(x_0, y_0, \delta_0) \cap supp(x_l, v_l\delta)$

  $= \emptyset \vee \neg supp(x_0, y_0, \delta_0) \cap supp\left(x_x, v_v, \delta'\right) = \emptyset.$

- $FLPcontain(FL, \cdot FP) : \begin{vmatrix} x_l & y_l & 1 \\ x_0 & y_r & 1 \\ y_0 & 1 & \end{vmatrix} = 0 \wedge$

  $supp(\min(x_l, x_x), \delta) < supp(x_0, \delta_0) < supp(\max(x_l, x_x), \delta) \vee$

  $(supp(\min(y_l, v_x), \delta) < supp(x_0, \delta_0) < supp(\max(y_l, v_x), \delta).$

- *FLPdisjoint* (*FL*, *FP*): ¬ (*FLPmeet* (*FL*, *FP*) ∨ *FLPcontain* (*FL*, *FP*)) ⇔ *FP* stays in the left of *FL* ⇔ $\begin{vmatrix} x_l & y_l & 1 \\ x_r & y_r & 1 \\ x_0 & y_0 & 1 \end{vmatrix} > 0$ represented as *FLPleft* (*FP*, δ_0, *FL*, δ, δ') ∨ *FP* stays in the right of *FL* ⇔ $\begin{vmatrix} x_l & y_l & 1 \\ x_r & y_r & 1 \\ x_0 & y_0 & 1 \end{vmatrix} < 0$ represented as *FLPright* (*FP*, δ_0, *FL*, δ, δ').

The fuzzy relations between two fuzzy lines encompass six distinct cases: fuzzy intersection (represented as *FLLintersect*), fuzzy equal (represented as *FLLequal*), fuzzy contain (represented as *FLLcontain*), fuzzy overlap (represented as *FLLoverlap*), fuzzy meet (represented as *FLLmeet*), and fuzzy disjoint (represented as *FLLdisjoint*).

Definition 4.9 (*Fuzzy relations of fuzzy lines*) For two fuzzy lines $FL_1 = (x_{1l}, y_{1l}, \delta_1, x_{1r}, y_{1r}, \delta_1')$ and $FL_2 = (x_{2l}, y_{2l}, \delta_2, x_{2r}, y_{2r}, \delta_2')$, then we have:

- *FLLintersect* (FL_1, FL_2): ((*FLPleft* (x_{1l} y_{1l}, δ_1, FL_2) ∧ *FLPright* (x_{1r}, y_{1r}, δ_1', FL_2)) ∨ *FLPleft* (x_{1r}, y_{1r}, δ_1', FL_2) ∧ *FLPright* (x_{1l}, y_{1l}, δ_1, FL_2)) ∧ ((*FLPleft* (x_{2l}, y_{2l}, δ_2, FL_1) ∧ *FLPright* (x_{2r}, y_{2r}, δ_2', FL_1)) ∨ (*FLPleft* (x_{2r}, y_{2r}, δ_2', FL_1) ∧ *FLPright* (x_{2l}, y_{2l}, δ_2, FL_1))).
- *FLLequal* (FL_1, FL_2): *FLPmeet* (FL_2, x_{1l}, y_{1l}, δ_1) ∧ *FLPmeet* (FL_2, x_{1r}, y_{1r}, δ_1').
- *FLLcontain* (FL_1, FL_2): (*FLPcontain* (FL_1, x_{2l}, y_{2l}, δ_2) ∧ (¬*FLPdisjoint* (FL_1, x_{2r}, y_{2r}, δ_2'))) ∨ (*FLPcontain* (FL_1, x_{2r}, y_{2r}, δ_2') ∧ (¬*FLPdisjoint* (FL_1, x_{2l}, y_{2l}, δ_2))).
- *FLLoverlap* (FL_1, FL_2): (*FLPcontain* (FL_1, x_{2l}, y_{2l}, δ_2) ∨ *FLPcontain* (FL_1, x_{2r}, y_{2r}, δ_2')) ∧ (*FLPcontain* (FL_2, x_{1l}, y_{1l}, δ_1) ∨ *FLPcontain* (FL_2, x_{1r}, y_{1r}, δ_1')).
- *FLLmeet* (FL_1, FL_2): ((*FLPcontain* (FL_1, x_{2l}, y_{2l}, δ_2) ∨ *FLPcontain* (FL_1, x_{2r}, y_{2r}, δ_2') ∨ *FLPcontain* (FL_2, x_{1l}, y_{1l}, δ_1) ∨ *FLPcontain* (FL_2, x_{1r}, y_{1r}, δ_1')) ∧ $(y_{1r} - y_{1l})/(x_{1r} - x_{1l}) \neq (y_{2r} - y_{2l})/(x_{2r} - x_{2l})$) ∨ ((*FLPmeet* ($FL_1$, x_{2l}, y_{2l}, δ_2) ∨ *FLPmeet* (FL_1, x_{2r}, y_{2r}, δ_2') ∨ *FLPmeet* (FL_2, x_{1l}, y_{1l}, δ_1) ∨ *FLPmeet* (FL_2, x_{1r}, y_{1r}, δ_1')) ∧ ¬ (*FLLequal* (FL_1, FL_2) ∨ *FLLcontain* (FL_1, FL_2) ∨ *FLLcontain* (FL_2, FL_1))).
- *FLLdisjoint* (FL_1, FL_2): *FLPdisjoint* (FL_2, x_{1l}, y_{1l}, δ_1) ∧ *FLPdisjoint* (FL_2, x_{1r}, y_{1r}, δ_1') ∧ *FLPdisjoint* (FL_1, x_{2l}, y_{2l}, δ_2) ∧ *FLPdisjoint* (FL_1, x_{2r}, y_{2r}, δ_2') ∧ ¬ (*FLLintersect* (FL_1, FL_2) ∨ *FLLcontain* (FL_1, FL_2) ∨ *FLLcontain* (FL_2, FL_1)).

A general definition characterizes a crisp region as a collection of disjoint, connected components, referred to as faces, which may include disjoint holes within the Euclidean space. Drawing parallels with the extension of crisp regions to fuzzy regions, we aim to develop fuzzy regions grounded in the point set paradigm and fuzzy concepts. For the sake of clarity, we limit the discussion to two-dimensional regions that do not contain holes.

A fuzzy region is characterized by its indeterminate boundaries. Such regions can be effectively represented using MBR, which allows researchers to utilize two fuzzy points to delineate these fuzzy regions.

Definition 4.10 (*Fuzzy region*) For a fuzzy region (denoted as *FR*), then we have $FR = (x_{min}, y_{min}, \delta, x_{max}, y_{max}, \delta')$, where:

- x_{min} and y_{min} represent the minimum projection values of the fuzzy region onto *x-axis* and *y-axis*, respectively (lower left ending point).
- x_{max} and y_{max} represent the maximum projection values of the fuzzy region onto *x-axis* and *y-axis*, respectively (upper right ending point).
- and δ' represent the membership degrees of the above two representing points, where $0 < \delta \leq 1$ and $0 < \delta' \leq 1$.

Analogous to the concept of a fuzzy line, the degree of membership within a fuzzy region is determined based on the membership degrees of its two representing endpoints.

Theorem 4.5 *For a fuzzy region* $FR = (x_{min}, y_{min}, \delta, x_{max}, y_{max}, \delta')$, *where* $0 < \delta \leq 1$ *and* $0 < \delta' \leq 1$, *the membership degree of the fuzzy line* δ_r *is*:

- if $\delta = 1$ (the lower left psoint is crisp), $\delta_r = \delta'$.
- if $\delta' = 1$ (the upper right point is crisp), $\delta_r = \delta$.
- if $0 < \delta < 1$ and $0 < \delta' < 1$, $\delta_r = \delta \times \delta'$.

The proof of Theorem 4.5 can be elucidated in a manner analogous to that of Theorem 4.4.

It is acknowledged that the two representing fuzzy points under consideration are presumed to be independent. Consequently, the variables δ and δ' are also regarded as independent. For instance, in the context of a fuzzy region, the membership degree of the fuzzy region $FR = (5, 9, 0.8, 8, 6, 0.8)$ can be calculated as $0.8 \times 0.8 = 0.64$.

The fuzzy relations between a fuzzy point and a fuzzy region can be categorized into three distinct cases: fuzzy disjoint (represented as *FRPdisjoint*), fuzzy meet (represented as *FRPmeet*), and fuzzy contain (represented as *FRPcontain*).

Definition 4.11 (*Fuzzy relations of fuzzy point and fuzzy region*) For a fuzzy point $FP = (x_0, y_0, \delta_0)$ and a fuzzy region $FR = (x_{min}, y_{min}, \delta, x_{max}, y_{max}, \delta')$, we denote four fuzzy line of the fuzzy region: $FL_1 = (x_{min}, y_{max}, \delta_1, x_{min}, y_{min}, \delta_1{}')$, $FL_2 = (x_{min}, y_{min}, \delta_2, x_{max}, y_{min}, \delta_2{}')$, $FL_3 = (x_{max}, y_{min}, \delta_3, x_{max}, y_{max}, \delta_3{}')$, $FL_4 = (x_{max}, y_{max}, \delta_4, x_{min}, y_{max}, \delta_4{}')$. Then we have:

- *FRPdisjoint* (*FP*, *FR*): $\neg$ (*FLPleft* (FL_1, *FP*) $\wedge$ *FLPleft* (FL_2, *FP*,) $\wedge$ *FLPleft* (FL_3, *FP*) $\wedge$ *FLPleft* (FL_4, *FP*)).
- *FRPmeet* (*FP*, *FR*): *FLPmeet* (FL_1, *FP*) $\vee$ *FLPmeet* (FL_2, *FP*) $\vee$ *FLPmeet* (FL_3, *FP*) $\vee$ *FLPmeet* (FL_4, *FP*) $\vee$ *FLPcontain* (FL_1, *FP*) $\vee$ *FLPcontain* (FL_2, *FP*) $\vee$ *FLPcontain* (FL_3, *FP*) $\vee$ *FLPcontain* (FL_4, *FP*).

- *FRPcontain* (*FP*, *FR*): *FLPleft* (FL_1, *FP*) ∧ *FLPleft* (FL_2, *FP*) ∧ *FLPleft* (FL_3, *FP*) ∧ *FLPleft* (FL_4, *FP*).
- The fuzzy relations between a fuzzy line and a fuzzy region can be categorized into four distinct cases: fuzzy contain (represented as *FRLcontain*), fuzzy intersect (represented as *FRLintersect*), fuzzy meet (represented as *FRLmeet*), and fuzzy disjoint (represented as *FRLdisjoint*).

Definition 4.12 (*Fuzzy relations of fuzzy line and fuzzy region*) For a fuzzy line *FL* $= (x_l, y_l, \delta, x_r, y_r, \delta')$ and a fuzzy region $FR = (x_{min}, y_{min}, \delta, x_{max}, y_{max}, \delta')$, we denote four fuzzy line of the fuzzy region: $FL_1 = (x_{min}, y_{max}, \delta_1, x_{min}, y_{min}, \delta_1{}')$, $FL_2 = (x_{min}, y_{min}, \delta_2, x_{max}, y_{min}, \delta_2{}')$, $FL_3 = (x_{max}, y_{min}, \delta_3, x_{max}, y_{max}, \delta_3{}')$, $FL_4 = (x_{max}, y_{max}, \delta_4, x_{min}, y_{max}, \delta_4{}')$. Then we have:

- *FRLcontain* (*FR*, *FL*): (*FRPcontain* (*FR*, x_l, y_l, δ) ∧ ¬ *FRPdisjoint* (*FR*, x_r, y_r, δ')) ∨ (*FRPcontain* (*FR*, x_r, y_r, δ') ∧ ¬ *FRPdisjoint* (*FR*, x_l, y_l, δ)).
- *FRLintersect* (*FR*, *FL*): *FLLintersect* (*FL*, FL_1) ∨ *FLLintersect* (*FL*, FL_2) ∨ *FLLintersect* (*FL*, FL_3) ∨ *FLLintersect* (*FL*, FL_4) ∨ ((*FRPcontain* (*FR*, x_l, y_l, δ) ∧ *FRPdisjoint* (*FR*, x_r, y_r, δ')) ∨ (*FRPcontain* (*FR*, x_r, y_r, δ') ∧ *FRPdisjoint* (*FR*, x_l, y_l, δ))).
- *FRLmeet* (*FR*, *FL*): ((¬ *FRPcontain* (*FR*, x_l, y_l, δ) ∧ ¬ *FRPcontain* (*FR*, x_r, y_r, δ')) ∧ (*FLLmeet* (*FL*, FL_1) ∨ *FLLmeet* (*FL*, FL_2) ∨ *FLLmeet* (*FL*, FL_3) ∨ *FLLmeet* (*FL*, FL_4) ∨ *FLLoverlap* (*FL*, FL_1) ∨ *FLLoverlap* (*FL*, FL_2) ∨ *FLLoverlap* (*FL*, FL_3) ∨ *FLLoverlap* (*FL*, FL_4) ∨ *FLLcontain* (*FL*, FL_1) ∨ *FLLcontain* (FL_1, *FL*) ∨ *FLLcontain* (*FL*, FL_2) ∨ *FLLcontain* (FL_2, *FL*) ∨ *FLLcontain* (*FL*, FL_3) ∨ *FLLcontain* (FL_3, *FL*) ∨ *FLLcontain* (*FL*, FL_4) ∨ *FLLcontain* (FL_4, *FL*)).
- *FRLdisjoint* (*FR*, *FL*): (*FRPdisjoint* (*FR*, x_l, y_l, δ) ∨ *FRPdisjoint* (*FR*, x_r, y_r, δ')) ∧ ¬ *FRLintersect* (*FR*, *FL*) ∧ ¬ *FRLmeet* (*FR*, *FL*).

The fuzzy relations between two fuzzy regions can be categorized into five distinct cases: fuzzy equal (represented as *FRRequal*), fuzzy contain (represented as *FRRcontain*), fuzzy overlap (represented as *FRRoverlap*), fuzzy meet (represented as *FRRmeet*), and fuzzy disjoint (represented as *FRRdisjoint*).

Definition 4.13 (*Fuzzy relations of fuzzy regions*) For two fuzzy regions $FR_1 = (x_{1min}, y_{1min}, \delta_1, x_{1max}, y_{1max}, \delta_1{}')$ and $FR_2 = (x_{2min}, y_{2min}, \delta_2, x_{2max}, y_{2max}, \delta_2{}')$, we denote four fuzzy line of the fuzzy region: $FL_{11} = (x_{1min}, y_{1max}, \delta_{11}, x_{1min}, y_{1min}, \delta_{11}{}')$, $FL_{12} = (x_{1min}, y_{1min}, \delta_{12}, x_{1max}, y_{1min}, \delta_{12}{}')$, $FL_{13} = (x_{1max}, y_{1min}, \delta_{13}, x_{1max}, y_{1max}, \delta_{13}{}')$, $FL_{14} = (x_{1max}, y_{1max}, \delta_{14}, x_{1min}, y_{1max}, \delta_{14}{}')$; $FL_{21} = (x_{2min}, y_{2max}, \delta_{21}, x_{2min}, y_{2min}, \delta_{21}{}')$, $FL_{22} = (x_{2min}, y_{2min}, \delta_{22}, x_{2max}, y_{2min}, \delta_{22}{}')$, $FL_{23} = (x_{2max}, y_{2min}, \delta_{23}, x_{2max}, y_{2max}, \delta_{23}{}')$, $FL_{24} = (x_{2max}, y_{2max}, \delta_{24}, x_{2min}, y_{2max}, \delta_{24}{}')$. Then we have:

- *FRRequal* (FR_1, FR_2): *FPPequal* (x_{1min}, y_{1min}, δ_{12}, x_{2min}, y_{2min}, δ_{22}) ∧ *FPPequal* (x_{1max}, y_{1min}, δ_{13}, x_{2max}, y_{2min}, δ_{23}) ∧ *FPPequal* (x_{1max}, y_{1max}, δ_{14}, x_{2max}, y_{2max}, δ_{24}) ∧ *FPPequal* (x_{1min}, y_{1max}, δ_{11}, x_{2min}, y_{2max}, δ_{21}).
- *FRRcontain* (FR_1, FR_2): ¬ *FRRequal* (FR_1, FR_2) ∧ (*FRPcontain* (FR_1, x_{min}, y_{min}, δ_{22}) ∨ *FRPmeet* (FR_1, x_{min}, y_{min}, δ_{22})) ∧ (*FRPcontain* (FR_1, x_{2max}, y_{2min}, δ_{23}) ∨ *FRPmeet* (FR_1, x_{2max}, y_{2min}, δ_{23})) ∧ (*FRPcontain* (FR_1, x_{2max}, y_{2max}, δ_{24})

$\vee$ *FRPmeet* (FR_1, x_{2max}, y_{2max}, δ_{24})) $\wedge$ (*FRPcontain* (FR_1, x_{2min}, y_{2max}, δ_{21}) $\vee$ *FRPmeet* (FR_1, x_{2min}, y_{2max}, δ_{21})).

- *FRRoverlap* (FR_1, FR_2): *FRLintersect* (FR_1, FL_{21}) $\vee$ *FRLintersect* (FR_1, FL_{22}) $\vee$ *FRLintersect* (FR_1, FL_{23}) $\vee$ *FRLintersect* (FR_1, FL_{24}).
- *FRRmeet* (FR_1, FR_2): (*FRPmeet* (FR_2, x_{1min}, y_{1min}, δ_{12}) $\vee$ *FRPmeet* (FR_2, x_{1max}, y_{1min}, δ_{13}) $\vee$ *FRPmeet* (FR_2, x_{1max}, y_{1max}, δ_{14}) $\vee$ *FRPmeet* (FR_2, x_{1min}, y_{1max}, δ_{11}) $\vee$ *FRPmeet* (FR_1, x_{2min}, y_{2max}, δ_{21}) $\vee$ *FRPmeet* (FR_1, x_{2min}, y_{2max}, δ_{21}) $\vee$ *FRPmeet* (FR_1, x_{2min}, y_{2max}, δ_{21}) $\vee$ *FRPmeet* (FR_1, x_{2min}, y_{2max}, δ_{21})) $\wedge$ $\neg$ (*FRRequal* (FR_1, FR_2) $\vee$ *FRRoverlap* (FR_1, FR_2) $\vee$ *FRRcontain* (FR_1, FR_2) $\vee$ *FRRcontain* (FR_2, FR_1)).
- *FRRdisjoint* (FR_1, FR_2): (*FRLdisjoint* (FR_1, FL_{21}) $\wedge$ *FRLdisjoint* (FR_1, FL_{22}) $\wedge$ *FRLdisjoint* (FR_1, FL_{23}) $\wedge$ *FRLdisjoint* (FR_1, FL_{24}) $\wedge$ $\neg$ *FRRcontain* (FR_1, FR_2)) $\vee$ (*FRLdisjoint* (FR_2, FL_{11}) $\wedge$ *FRLdisjoint* (FR_2, FL_{12}) $\wedge$ *FRLdisjoint* (FR_2, FL_{13}) $\wedge$ *FRLdisjoint* (FR_2, FL_{14}) $\wedge$ $\neg$ *FRRcontain* (FR_2, FR_1)).

Next, we delineate the relations between fuzzy spatial topological relations and spatial topological relations, drawing upon the findings discussed above. These relations are categorized into six distinct cases: point and point, line and point, line and line, region and point, region and line, and region and region.

Figure 4.4 illustrates the fuzzy relations between two fuzzy points, along with their associated crisp relations. The two circles depicted in Fig. 4.4 encompass all potential points associated with the fuzzy points. The grey points signify the possibility of their existence within one fuzzy point, whereas the dark points indicate their potential presence in both fuzzy points. The relation denoted as *FPPequal* represents the scenario in which there are possible equal points, while *FPPdisjoint* denotes the relation in which no possible equal points exist.

Figure 4.5 illustrates the fuzzy relations between a fuzzy point and a fuzzy line, along with their respective crisp relationships. The points enclosed within the circle in Fig. 4.5 denote all potential locations of the fuzzy point, while the grey line signifies the possible configurations of the fuzzy line. Figure 4.6 presents six fuzzy relations between two fuzzy lines, accompanied by their corresponding crisp relations. The two grey lines represent all possible lines associated with the two fuzzy lines.

In the context of fuzzy relations involving a fuzzy point and a fuzzy line, the relation denoted as *FLPmeet* is characterized by the existence of a potential meeting between a possible point and a possible line. Conversely, the relation referred to as *FLPcontain* arises when there exists a possible line that encompasses possible points, while simultaneously ensuring that no possible point intersects with a possible line. Lastly, the relation identified as *FLPdisjoint* is denoted by the condition that all possible points and all possible lines are mutually exclusive.

In the context of fuzzy relations between two fuzzy lines, the relation denoted as *FLLintersect* is defined as the scenario in which a potential line from one fuzzy line intersects with a potential line from the other fuzzy line. The relation *FLLequal* is characterized by the existence of two potential lines from the respective fuzzy lines that are equal. The relation *FLLcontain* is established when the minimum and maximum endpoints of a potential line from one fuzzy line are encompassed within

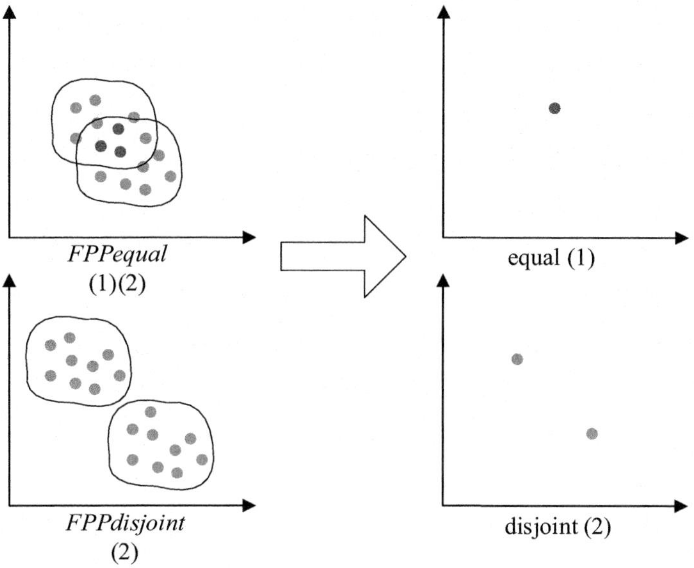

Fig. 4.4 Fuzzy relations of fuzzy points

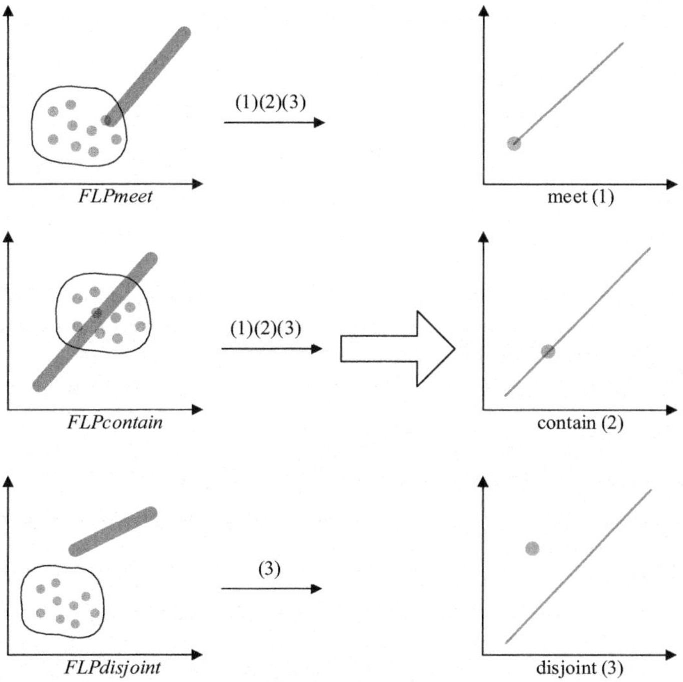

Fig. 4.5 Fuzzy relations of a fuzzy point and a fuzzy line

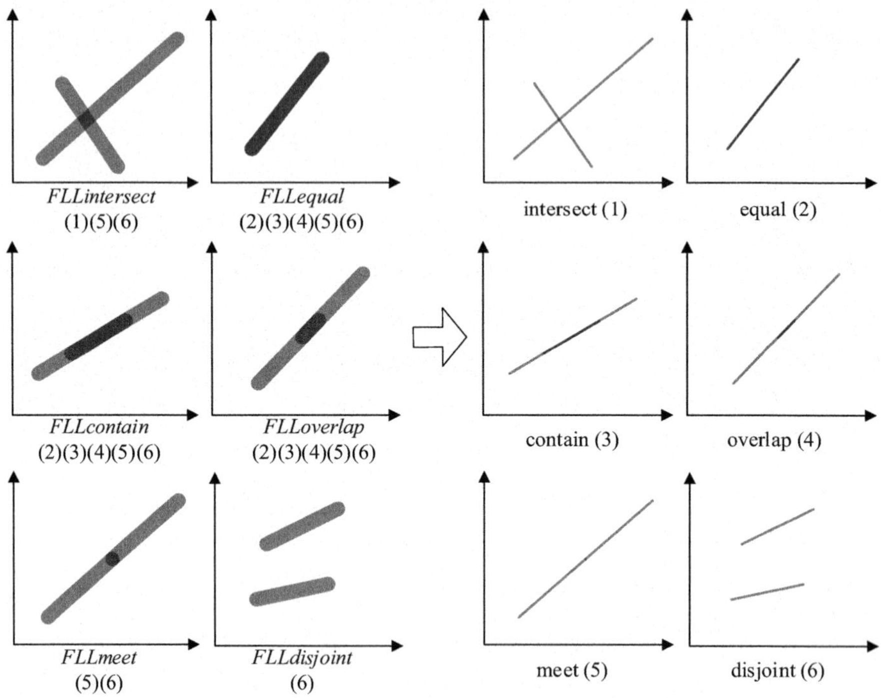

Fig. 4.6 Fuzzy relations of two fuzzy lines

a potential line of the other fuzzy line. The relation *FLLoverlap* occurs when the maximum endpoint of a potential line from one fuzzy line is contained within a potential line of the other fuzzy line, while simultaneously, the minimum endpoint of a potential line from the second fuzzy line is contained within a potential line of the first fuzzy line. The relation *FLLmeet* is defined by the presence of an endpoint from a potential line of one fuzzy line coinciding with an endpoint from a potential line of the other fuzzy line. Lastly, the relation *FLLdisjoint* is characterized by the condition that the two potential lines from the respective fuzzy lines do not intersect at any point.

Figure 4.7 illustrates three fuzzy relations involving a fuzzy point and a fuzzy region. The circle area denotes all potential locations of the fuzzy point, while the slashed region signifies the possible areas of the fuzzy region. Figure 4.8 depicts the fuzzy relations between a fuzzy line and a fuzzy region, where the grey lines represent all possible lines of the fuzzy line, and the dashed areas indicate the possible regions of the fuzzy region. Figure 4.9 presents the fuzzy relations between two fuzzy regions, with the two dashed areas representing the potential regions of each fuzzy region. Collectively, the fuzzy relations depicted in Figs. 4.7, 4.8, and 4.9 also reflect their corresponding crisp relations.

In the context of fuzzy relations involving a fuzzy point and a fuzzy region, the relationship is classified as *FRPdisjoint* if every potential point associated with the

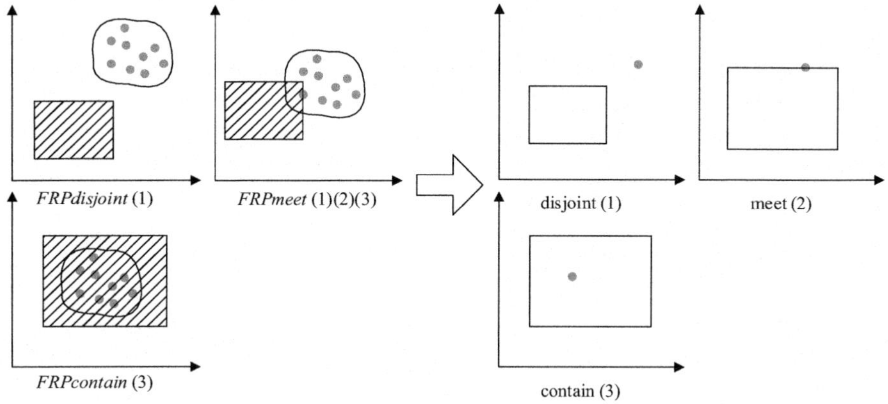

Fig. 4.7 Fuzzy relations of a fuzzy point and a fuzzy region

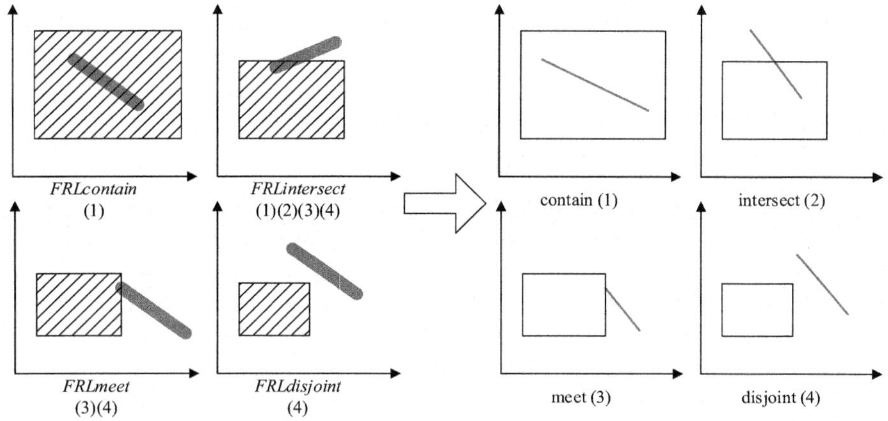

Fig. 4.8 Fuzzy relations of a fuzzy line and a fuzzy region

fuzzy point is disjoint from all potential regions associated with the fuzzy region. Conversely, the relationship is termed *FRPmeet* if there exists at least one potential point of the fuzzy point that intersects with a potential region of the fuzzy region, while not all potential points of the fuzzy point are confined to either the interior or exterior of all potential regions of the fuzzy region. Furthermore, the relationship is denoted as *FRPcontain* if every potential point of the fuzzy point is encompassed within all potential regions of the fuzzy region.

In the context of fuzzy relations involving a fuzzy line and a fuzzy region, *FRLcontain* is used to describe a situation in which all potential lines of the fuzzy line are encompassed within all potential regions of the fuzzy region. Conversely, *FRLdisjoint* refers to a scenario where all potential lines of the fuzzy line and all potential regions of the fuzzy region are mutually exclusive. *FRLmeet* is applicable when

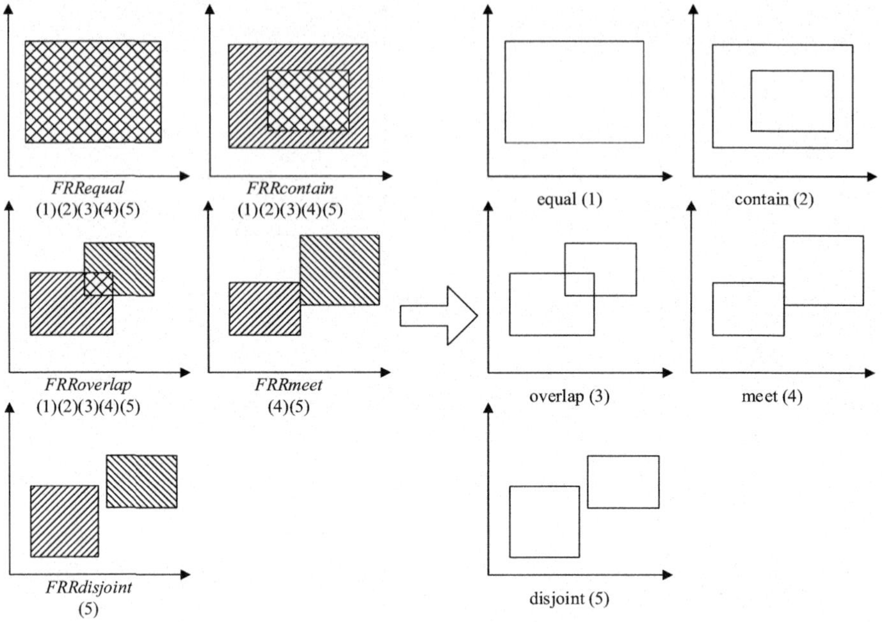

Fig. 4.9 Fuzzy relations of two fuzzy regions

there exists at least one potential line of the fuzzy line that intersects with a potential region of the fuzzy region, while the relationship does not fall under the categories of *FRLcontain* or *FRLdisjoint*. Lastly, *FRLintersect* is defined as the condition in which a potential line of the fuzzy line intersects with a potential region of the fuzzy region, and this relationship is not classified as *FRLcontain*, *FRLdisjoint*, or *FRLmeet*.

In the context of fuzzy relations between two fuzzy regions, several classifications can be identified. A fuzzy relation is *FRRdisjoint* if all potential regions of one fuzzy region are entirely disjoint from all potential regions of the other fuzzy region. Conversely, it is classified as *FRRcontain* if all potential regions of one fuzzy region are entirely contained within all potential regions of the other fuzzy region. The relation is designated as *FRRoverlap* when there exists at least one potential region of one fuzzy region that overlaps with a potential region of the other fuzzy region, provided that the relation is neither *FRRdisjoint* nor *FRRcontain*. Furthermore, the relation is referred to as *FRRmeet* if there exists a potential region of one fuzzy region that meets a potential region of the other fuzzy region, and the relation does not fall under the categories of *FRRdisjoint*, *FRRcontain*, or *FRRoverlap*. Finally, the relation is classified as *FRRequal* if a potential region of one fuzzy region is equal to a potential region of the other fuzzy region, and it is not categorized as *FRRdisjoint*, *FRRcontain*, *FRRoverlap*, or *FRRmeet*.

The corresponding crisp relations associated with fuzzy relations among various spatial entities (point and point, line and point, line and line, region and point, region and line, region and region) are illustrated in Figs. 4.7, 4.8, and 4.9. To facilitate an intuitive understanding of the correspondences between fuzzy spatial topological relations and their crisp counterparts, Fig. 4.10 presents these relationships. In Fig. 4.10, a cross symbol indicates the absence of correspondences, a tick symbol signifies the presence of correspondences where the fuzzy relation has a corresponding crisp relation, and a blank space denotes the existence of correspondences without a corresponding crisp relation for the fuzzy relation.

In the followings, we delineate the methodology for determining fuzzy spatial topological relations. Initially, they introduce a comprehensive algorithm designed for the identification of fuzzy relations, followed by an illustrative example to elucidate its application.

There exists a total of 23 distinct fuzzy relations, namely: *FPPdisjoint*, *FPPdisjoint*, *FLPmeet*, *FLPcontain*, *FLPdisjoint*, *FLLequal*, *FLLdisjoint*, *FLLmeet*, *FLLcontain*, *FLLoverlap*, *FLLintersect*, *FRPdisjoint*, *FRPmeet*, *FRPcontain*, *FRLcontain*, *FRLintersect*, *FRLmeet*, *FRLdisjoint*, *FRRequal*, *FRRcontain*, *FRRoverlap*, *FRRmeet*, and *FRRdisjoint*. Each of these relations necessitates the development of a specific algorithm for the determination of fuzzy relations. Given the presence of commonalities among these relations, we propose the formulation of a general algorithm. Subsequently, this general algorithm can be adapted to accommodate the requirements of any particular fuzzy relation.

			FPPequal			*FPPdisjoint*			*FLPmeet*			*FLPcontain*			*FLPdisjoint*			*FLLequal*		
1	2	3	√	√	×		√	×	×	√	×	×	√	×	×	√	×	√	√	√
4	5	6	×	×	×	×	×	×	×	√	√	×	√	√	×				√	√
equal (1)			*FLLdisjoint*			*FLLmeet*			*FLLcontain*			*FLLoverlap*			*FLLintersect*			*FRPdisjoint*		
disjoint (2)				√			√		√	√	√	√	√	√		√		×	√	×
							√			√	√		√	√	√	√		×		
overlap (3)			*FRPmeet*			*FRPcontain*			*FRLcontain*			*FRLintersect*			*FRLmeet*			*FRLdisjoint*		
intersect (4)			×	√	×	×		×	×		×	×	√	×	×	√	×	×	√	×
			×	√	√	×		√			√	√	√	√		√				
meet (5)			*FRRequal*			*FRRcontain*			*FRRoverlap*			*FRRmeet*			*FRRdisjoint*					
contain (6)			√	√	√	√	√	√	√	√	√		√			√				
			×	√	√	×	√	√	×	√	√	×	√		×					

Fig. 4.10 Correspondences between fuzzy and crisp spatial relations

Algorithm 4.2 serves as a comprehensive method for identifying fuzzy relations. It incorporates two iterative loops to facilitate the comparison of all potential fuzzy relations. The computation of the relation is conducted in a cumulative manner. The resultant value is subsequently normalized by dividing it by the total of all membership degrees, as each target membership degree is derived from the combination of two membership degrees associated with two distinct fuzzy points. Consequently, it is essential to convert the membership degrees into relative values.

Algorithm 4.2 *Frelation Y, Z*

```
for (k = 1; k ≤ X; k++)
let δ_k = 0
end for
for (m = 1; m≤ i; i++)
   for (n = 1; n≤ m; n++)
      for (r = 1; r ≤ X; r++)
      if F_r relation (Y, Z)
      δ_r = δ_r + δ_X
      end for
   end for
end for
if true (F_T relation)
   return δ_T /∑_{i=1}^{X} δ_i
```

For Algorithm 4.2, it is essential to clarify several key components: Y and Z represent two fuzzy spatial entities, which can be further elucidated through their corresponding representative points. The variable X denotes the quantity of fuzzy relations that exist between Y and Z. $F_r relation\ (Y, Z)$ evaluates these fuzzy relations X times, in accordance with Definitions 4.6, 4.8, 4.9, 4.11, 4.12, or 4.13. T signifies the count of instances that satisfy the fuzzy relation, which ranges from 1 to X. The resultant value is subsequently normalized by dividing it by the total possible membership degrees.

Based on Algorithm 4.2, it is possible to extend any algorithm designed for the determination of fuzzy relations. Due to constraints on space, we are unable to provide a comprehensive overview of all algorithms for establishing fuzzy relations. Instead, we focus on the fuzzy relations between two fuzzy regions as outlined by this general algorithm, as this represents the most complex case. Other algorithms can similarly be extended in an analogous manner.

Algorithm 4.3 *FRRrelation* $FR_1 = (x_{1imin}, y_{1imin}, \delta_{1i}, x_{1imax}, y_{1imax}, \delta_{1i}')$, $FR_2 = (x_{2jmin}, y_{2jmin}, \delta_{2j}, x_{2jmax}, y_{2jmax}, \delta_{2j}')$

01 **for** $(k = 1; k \leq 5; k{+}{+})$
02 **let** $\delta_k = 0$
03 **end for**
04 **for** $(m = 1; m \leq i; i{+}{+})$
05   **for** $(n = 1; n \leq m; n{+}{+})$
06   **if** *FRRequal* (FR_1, FR_2)
07         $\delta_1 = \delta_1 + \delta_{1i} \times \delta_{1i}' \times \delta_{2j} \times \delta_{2j}'$
08   **if** *FRRcontain* (FR_1, FR_2)
09         $\delta_2 = \delta_2 + \delta_{1i} \times \delta_{1i}' \times \delta_{2j} \times \delta_{2j}'$
10   **if** *FRRoverlap* (FR_1, FR_2)
11         $\delta_3 = \delta_3 + \delta_{1i} \times \delta_{1i}' \times \delta_{2j} \times \delta_{2j}'$
12   **if** *FRRmeet* (FR_1, FR_2)
13         $\delta_4 = \delta_4 + \delta_{1i} \times \delta_{1i}' \times \delta_{2j} \times \delta_{2j}'$
14   **if** *FRRdisjoint* (FR_1, FR_2)
15         $\delta_5 = \delta_5 + \delta_{1i} \times \delta_{1i}' \times \delta_{2j} \times \delta_{2j}$
16   **end for**
17 **end for**
18 **if** true (equal) return $\delta_1/(\delta_1 + \delta_2 + \delta_3 + \delta_4 + \delta_5)$
19 **if** true (contain) return $\delta_2/(\delta_1 + \delta_2 + \delta_3 + \delta_4 + \delta_5)$
20 **if** true (overlap) return $\delta_3/(\delta_1 + \delta_2 + \delta_3 + \delta_4 + \delta_5)$
21 **if** true (meet) return $\delta_4/(\delta_1 + \delta_2 + \delta_3 + \delta_4 + \delta_5)$
22 **if** true (disjoint) return $\delta_5/(\delta_1 + \delta_2 + \delta_3 + \delta_4 + \delta_5)$

In Algorithm 4.3, Y and Z are denoted as FR_1 and FR_2, respectively. Their corresponding representative points are defined as $(x_{1imin}, y_{1imin}, \delta_{1i}, x_{1imax}, y_{1imax}, \delta_{1i}')$ and $(x_{2jmin}, y_{2jmin}, \delta_{2j}, x_{2jmax}, y_{2jmax}, \delta_{2j}')$. The algorithm considers five distinct fuzzy relations between the two fuzzy regions: *FRRequal*, *FRRcontain*, *FRRoverlap*, *FRRmeet*, and *FRRdisjoint*, denoted as $X = 5$. These five types of outcomes are enumerated to facilitate the assessment of the fuzzy relations, as the algorithm itself is unable to determine which specific fuzzy relation the two regions satisfy. Ultimately, the resultant value is normalized by dividing it by the five membership degrees.

Next, we present an illustrative example to elucidate the methodology for ascertaining fuzzy relations through the proposed algorithm. For the sake of clarity, we utilize the scenario of two fuzzy regions.

Consider two fuzzy regions $FR_1 = \{(1, 2, 0.6, 6, 7, 0.7), (2, 4, 0.4, 7, 8, 0.3)\}$ and $FR_2 = \{(6, 3, 0.8, 9, 10, 0.6), (6, 1, 0.2, 9, 3, 0.4)\}$. According to line 10 in algorithm *FRRrelation*, we can obtain the fuzzy relation of FR_1 and FR_2 is *FRRoverlap*. The *meet* pair are {(1, 2, 0.6, 6, 7, 0.7), (6, 3, 0.8, 9, 10, 0.6)}, {(1, 2, 0.6, 6, 7, 0.7), (6, 1, 0.2, 9, 3, 0.4)}; the *overlap* pair is {(2, 4, 0.4, 7, 8, 0.3), (6, 3, 0.8, 9, 10, 0.6)}; the *disjoint* pair is {(2, 4, 0.4, 7, 8, 0.3), (6, 1, 0.2, 9, 3, 0.4)}. Possibility of each relation is: $\delta_4 = 0 + 0.6 \times 0.7 \times 0.8 \times 0.6 + 0.6 \times 0.7 \times 0.2 \times 0.4 = 0.2352$; $\delta_3 = 0 + 0.4 \times 0.3 \times 0.8 \times 0.6 = 0.0576$; $\delta_5 = 0 + 0.4 \times 0.3 \times 0.2 \times 0.4 = 0.0096$. Finally, we can obtain the possibility of each relation: $\delta_4/(\delta_1 + \delta_2 + \delta_3 + \delta_4 + \delta_5) \approx 0.778$; $\delta_3/(\delta_1 + \delta_2 + \delta_3 + \delta_4 + \delta_5) \approx 0.190$; $\delta_5/(\delta_1 + \delta_2 + \delta_3 + \delta_4 + \delta_5) = 0.032$. Therefore, the likelihood that the relationship between the two fuzzy regions is classified as *meet* is approximately 0.778, as *overlap* is approximately 0.190, and as *disjoint* is approximately 0.032.

4.5 Fuzzy Topological Relation Model in XML

In this subsection, we present a model for topological relations aimed at discerning the topological relations between two fuzzy spatiotemporal regions over time in XML. We begin by defining fuzzy spatiotemporal data and fuzzy spatiotemporal XML documents, along with their interpretations [62, 63].

Definition 4.14 Fuzzy spatiotemporal data (denoted as *FSP*) is a 5-tuple, *FSP* = (*OID*, *ATTR*, *FP*, *FM*, *FT*), where.

- *OID* represents the evolving historical record of fuzzy spatiotemporal data.
- *ATTR* represents the fuzzy attributes of fuzzy spatiotemporal data.
- *FP* represents the fuzzy position of fuzzy spatiotemporal data.
- *FM* represents the fuzzy motion of fuzzy spatiotemporal data.
- *FT* represents the fuzzy time of fuzzy spatiotemporal data.

It is evident that conventional XML schema definitions require enhancement to effectively support fuzzy spatiotemporal data. Consequently, we will extend XML schema to integrate fuzzy spatiotemporal information from four distinct perspectives.

XML Schema (part1-evolution.xsd)

```
<?xml version = “1.0” encoding = “UTF-8”?>
<xs:schema xmlns:xs = “http://www.w3.org/2001/XMLSchema” elementFormDefault = “qualified”>
<xs:element name = “OID”>
<xs:complexType>
<xs:sequence>
<xs:element name = “time”/>
<xs:element name = “pre”/>
<xs:element name = “suc”/>
<xs:element name = “type”/>
</xs:sequence>
</xs:complexType>
</xs:element>
<xs:element name = “time”>
<xs:simpleType>
<xs:restriction base = “xs:dateTime”>
<xs:minInclusive value = “1000-01-01T00:00:00"/>
<xs:maxInclusive value = “9999-12-30T23:59:59”/>
</xs:restriction>
</xs:simpleType>
</xs:element>
<xs:element name = “type”>
<xs:simpleType>
<xs:restriction base = “xs:string”>
<xs:enumeration value = “c”/>
<xs:enumeration value = “s”/>
<xs:enumeration value = “m”/>
<xs:enumeration value = “e”/>
</xs:restriction>
</xs:simpleType>
</xs:element>
</xs:schema>
```

XML Schema (part2-general.xsd)

```
<?xml version = “1.0” encoding = “UTF-8”?>
<xs:schema xmlns:xs = “http://www.w3.org/2001/XMLSchema” elementFormDefault = “qualified”>
<xs:element name = “General”>
<xs:complexType>
<xs:sequence>
<xs:element name = “time”/>
<xs:element name = “UA”/>
</xs:sequence>
</xs:complexType>
</xs:element>
<xs:element name = “time”>
<xs:complexType>
<xs:sequence>
<xs:element name = “t-start”/>
<xs:element name = “t-end”/>
</xs:sequence>
</xs:complexType>
</xs:element>
<xs:element name = “t-start”>
<xs:simpleType>
<xs:restriction base = “xs:dateTime”>
<xs:minInclusive value = “1000-01-01T00:00:00”/>
<xs:maxInclusive value = “9999-12-30T23:59:59”/>
</xs:restriction>
</xs:simpleType>
</xs:element>
<xs:element name = “t-end”>
<xs:simpleType>
<xs:restriction base = “xs:dateTime”>
<xs:minInclusive value = “1000-01-01T00:00:00”/>
<xs:maxInclusive value = “9999-12-30T23:59:59”/>
</xs:restriction>
</xs:simpleType>
</xs:element>
<xs:complexType name = “UAType”>
<xs:sequence>
<xs:element name = “per”/>
</xs:sequence>
</xs:complexType>
<xs:element name = “per”>
<xs:complexType>
<xs:sequence>
<xs:element name = “element” minOccurs = “0” maxOccurs = “unbounded"/>
</xs:sequence>
</xs:complexType>
</xs:element>
<xs:complexType name = “extendUAType”>
<xs:complexContent>
<xs:extension base = “UAType”>
<xs:attribute name = “patt”/>
</xs:extension>
</xs:complexContent>
</xs:complexType>
<xs:attribute name = “patt”>
<xs:simpleType>
<xs:restriction base = “xs:string”>
<xs:enumeration value = “disjunctive”/>
<xs:enumeration value = “conjunctive”/>
</xs:restriction>
</xs:simpleType>
</xs:attribute>
<xs:complexType name = “elementType”/>
<xs:complexType name = “extendelementType”>
<xs:complexContent>
<xs:extension base = “elementType”>
<xs:attribute name = “pro” type = “xs:float” use = “optional” default = “1.0"/>
</xs:extension>
</xs:complexContent>
</xs:complexType>
</xs:schema>
```

XML Schema (part3-position.xsd)

```
<?xml version = “1.0” encoding = “UTF-8”?>
<xs:schema xmlns:xs = “http://www.w3.org/2001/XMLSchema” elementFormDefault = “qualified”>
<xs:element name = “position”>
<xs:complexType>
<xs:sequence>
<xs:element name = “time"/>
<xs:element name = “val"/>
</xs:sequence>
</xs:complexType>
</xs:element>
<xs:element name = “time”>
<xs:simpleType>
<xs:restriction base = “xs:dateTime”>
<xs:minInclusive value = “1000-01-01T00:00:00”/>
<xs:maxInclusive value = “9999-12-30T23:59:59”/>
</xs:restriction>
</xs:simpleType>
</xs:element>
<xs:complexType name = “valType”>
<xs:sequence>
<xs:element name = “xstart” type = “xs:int"/>
<xs:element name = “ystart” type = “xs:int"/>
<xs:element name = “xend” type = “xs:int"/>
<xs:element name = “yend” type = “xs:int"/>
</xs:sequence>
</xs:complexType>
<xs:complexType name = “extendvalType”>
<xs:complexContent>
<xs:extension base = “valType”>
<xs:attribute name = “pro” type = “xs:float” use = “optional” default = “1.0”/>
</xs:extension>
</xs:complexContent>
</xs:complexType>
</xs:schema>
```

XML Schema (part4-position.xsd)

```
<?xml version = “1.0” encoding = “UTF-8”?>
<xs:schema xmlns:xs = “http://www.w3.org/
2001/XMLSchema” elementFormDefault =
“qualified”>
<xs:element name = “motion”>
<xs:complexType>
<xs:sequence>
<xs:element name = “time”/>
<xs:element name = “tend”/>
<xs:element name = “range”/>
</xs:sequence>
</xs:complexType>
</xs:element>
<xs:element name = “time”>
<xs:simpleType>
<xs:restriction base = “xs:dateTime”>
<xs:minInclusive value = “1000-01-
01T00:00:00”/>
<xs:maxInclusive value = “9999-12-
30T23:59:59”/>
</xs:restriction>
</xs:simpleType>
</xs:element>
<xs:complexType name = “tendType”>
<xs:sequence>
<xs:element name = “xaxis”/>
<xs:element name = “yaxis”/>
</xs:sequence>
</xs:complexType>
<xs:complexType name = “extendtendType”>
<xs:complexContent>
<xs:extension base = “tendType”>
<xs:attribute name = “pro” type = “xs:float” use
= “optional” default = “1.0"/>
</xs:extension>
</xs:complexContent>
</xs:complexType>
<xs:element name = “xaxis”>
<xs:simpleType>
<xs:restriction base = “xs:string”>
<xs:enumeration value = “L”/>
<xs:enumeration value = “R”/>
<xs:enumeration value = “B”/>
</xs:restriction>
</xs:simpleType>
</xs:element>
<xs:element name = “yaxis”>
<xs:simpleType>
<xs:restriction base = “xs:string”>
<xs:enumeration value = “U”/>
<xs:enumeration value = “D”/>
<xs:enumeration value = “A”/>
</xs:restriction>
</xs:simpleType>
</xs:element>
<xs:complexType name = “rangeType”>
<xs:sequence>
<xs:elementname = “xgra” type = “xs:int”/>
<xs:element name = “ygra” type = “xs:int”/>
</xs:sequence>
</xs:complexType>
<xs:complexType name = “exten-
drangeType”>
<xs:complexContent>
<xs:extension base = “rangeType”>
<xs:attribute name = “pro” type = “xs:float”
use = “optional” default = “1.0”/>
</xs:extension>
</xs:complexContent>
</xs:complexType>
</xs:schema>
```

Definition 4.15 For the fuzzy spatiotemporal XML document, we have $F = (V, \psi, T, \varpi, \wp, \tau, \xi, \delta, \pi)$, including:

- $V = \{V_1, V_2, \ldots, V_n\}$ is a finite set of vertices.
- $\psi \subset \{(V_i, V_j) \mid V_i, V_j \in V\}$, (V, ψ) is a directed tree.
- $T \in \psi \subset \{(V_i, V_j) \mid V_i, V_j \in V\}$ specifies time of V_j.
- ϖ represents the nesting depth of V in the spatiotemporal data tree of the document. The nesting depth ϖ of root node is 1, and the nesting depth ϖ of each following level adds 1.
- $\wp$ maintains the order information in the fuzzy XML data tree. It is derived by quantifying the number of words from root of the fuzzy data tree to the respective start and end points of the element. Here, preorder traversal is employed.

- τ is a set of labels.
- For the node $v \in V$ and the label $\nabla \in \tau$, $\xi(v, \nabla)$ specifies v exists with label ∇.
- $\delta \in \tau$ is a mapping from the element node $v \in V$ to membership degree functions. It defines the possibility of the element node exists.
- $\pi \in \tau$ is a mapping from the attribute node $v \in V$ to possibility distribution functions. It defines the possibility of the attribute node exists.

Due to its extensible and structured nature, XML is well-suited for the representation of fuzzy information. Two distinct types of fuzziness can be discerned within an XML document [64]: fuzziness pertaining to elements and fuzziness associated with the attribute values of those elements. The former is characterized by the use of membership degrees linked to elements, while the latter is represented through possibility distributions. In the context of attribute values, it is common for certain data items to possess multiple potential values, which may be entirely unknown and can be articulated using a possibility distribution. Conversely, some data items are recognized as having a singular, definitive value. Consequently, fuzzy data represented by a possibility distribution can be interpreted in two ways: as fuzzy disjunctive data or fuzzy conjunctive data. Furthermore, possibility distributions can also be employed to convey fuzzy values associated with elements.

Spatiotemporal data can be naturally represented as rooted, ordered, and directed trees in XML, where the nodes correspond to elements and the edges signify the relationships between these nodes. A fuzzy XML document is similarly characterized as a data tree. Within fuzzy XML documents, there exist two distinct types of nodes: crisp nodes and fuzzy nodes. Crisp nodes represent standard XML nodes, while fuzzy nodes encapsulate a fuzzy perspective regarding the subsets of child nodes, indicating the likelihood that a particular element node is present. It is more appropriate to conceptualize fuzzy spatiotemporal data as a structured "tuple" of values. In this context, the existential possibility associated with an element should be interpreted as the likelihood that the state of the world encompasses this element.

Definition 4.16 For the fuzzy XML spatiotemporal data model $FSP = (OID, ATTR, FP, FM, FT)$ and the fuzzy spatiotemporal data tree $F = (V, \psi, T, \varpi, \wp, \tau, \xi, \delta, \pi)$, we have:

- $OID \sim V_2\psi, T, \varpi, \wp, \tau, \xi.$
- $ATTR \sim V_2\psi, \cdot T, \varpi, \wp, \tau, \xi, \pi.$
- $FP \sim V, \psi, T, \varpi, \delta, \tau, \xi, \delta.$
- $FM \sim V, \psi, T, \varpi, \wp, \tau, \xi, \delta.$
- $FT \sim T, \tau, \xi, \delta.$

We will elucidate the concept of the membership degree associated with an element. The existential membership degree of an element is considered as the likelihood that the current state of the world encompasses this element along with the sub-tree that is rooted at it. For any given element, the nodes within its sub-tree are contingent upon the chain connecting the root to each node. Each possibility within the source fuzzy spatiotemporal XML document is conditioned on the existence of the parent element. In essence, this possibility is relative, predicated on the assumption that the existence of the parent element is certain, with a probability of 1.0. To derive the absolute possibility of an element, it is necessary to take into account the relative possibility associated with the parent element. Generally, the absolute possibility of an element can be calculated by multiplying the relative possibilities identified in the source fuzzy spatiotemporal XML document along the designated path. Consequently, relative possibilities are typically assumed to be 1.0, indicating that they are equal to 1.0 in cases where certainty is established.

Let U be a universe of discourse and F be a fuzzy set in U. A membership function μ_F: $U \rightarrow [0, 1]$ is defined for F, where $\mu_F(u)$, for each $u \in U$, denotes the membership degree of u in the fuzzy set F. Thus, the fuzzy set F is described as follows:

$$F = \{\mu_E(u_1)/u_1, \mu_F(u_2)/u_2, \ldots, \mu_F(u_n)/u_n\}$$

When U is not a discrete set, the fuzzy set F can be represented by

$$F = \int_{u \in U} \mu_F(u)/u.$$

It is important to note that μ_F denotes the membership function associated with fuzzy set F, while $\mu_F(u)$ signifies the degree of membership of element u within fuzzy set F. Following this, we will outline the methodology for the automatic generation of rational membership functions and suitable decision rules derived from the training data. This methodology can be categorized into six primary steps [65].

- Cluster and fuzzify the output data;
- Construct initial membership functions for input attributes;
- Construct the initial decision table;
- Simplify the initial decision table;
- Rebuild membership functions in the simplification process;
- Derive decision rules from the decision table.

In relation to fuzzy spatiotemporal XML data, as previously discussed, our focus is on the membership degree associated with a given element. The membership functions employed are typically defined by users based on specific practical applications. Various methods for determining these functions exist, including triangular functions, trapezoidal functions (both R-functions and L-functions), Gaussian functions, and similarity relations, among others.

Definition 4.17 Fuzzy spatiotemporal region F is a 4-tuple, $F = (C, I, E, T)$, where

- C is the core of fuzzy spatiotemporal region.
- I is the indeterminate boundary of fuzzy spatiotemporal region with an *α-cut* level (denoted by A_α).
- E is the exterior of fuzzy spatiotemporal region.

- T is the valid time of fuzzy spatiotemporal region.

The notion of an *α-cut* level region servers to approximate the indeterminate boundaries of a fuzzy region and is defined as follows:

$$A_\alpha = \{(x, y, \omega)|\mu_A(x, y, \omega) \geq \alpha\}(0 < \alpha < 1)$$

The concept of A_α represents the indeterminate boundary of a fuzzy spatiotemporal region (I), characterized by all points possessing membership degrees equal to α. To facilitate the annotation of XML documents with truth degrees, we propose the implementation of a possibility attribute, denoted as *Poss*, in conjunction with the element *Val*. This attribute is designed to indicate the likelihood of a specific element's existence within XML documents, with values ranging from 0.0 to 1.0. Truth degrees may be articulated through either internal or external mechanisms. However, it is imperative to ensure that all users possess a valid schema that enables the identification of truth degrees within XML documents. Building upon this foundation, future research can explore fuzzy regions, particularly focusing on fuzzy topological relations among these regions, which can be defined by aggregating the *α-cut* levels of fuzzy regions. Furthermore, it establishes a basis for our investigation into the determination of topological relations of fuzzy spatiotemporal regions over time in XML.

Assuming the existence of two fuzzy spatiotemporal regions, designated as A and B, with T_1 and T_2 representing the valid time intervals for regions A and B, respectively, we propose to formalize the topological relationships by determining the intersection of the parts (core, indeterminate boundary, and exterior) of these regions over the temporal overlap $T_1 \cap T_2$.

In every scenario, the evolving fuzzy regions A and B maintain identical topological relationships; however, their spatiotemporal topological relations are distinctly different. Our focus is solely on the evolving fuzzy regions within the same temporal interval, as illustrated in Fig. 4.11a, rendering the situation depicted in Fig. 4.11b irrelevant. Therefore, it is necessary to establish a definition for "$\cap_T$" in order to streamline the problem and discard irrelevant topological relations.

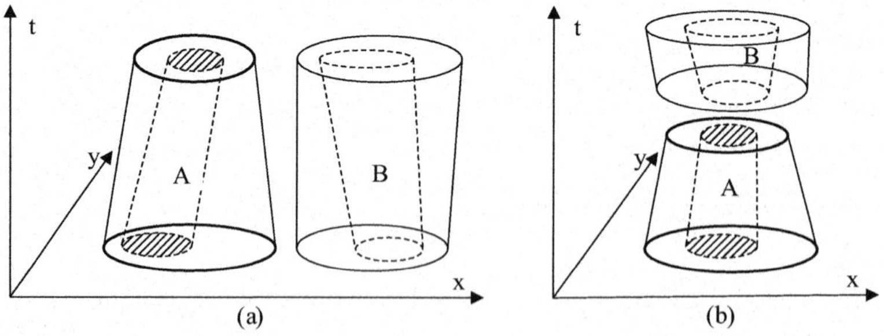

Fig. 4.11 Different spatiotemporal relations over time with the same topological relation

Definition 4.18 For two evolving fuzzy spatiotemporal regions A and B, T_A is the time interval of the fuzzy region A and T_B is the time interval of the fuzzy region B, we have:

$$A \cap_T B := \xi_A \cap_{TA \cap TB} \zeta_B \wedge C(A)(T_A \cap T_B) \neq \emptyset \wedge C(B)(T_A \cap T_B) \neq \emptyset$$

where $\xi_A = E(A) \vee C(A) \vee I(A)$, $\zeta_B = E(B) \vee C(B) \vee I(B)$.

Definition 4.19 For two evolving fuzzy spatiotemporal regions A and B, T_A is the time interval of the fuzzy region A and T_B is the time interval of the fuzzy region B, then fuzzy topological relation $R = (E(A) \cap_{TA \cap TB} E(B), E(A) \cap_{TA \cap TB} C(B), E(A) \cap_{TA \cap TB} I(B), C(A) \cap_{TA \cap TB} E(B), C(A) \cap_{TA \cap TB} C(B), C(A) \cap_{TA \cap TB} I(B), I(A) \cap_{TA \cap TB} E(B), I(A) \cap_{TA \cap TB} C(B), I(A) \cap_{TA \cap TB} I(B))$, which can be determined by (denoted as $q\ (\alpha i)$).

$$\sum_{i=1}^{n} \sum_{j=1}^{m} m_j(A_{\alpha_j}) m(B_{\alpha_j}) \tau(A_{\alpha_j} \cdot B_{o_j})$$

The fundamental probability assignment, denoted as $m\ (A_{\alpha i})$, can be understood as the likelihood that $A_{\alpha i}$ accurately represents the fuzzy spatiotemporal region A. This is expressed mathematically as $m\ (A_{\alpha i}) = \alpha_i - \alpha_{i+1}$ for $1 \leq i \leq n$, where n is a natural number ($n \in N$) and the sequence of values satisfies the condition $1 = \alpha_1 > \alpha_2 > \ldots > \alpha_n > \alpha_{n+1} = 0$. $\tau\ (A_{\alpha i}, B_{\alpha j})$ represents a specific topological relationship value, which yields a membership degree indicative of the fuzzy topological relations between the two fuzzy spatiotemporal regions in question. The value of τ is defined as the representation of the topological relationship between two fuzzy spatiotemporal regions at a given *α-cut* level.

Topological relationships among fuzzy regions characterize the spatial intersections or associations between these regions within a given space. The 9-Intersection model [11] provides a framework for analyzing these topological relations by examining the intersections among the interior, boundary, and exterior components of the regions in question.

Fuzzy spatiotemporal regions can be conceptualized as comprising core parts, indeterminate boundary parts, and exterior parts [66]. The intersection matrix between two fuzzy spatiotemporal regions, A and B, over a temporal dimension can be illustrated in Table 4.2, analogous to the construction of the topological model known as the 9-Intersection [24]. The intersection matrix pertaining to the two fuzzy spatiotemporal regions A and B over time can be articulated through the following expression:

$$R(A, \beta) = \begin{bmatrix} \partial A \cap_T \partial B & \partial A \cap_T B^0 & \partial A \cap_T B^- \\ A^0 \cap_T \partial B & A^0 \cap_T B^0 & A^0 \cap_T B^- \\ A^- \cap_T \partial B & A^- \cap_T B^0 & A^- \cap_T B^- \end{bmatrix}$$

Table 4.2 Intersection matrix

$\cap_T$	∂B	B^0	B^-
∂A	$\partial A \cap_T \partial B$	$\partial A \cap_T B^0$	$\partial A \cap_T B^-$
A^0	$A^0 \cap_T \partial B$	$A^0 \cap_T B^0$	$A^0 \cap_T B^-$
A^-	$A^- \cap_T \partial B$	$A^- \cap_T B^0$	$A^- \cap_T B^-$

Definition 4.20 For two evolving fuzzy spatiotemporal regions A and B, T_A is the time interval of the fuzzy region A and T_B is the time interval of the fuzzy region B, $\xi_A = E(A) \vee C(A) \vee I(A)$, $\zeta_B = E(B) \vee C(B) \vee I(B)$, R is fuzzy topological relation of A and B. Supposing e_i is the element of R where $1 \leq i \leq 9$, $\Re(.)$ is the function of mapping the values of *empty* ($\emptyset$) and *non-empty* ($\neg\emptyset$) onto the integers, *dim* denote the dimension, then we have:

- $\Re(.) = 0, \text{ if } e_i = \emptyset.$
- $\Re(.) = 1^{>}, \text{ if } e_A = \neg\emptyset \wedge ((\dim(\xi_A) = 2 \wedge \dim(\xi_B) = 2)$

  $\vee(\dim(\xi_A) = 2 \wedge \dim \cdot(\xi_B) = 1)) \wedge \xi_A \supset \zeta_B.$
- $\Re(.) = 1^{<}, \text{ if } e_n = \neg\emptyset \wedge ((\dim(\xi_A) = 2 \wedge \dim(\zeta_B) = 2)$

  $\vee(\dim(\xi_A) = 1 \wedge \dim(\zeta_B) = 2)) \wedge \xi_A \subset \zeta_B.$
- $\Re(.) = 1^{*}, \text{ if } e_A = \neg\emptyset \wedge \dim \cdot(\xi_A \cap \zeta_B) = 2 \wedge \xi_A \cap \zeta_B \neq \xi_A \wedge \xi_A \cap \zeta_B \neq \zeta_B.$
- $\Re(.) = 1+, \text{ if } \forall t_{t_j}, t_j, \in T_A \cap T_B, i \neq j, \xi_A(t_j) \cap \zeta_B(t_j) = \xi_A(t_j) \cap \zeta_B(t_j) = 1.$
- $\Re(.) = 1_{\lambda}, \text{ if } \exists t_1, t_2, \ldots, t_j, \ldots, t_j \ldots, t_{n_n} \in T_A \cap T_B,$

  $$t_l < t_2 < \cdots < t_i < \cdots < t_i < \cdots < t_n, \xi_A(t_i) \cap \zeta_B(t_j) \neq \xi_A(t_j) \cap \zeta_B(t_j) \wedge \delta_l$$

  $$= \xi_A(t_1) \cap \zeta_B(t_1) \wedge \delta_2 = \xi_A(t_2) \cap \zeta_B(t_2) \wedge \cdots \wedge \delta_n$$

  $$\xi_A(t_n) \cap \zeta_B(t_n) \wedge \lambda = \delta_l \times 2^0 + \delta_2 \times 2^1 + \cdots + \delta_n \times 2^{n-1} + n.$$

As outlined in Definition 4.20, the identification of topological relationships between two fuzzy spatiotemporal regions over time is a straightforward process. To exemplify our model of topological relations, we will present the following example.

Example 4.1 Figure 4.12 [66] illustrates three distinct fuzzy spatiotemporal relations between entities A and B. The relation depicted in Fig. 4.12a is characterized by its validity throughout the entire time interval. In contrast, the relations represented in Fig. 4.12b and c do not maintain validity across the entire time span. Notably, while both relations in Fig. 4.12b and c are not consistently valid, they exhibit differing patterns: the relation in Fig. 4.12b transitions from valid to invalid, whereas the relation in Fig. 4.12c shifts from invalid to valid. This distinction indicates that their fuzzy spatiotemporal relations evolve differently over time. However, prior methodologies may not effectively differentiate these relations. Utilizing Definition 4.20 facilitates the identification of these scenarios. For instance, the intersection of the indeterminate boundary for the relation in Fig. 4.12a is represented as $\Re(.) = 1_{+}$, indicating a persistent non-empty intersection. Conversely, for Fig. 4.12b, the

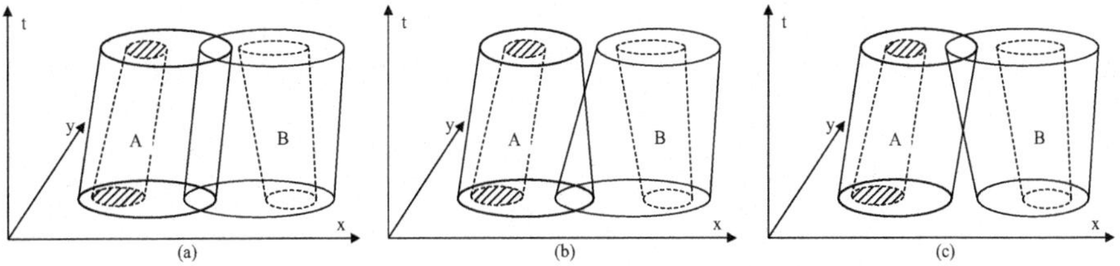

Fig. 4.12 Always and sometimes spatiotemporal topological relations

intersection is expressed as $\Re(.) = 1_3$ ($\lambda = 1 \times 2^0 + 0 \times 2^1 + 2 = 3$), reflecting a transition from a non-empty to an empty intersection over time. For Fig. 4.12c, the intersection is denoted as $\Re(.) = 1_4$ ($\lambda = 0 \times 2^0 + 1 \times 2^1 + 2 = 4$), indicating a change from an empty to a non-empty intersection over time. Similar analyses can be conducted for the remaining relations. Ultimately, the fuzzy topological relations for the three scenarios depicted in Fig. 4.12 are represented as follows: $R = (1_+, 0, 1_+, 0, 0, 1_+, 1_+, 1_+, 1_+)$; $R = (1_3, 0, 1_+, 0, 0, 1_+, 1_+, 1_+, 1_+)$; $R = (1_4, 0, 1_+, 0, 0, 1_+, 1_+, 1_+, 1_+)$.

As delineated in Definition 4.20, each relational value of e_i within the set R can be expressed through a combination of superscripts and subscripts, thereby allowing for the representation of fuzzy topological relations between two fuzzy regions over time as a unique R. Conversely, this unique R is capable of identifying a singular instance of fuzzy topological relations over time. In essence, there exists a one-to-one mutual mapping between the fuzzy topological relations over time and the set R.

In contrast to static objects, spatiotemporal objects possess dynamic characteristics, including their positions, shapes, and movements over time. To effectively represent these characteristics, abstract data types have been developed. The integration of temporal and spatial information is achieved by documenting the spatial state over time, thereby establishing a novel spatiotemporal framework. The various types of dynamic features can be conceptualized as a mapping from a specific time t to a corresponding spatial object s. More broadly, this relationship can be defined as a function that maps time to a spatial region:

$$\tau(\alpha) = time \xrightarrow{fuzzy} \alpha$$

Type $\tau(\alpha)$ represents the set of all functions mapping time to the fuzzy spatiotemporal data type α, which pertains to regions. It is important to recognize that fuzziness is present not only in the dimensions of time and α, but also within the developmental functions that approximate the relationship between time and a spatial region. This chapter primarily focuses on two-dimensional spatiotemporal data, as zero-dimensional and one-dimensional spatiotemporal data can be similarly extended.

Utilizing dimensional extensibility, a total of 92 fuzzy topological relations can be established between two fuzzy regions over time [66]. Based on these relations, the fuzzy topological relation R for each individual relation can be derived in accordance with Definition 4.20, as illustrated in Fig. 4.13. It is evident that the fuzzy topological relations over time and R can be mutually mapped in a one-to-one correspondence. It is important to highlight three primary distinctions between our methodology and existing approaches. First, in the study conducted by Bejaoui et al. [36], the 242 relations between two fuzzy regions are further categorized into multiple relations based on their levels of vagueness. In contrast, our method addresses this scenario by implementing a quantifier Q (αi) in the subsequent step. Additionally, the research by Tang [37] utilizes a 4×4 intersection matrix to establish 152 relations, whereas our approach is grounded in the extended 9-intersection model developed in our prior work [66]. Second, our focus is on the identification of continuous topological relations among fuzzy spatiotemporal regions over time. Consequently, we concentrate exclusively on the evolving fuzzy regions within the same temporal interval, rendering different spatiotemporal relations over time that share the same topological relation as irrelevant and thus excluded from consideration. Finally, the extended topological relations in existing frameworks typically account for a maximum of two distinct circumstances over time. In contrast, our model for identifying continuous topological relations over time is capable of representing intricate fuzzy topological relations that may undergo multiple changes throughout the temporal continuum. This capability can be further extended in accordance with Definition 4.20 in a similar manner.

(1)$(0,0,1_+,0,0,1_+,1_+,1_+,1_+)$	(2)$(1_4,0,1_+,0,0,1_+,1_+,1_+,1_+)$	(3)$(1_+,0,1_+,0,0,1_+,1_+,1_+,1_+)$	(4)$(1_4,1_4,1_+,0,0,1_+,1_+,1_+,1_+)$
(5)$(1_+,1_+,1_+,0,0,1_+,1_+,1_+,1_+)$	(6)$(1_4,0,1_+,1_4,0,1_+,1_+,1_+,1_+)$	(7)$(1_+,0,1_+,1_+,0,1_+,1_+,1_+,1_+)$	(8)$(1_+,0,1_+,0,0,1_+,1_+,0,1_+)$
(9)$(1_+,1_4,1_+,0,0,1_+,1_+,1_4,1_+)$	(10)$(1_+,1_5,1_+,0,0,1_+,1_+,1_4,1_+)$	(11)$(1_3,1_5,1_+,0,0,1_+,1_+,1_4,1_+)$	(12)$(1_+,0,1_+,0,0,0,1_+,1_+,1_+)$
(13)$(1_+,0,1_+,1_4,0,1_4,1_+,1_+,1_+)$	(14)$(1_+,0,1_+,1_5,0,1_4,1_+,1_+,1_+)$	(15)$(1_3,0,1_+,1_5,0,1_4,1_+,1_+,1_+)$	(16)$(0,0,1_+,0,0,1_+,0,0,1_+)$
(17)$(0,0,0,0,0,0,1_+,1_+,1_+)$	(18)$(1_+,1_+,1_+,1_+,0,1_+,1_+,1_+,1_+)$	(19)$(1_+,1_+,1_+,1_3,0,1_3,1_+,1_+,1_+)$	(20)$(1_+,1_+,1_+,1_3,0,1_+,1_+,1_+,1_+)$
(21)$(1_+,1_3,1_+,1_+,0,1_+,1_+,1_3,1_+)$	(22)$(1_+,1_3,1_+,1_+,0,1_+,1_+,1_+,1_+)$	(23)$(1_+,0,1_+,1_+,0,1_+,1_+,0,1_+)$	(24)$(1_+,0,1_+,1_3,0,1_3,1_+,0,1_+)$
(25)$(1_+,0,1_+,1_3,0,1_+,1_+,0,1_+)$	(26)$(1_+,1_+,1_+,0,0,0,1_+,1_+,1_+)$	(27)$(1_+,1_3,1_+,0,0,0,1_+,1_3,1_+)$	(28)$(1_+,1_3,1_+,0,0,0,1_+,1_+,1_+)$
(29)$(0,0,1_+,1_+,0,1_+,0,0,1_+)$	(30)$(0,0,1_+,1_3,0,1_3,0,0,1_+)$	(31)$(0,0,1_+,1_3,0,1_+,0,0,1_+)$	(32)$(0,1_+,0,0,0,0,1_+,1_+,1_+)$
(33)$(0,1_3,0,0,0,0,1_+,1_3,1_+)$	(34)$(0,1_3,0,0,0,0,1_+,1_+,1_+)$	(35)$(1_+,1_+,1_+,1_+,1_+^*,1_+,1_+,1_+,1_+)$	(36)$(1_+,1_+,1_+,1_3,1_+^*,1_3,1_+,1_+,1_+)$
(37)$(1_+,1_3,1_+,1_+,1_+^*,1_+,1_+,1_3,1_+)$	(38)$(1_+,1_3,1_+,1_3,1_+^*,1_3,1_+,1_3,1_+)$	(39)$(1_+,0,1_+,1_+,1_+^*,1_+,1_+,0,1_+)$	(40)$(1_+,0,1_+,1_3,1_+^*,1_3,1_+,0,1_+)$
(41)$(1_+,1_+,1_+,0,1_+^*,0,1_+,1_+,1_+)$	(42)$(1_+,1_3,1_+,0,1_+^*,0,1_+,1_3,1_+)$	(43)$(1_+,0,1_+,1_+,1_+^>,1_+,1_+,0,1_+)$	(44)$(1_+,0,1_+,1_3,1_+^>,1_3,1_+,0,1_+)$
(45)$(1_+,1_+,1_+,0,1_+^<,0,1_+,1_+,1_+)$	(46)$(1_+,1_3,1_+,0,1_+^<,0,1_+,1_3,1_+)$	(47)$(1_+,0,1_+,0,0,0,1_+,0,1_+)$	(48)$(1_+,0,1_+,0,1_+^*,0,1_+,0,1_+)$
(49)$(1_+,0,1_+,0,1_3^*,0,1_+,0,1_+)$	(50)$(0,0,1_+,1_+,1_+^*,1_+,0,0,1_+)$	(51)$(0,0,1_+,1_3,1_+^*,1_3,0,0,1_+)$	(52)$(0,0,1_+,1_+,1_3^*,1_+,0,0,1_+)$
(53)$(0,0,1_+,1_3,1_3^*,1_3,0,0,1_+)$	(54)$(0,1_+,0,0,1_+^*,0,1_+,1_+,1_+)$	(55)$(0,1_3,0,0,1_+^*,0,1_+,1_3,1_+)$	(56)$(0,1_+,0,0,1_3^*,0,1_+,1_+,1_+)$
(57)$(0,1_3,0,0,1_3^*,0,1_+,1_3,1_+)$	(58)$(0,0,1_+,1_+^>,1_+^>,1_+,0,0,1_+)$	(59)$(0,1_+^<,0,0,1_+^<,0,1_+,1_+,1_+)$	(60)$(0,0,1_+,1_+,1_+^>,1_+,0,0,1_+)$
(61)$(0,0,1_+,1_+,1_5^*,1_+,0,0,1_+)$	(62)$(0,0,1_+,1_+,1_3^*,1_+,0,0,1_+)$	(63)$(0,0,1_+,1_3,1_3^*,1_+,0,0,1_+)$	(64)$(0,1_+,0,0,1_+^<,0,1_+,1_+,1_+)$
(65)$(0,1_+,0,0,1_5^*,0,1_+,1_+,1_+)$	(66)$(0,1_+,0,0,1_3^*,0,1_+,1_+,1_+)$	(67)$(0,1_3,0,0,1_3^*,0,1_+,1_+,1_+)$	(68)$(1_+,0,1_+,0,1_+^>,0,1_+,0,1_+)$
(69)$(1_+,0,1_+,1_4,1_+^>,1_4,1_+,0,1_+)$	(70)$(0,0,1_+,0,0,0,0,0,1_+)$	(71)$(0,0,1_+,0,1_+^*,0,0,0,1_+)$	(72)$(0,0,1_+,0,1_3^*,0,0,0,1_+)$
(73)$(0,0,1_+,1_+^>,1_+^>,0,0,0,1_+)$	(74)$(0,0,1_+,0,1_+^>,0,0,0,1_+)$	(75)$(1_+,0,1_+,0,1_+^<,0,1_+,0,1_+)$	(76)$(1_+,1_4,1_+,0,1_+^<,0,1_+,1_4,1_+)$
(77)$(0,0,1_+,0,1_+^<,0,0,0,1_+)$	(78)$(0,0,0,0,0,0,1_+,0,1_+)$	(79)$(0,0,0,0,1_+^*,0,1_+,0,1_+)$	(80)$(0,0,0,0,1_3^*,0,1_+,0,1_+)$
(81)$(0,0,0,0,1_+^>,0,1_+,0,1_+)$	(82)$(0,1_+,0,0,1_+^<,0,1_+,0,1_+)$	(83)$(0,0,0,0,1_+^<,0,1_+,0,1_+)$	(84)$(1_+,0,1_+,0,1_+,0,1_+,0,1_+)$
(85)$(0,0,1_+,0,1_+,0,0,0,1_+)$	(86)$(0,0,0,0,1_+,0,1_+,0,1_+)$	(87)$(1_+,0,0,0,0,0,0,0,1_+)$	(88)$(1_+,0,0,0,1_+^*,0,0,0,1_+)$
(89)$(1_+,0,0,0,1_3^*,0,0,0,1_+)$	(90)$(1_+,0,0,0,1_+^>,0,0,0,1_+)$	(91)$(1_+,0,0,0,1_+^<,0,0,0,1_+)$	(92)$(1_+,0,0,0,1_+,0,0,0,1_+)$

Fig. 4.13 Fuzzy topological relation R of each relation presented in [66]

In the context of fuzzy topological relations between two fuzzy regions over time, the temporal dimension can be conceptualized as an additional geometric dimension. This perspective is supported by the notion that spatiotemporal space is homeomorphic and topologically equivalent to three-dimensional geometric space [52]. Consequently, fuzzy topological relations serve as a framework for articulating the spatiotemporal interactions between two fuzzy regions as they evolve. As delineated in Definition 4.20, our model adeptly addresses the challenges posed by fuzziness in topological relations, particularly in instances where the topological relationship may fluctuate as the fuzzy regions undergo temporal changes. Thus, it follows that spatiotemporal space maintains a homeomorphic and topologically equivalent relationship with our extended three-dimensional geometric space.

Regarding the fuzzy topological relationships between two fuzzy regions over time, it is posited that these relationships are inherently fuzzy. For instance, consider two fuzzy spatiotemporal regions, *A* and *B*, which may exhibit an overlap characterized by a membership degree of 0.4 while simultaneously demonstrating a meeting with a membership degree of 0.6 at a specific moment in time. To identify such scenarios, a quantifier has been developed as follows:

$$Q(ai) = \begin{cases} \text{"no"}, & \text{if } q(\alpha i) \leq 0.05 \\ \text{"slightly"}, & \text{if } 0.05 < q(\alpha i) \leq 0.3 \\ \text{"somewhat"}, & \text{if } 0.3 < q(\alpha i) \leq 0.7 \\ \text{"mostly"}, & \text{if } 0.7 < q(\alpha i) \leq 0.95 \\ \text{"clearly"}, & if\ q(\alpha i) > 0.95 \end{cases}$$

The numbers in $Q\ (ai)$ are chosen from the interval [0, 1], which encompasses three broad categories in the central region and a narrow category at each extremity of the interval. The classification of these categories is largely subjective.

4.6 Determination of Fuzzy Topological Relation

The identification of topological relations constitutes a critical operation in the analysis of fuzzy spatiotemporal data. In this subsection, we propose an algorithm designed to ascertain fuzzy topological relations between two fuzzy regions over time. Initially, we examine the neighborhood relations between fuzzy regions over time, as this analysis can facilitate a reduction in unnecessary computational time in the subsequent algorithm. Following this, we present the proposed algorithm for determining fuzzy topological relations.

4.6.1 Neighborhood Relations

Given that fuzzy topological relations can evolve over time, it is essential to establish a metric for evaluating how far these two relations are. The concept of topological distance is introduced to quantify the extent of divergence between two fuzzy topological relations by determining the number of corresponding elements that exhibit differing values within the fuzzy topological relation R. This measure serves to assess how far these two fuzzy topological relations are.

Definition 4.21 For two fuzzy topological relations R_x and R_y among 92 fuzzy topological relations obtained by our topological relation model with $1 \leq x, y \leq 92$, e_i is the element of R where $1 \leq i \leq 9$, and $\Re(.)$ is the function of mapping the values of *empty* (∅) and *non-empty* (¬∅) onto the integers. The topological distance, denoted by TD (R_x, R_y), is the sum of the topological distance of their core parts and indeterminate boundary, which are denoted as TD_C (R_x, R_y) and TD_I (R_x, R_y), respectively. Then we have TD $(R_x, R_y) = TD_C$ $(R_x, R_y) + TD_I$ (R_x, R_y), where

- $TD_C(R_x, R_y) = \left|\Re_x(e_1) - \Re_y(e_1)\right| + \left|\Re_x(e_2) - \Re_y(e_2)\right| + \left|\Re_x(e_4) - \Re_y(e_4)\right| + \left|\Re_x(e_5) - \Re_y(e_5)\right|$.
- $TD_I(R_x, R_y) = \left|\Re_x(e_1) - \Re_y(e_1)\right| + \left|\Re_x(e_3) - \Re_y(e_3)\right| + \left|\Re_x(e_7) - \Re_y(e_7)\right| + \left|\Re_x(e_9) - \Re_y(e_9)\right|$.

Topological distance between any two fuzzy topological relations can be determined in accordance with Definition 4.21. For example, the topological distance between the 14th and 67th fuzzy topological relation R in Fig. 4.13 can be calculated according to Definition 4.21 as TD $(R_x, R_y) = TD$ $(R_{14}, R_{67}) = TD_C$ $(R_{14}, R_{67}) + TD_I$ $(R_{14}, R_{67}) = (|1 - 0| + |0 - 1| + |1 - 0| + |0 - 1|) + (|1 - 0| + |1 - 0| + |1 - 1| + |1 - 1|) = 6$. In a similar manner, topological distances between other fuzzy topological relation R can be computed in an analogously way.

Neighborhood relations are determined by the topological distance, which is quantified by the count of corresponding elements exhibiting differing values within the fuzzy topological relation R. Traditional neighborhood relations delineate the relations between topological relations. However, in the context of fuzzy topological relations involving two fuzzy regions over time, topological relations with a distance of 1 may be grouped into distinct clusters that lack connectivity. Consequently, it is necessary to broaden the concept of neighborhood relations to encompass not only the relationships between topological relations but also the interrelations among clusters of topological relations.

Definition 4.22 For two fuzzy topological relations R_1 and R_2, their fuzzy topological relation R is denoted as R_x and R_y with $1 \leq x, y \leq 92$, we have:

- R_1 and R_2 are neighborhood relation between topological relations if TD_I $(R_1, R_2) = 1$, where $R_x \neq R_y$.
- R_1 and R_2 are neighborhood relation between clusters of topological relations if TD_I $(R_x, R_y) = 2 \wedge R_x = max$ (R_i), where $R_x \neq R_y$ and M_i is all possible topological relation in the cluster.

It can be noted that the established neighborhood relations are governed by topological distance, comprising solely TD_I (R_1, R_2) without the inclusion of TD_C (R_1, R_2). This phenomenon can be attributed to the fact that neighborhood relations pertain to their indeterminate boundaries, while the core parts do not influence the fuzzy topological relations. It is important to highlight that fuzzy topological relations are associated with both their indeterminate boundaries and their core parts, whereas neighborhood relations are exclusively determined by their indeterminate boundaries. Based on fuzzy topological relations, it is particularly advantageous to forecast the most probable topological relations between two fuzzy regions over time, especially in the context of change, as this approach can significantly decrease the execution time required for determining fuzzy topological relations.

4.6.2 Determining Fuzzy Topological Relations in XML

In this subsection, we delineate the proposed methodology for determining fuzzy topological relations between fuzzy spatiotemporal regions over time in XML. The comprehensive determining definitions are presented as follows, drawing inspiration from the representation and determination of fuzzy spatial relations [67]. We implement the proposed framework to determe fuzzy topological relations in XML. In this chapter, we utilize MBR [68] to represent two-dimensional spatiotemporal objects. Accordingly, *xstart* and *xend* denote the representing points by MBR; $\$\xi_A$*/../xstart* and $\$\xi_A$*/../xend* refer to the corresponding element nodes of the representing points in XML. The notations employed will remain consistent throughout the subsequent definitions.

Definition 4.23 For two evolving fuzzy spatiotemporal regions *A* and *B*, T_A is the time interval of the fuzzy region *A* and T_B is the time interval of the fuzzy region *B*, $\xi_A \cap_T \zeta_B$ is spatiotemporal disjoint, denoted as *stDisjoint*, if $\$\xi_A/../xend < \$\zeta_B/../xstart$ or $\$\xi_A/../yend < \$\zeta_B/../ystart$ or $\$\xi_A/../xstart > \$\zeta_B/../xend$ or $\$\xi_A/../ystart > \$\zeta_B/../yend$, where $\xi_A = C(A) \vee I(A)$, $\zeta_B = C(B) \vee I(B)$.

Definition 4.24 For two evolving fuzzy spatiotemporal regions A and B, T_A is the time interval of the fuzzy region A and T_B is the time interval of the fuzzy region B, $\xi_A \cap_T \zeta_B$ is spatiotemporal meet, denoted as *stMeet*, if ((($\$\xi_A$/../*xend* = $\$\zeta_B$/../*xstart* and ($\$\xi_A$/../*ystart*< = $\$\zeta_B$/../*ystart*< = $\$\xi_A$/../*yend or* $\$\xi_A$/../*ystart*< = $\$\zeta_B$/../*yend*< = $\$\xi_A$/../*yend*)) or ($\$\xi_A$/../*yend* = $\$\zeta_B$/../*ystart* and ($\$\xi_A$/../*xstart*< = $\$\zeta_B$/../*xstart*< = $\$\xi_A$/../*xend* or $\$\xi_A$/../*xstart*< = $\$\zeta_B$/../*xend*< = $\$\xi_A$/../*xend*)) or ($\$\xi_A$/../*xstart* = $\$\zeta_B$/../*xend* and ($\$\xi_A$/../*ystart*< = $\$\zeta_B$/../*ystart*< = $\$\xi_A$/../*yend or* $\$\xi_A$/../*ystart*< = $\$\zeta_B$/../*yend*< = $\$\xi_A$/../*yend*)) or ($\$\xi_A$/../*ystart* = $\$\zeta_B$/../*yend* and ($\$\xi_A$/../*xstart*< = $\$\zeta_B$/../*xend*< = $\$\xi_A$/../*xend* or $\$\xi_A$/../*xstart*< = $\$\zeta_B$/../*xstart*< = $\$\xi_A$/../*xend*))), where $\xi_A = C(A) \vee I(A)$, $\zeta_B = C(B) \vee I(B)$.

Definition 4.25 For two evolving fuzzy spatiotemporal regions A and B, T_A is the time interval of the fuzzy region A and T_B is the time interval of the fuzzy region B, $\xi_A \cap_T \zeta_B$ is spatiotemporal overlap, denoted as *stOverlap*, if ((($\$\xi_A$/../*xstart*< = $\$\zeta_B$/../*xend* and $\$\xi_A$/../*ystart*< = $\$\zeta_B$/../*yend*) and $\$\xi_A$/../*xstart* > $\$\zeta_B$/../*xstart*) or (($\$\xi_A$/../*xend* > = $\$\zeta_B$/../*xstart* and $\$\xi_A$/../*ystart*< = $\$\zeta_B$/..*yend*) and $\$\xi_A$/../*xstart* > $\$\zeta_B$/../*xstart*) or (($\$\xi_A$/../*xend* > = $\$\zeta_B$/../*xstart* and $\$\xi_A$/../*ystart*< = $\$\zeta_B$/../*yend*) $\$\xi_A$/../*xend*<$\$\zeta_B$/../*xend*) or (($\$\xi_A$/../*xend* > = $\$\zeta_B$/../*start* and $\$\xi_A$/..*yend* > = $\$\zeta_B$/../*ystart*) and $\$\xi_A$/../*xend*<$\$\zeta_B$/../*xend*)), where $\xi_A = C(A) \vee I(A)$, $\zeta_B = C(B) \vee I(B)$.

Definition 4.26 For two evolving fuzzy spatiotemporal regions A and B, T_A is the time interval of the fuzzy region A and T_B is the time interval of the fuzzy region B, $\xi_A \cap_T \zeta_B$ is spatiotemporal contain, denoted as *stContain*, if ($\$\xi_A$/../*xstart*<$\$\zeta_B$/../*xstart* and $\$\xi_A$/../*ystart*<$\$\zeta_B$/../*ystart*) and ($\$\xi_A$/../*xend* > $\$\zeta_B$/../*xend* and $\$\xi_A$/../*yend* > $\$\zeta_B$/../*yend*), where $\xi_A = C(A) \vee I(A)$, $\zeta_B = C(B) \vee I(B)$.

Definition 4.27 For two evolving fuzzy spatiotemporal regions A and B, T_A is the time interval of the fuzzy region A and T_B is the time interval of the fuzzy region B, $\xi_A \cap_T \zeta_B$ is spatiotemporal disjoint, denoted as *stEqual*, if ($\$\xi_A$/../*xstart* = $\$\zeta_B$/../*xstart* and $\$\xi_A$/../*ystart* = $\$\zeta_B$/../*ystart*) and ($\$\xi_A$/../*xend* = $\$\zeta_B$/../*xend* and $\$\xi_A$/../*yend* = $\$\zeta_B$/../*yend*), where $\xi_A = C(A) \vee I(A)$, $\zeta_B = C(B) \vee I(B)$.

In light of the aforementioned definitions, we propose an algorithm designed to determine fuzzy topological relations. The corresponding pseudo-code is illustrated in the subsequent algorithm.

Algorithm 4.4 Determining fuzzy topological relations *FT* (*A*, *B*)

```
Input: two evolving fuzzy spatiotemporal regions A and B
Output: fuzzy topological relation of A and B
01.  if $A/.../position/@T ∩ $B/.../position/@T ≠ Ø
02.      determining I(A) ∩ T.start I(B) = stDisjoint, stMeet, stOverlap, stContain, stEqual
03.      determining C(A) ∩ T.start C(B) = stDisjoint or
04.          C(A) ∩ T.start C(B) = stDisjoint, stMeet, stOverlap or
05.          C(A) ∩ T.start C(B) = stDisjoint, stMeet, stOverlap, stContain, stEqual
06.      finding neighborhood relations of I(A) ∩ T.start I(B) and C(A) ∩ T.start C(B)
07.      //according to Definition 4.21 and Definition 4.22
08.      determining I(A) ∩ T.end I(B) and C(A) ∩ T.end C(B) = stDisjoint, stMeet or
09.          I(A) ∩ T.end I(B) and C(A) ∩ T.end C(B) = stDisjoint, stMeet, stOverlap or
10.          I(A) ∩ T.end I(B) and C(A) ∩ T.end C(B) = stMeet, stOverlap, stContain or
11.          I(A) ∩ T.end I(B) and C(A) ∩ T.end C(B) = stOverlap, stContain, stEqual or
12.          I(A) ∩ T.end I(B) and C(A) ∩ T.end C(B) = stContain, stEqual
13.      return relation FT (A, B) and
14.      $A/.../position/@T ∩ $B/.../position/@T and
15.      $A/.../position/Val/@Poss × $B/.../position/Val/@Poss
16.      //according to q (αi)
17.  end if
     Function stDisjoint
01.  if ξA ∩T ζB is disjoint
02.  //according to Definition 4.23
03.  end if
04.  return true ()
     Function stMeet
01.  if ξA ∩T ζB is meet
02.  //according to Definition 4.24
03.  end if
04.  return true ()
     Function stOverlap
01.  if ξA ∩T ζB is overlap
02.  //according to Definition 4.25
03.  end if
04.  return true ()
     Function stContain
01.  if ξA ∩T ζB is contain
02.  //according to Definition 4.26
03.  end if
04.  return true ()
     Function stEqual
01.  if ξA ∩T ζB is equal
02.  //according to Definition 4.27
03.  end if
04.  return true ()
```

In order to elucidate the procedural aspects of the aforementioned algorithm, we provide a detailed illustration through the example depicted in Fig. 4.12. Specifically, in Fig. 4.12b, I(A) $\cap_{\mathrm{T.start}}$ $I(B)$ is classified as *stDisjoint* by applying Definition 4.23 according to Line 1 of Algorithm 4.4; $C(A) \cap_{\mathrm{T.start}} C(B)$ is classified as *stDisjoint* by applying Definition 4.23 according to Line 2 of Algorithm 4.4; neighborhood relations of $I(A) \cap_{\text{T.start}} I(B)$ and $C(A) \cap_{\text{T.start}} C(B)$ are classified as *stMeet*, *stContain*, and *stMeet* according to Line 3 of Algorithm 4.4; $I(A) \cap_{\text{T.end}} I(B)$ is classified as *stOverlap* by applying Definition 4.25, and $C(A) \cap_{\text{T.end}} C(B)$ is classified as *stDisjoint* by applying Definition 4.23 according to Line 4 of Algorithm 4.4. Consequently, the fuzzy topological relation of the fuzzy spatiotemporal regions *A* and *B* can be obtained as the 2nd fuzzy topological relation in Fig. 4.13.

4.7 Experiments

Our model is designed for the quantitative identification of continuous topological relationships among fuzzy spatiotemporal regions over time in XML. To assess the effectiveness and preciseness of our approach, we conduct a comparative analysis of our proposed algorithm against existing methodologies. The experiments were executed using Microsoft Visual Studio 10.0 on a system equipped with a 1.8 GHz Pentium IV processor, 1 GB of RAM, and operating on Windows XP. For the purpose of our experimental evaluations, we utilized a real dataset derived from meteorological observations in Hong Kong from the year 2011 [69], which includes data on the daily mean percentage of cloud cover, as presented in Table 4.3.

In the initial set of experiments, we apply the proposed position schema to the application. Figure 4.14 illustrates the position schema pertaining to fuzzy spatiotemporal data. It is evident from the validation of Schema-Position.xsd, indicated at the bottom of Fig. 4.14, that our model is both valid and useful.

In the second set of experiments, we analyze the average execution time required to determine fuzzy topological relations over a continuous temporal framework in both XML and traditional databases. In this context, traditional database refers specifically to relational databases, and the methodology for implementing topological relationships within these databases is based on the work of [70]. Figure 4.15 illustrates the execution times as the number of comparison instances varies, with traditional database and XML denoting the processes of determining fuzzy topological relations in their respective systems. The results indicate that XML demonstrates superior performance compared to traditional databases as the number of comparisons increases. This advantage can be attributed to XML's significant extensibility and structural benefits.

In the third set of experiments, we analyze the execution time required to determine fuzzy topological relations by employing both a general approach [70] and our proposed algorithm within the context of XML. Figure 4.16 illustrates the execution time associated with various fuzzy topological relations. The results indicate that our proposed algorithm demonstrates superior performance compared to the general approach in XML across all examined fuzzy topological relations. Furthermore, it is noteworthy that the execution time vary among the fuzzy topological relations, influenced by their respective neighborhood relations.

In the fourth set of experiments, we analyze the execution time required to determine fuzzy topological relations over time, comparing TRTwig [43] with our

Table 4.3 Daily mean amount of cloud (%) at the Hong Kong observatory in 2011

Day	Jan	Feb	Mar	Apr	May	Jun	Jul	Aug	Sep	Oct	Nov	Dec
01	14	58	72	21	81	40	74	39	82	83	54	59
02	83	31	77	20	83	63	53	23	85	82	70	35
03	91	32	64	41	83	68	46	22	77	88	80	3
04	95	4	76	92	86	77	33	39	64	88	63	29
05	91	6	82	90	91	70	26	55	62	85	61	88
06	73	27	87	65	76	74	39	45	56	88	80	84
07	70	60	66	41	78	67	42	62	58	81	82	83
08	36	50	87	16	82	70	59	77	52	68	89	66
09	25	67	88	60	63	59	67	86	53	76	94	87
10	41	44	88	41	64	70	72	85	51	86	88	30
11	85	85	87	24	74	84	88	67	77	88	88	13
12	88	87	51	76	75	88	88	65	81	90	54	19
13	73	89	38	64	85	75	80	53	49	86	45	52
14	65	90	48	36	89	80	87	59	46	83	80	85
15	36	89	66	64	89	74	87	45	66	54	55	39
16	54	93	85	63	91	87	88	35	63	30	77	49
17	41	100	91	92	86	91	82	42	61	14	83	63
18	59	95	92	81	63	66	84	53	72	23	88	33
19	70	100	99	38	68	59	86	61	77	36	77	43
20	86	91	94	80	70	39	89	45	87	27	67	88
21	27	88	60	73	83	67	68	50	56	23	84	79
22	77	82	79	67	90	89	64	51	80	42	85	41
23	53	60	88	84	91	88	43	55	86	62	77	32
24	81	63	86	32	88	72	30	49	88	25	73	9
25	58	67	84	31	88	64	51	69	88	57	88	19
26	29	78	84	48	59	85	48	61	81	79	86	84
27	77	54	86	67	56	85	35	54	63	52	11	55
28	77	23	84	89	23	94	49	75	71	60	32	7
29	66		84	95	51	94	86	75	88	62	43	37
30	78		88	85	28	86	78	68	79	57	41	82
31	40		74		42		60	58		42		68

proposed method. As illustrated in Fig. 4.17, our method demonstrates superior performance compared to TRTwig when the number of comparisons is fewer than 24. This advantage can be attributed to the fact that the primary expense associated with TRTwig arises from the transformation process, whereas our approach utilizes neighborhood relations for the determination of these relations.

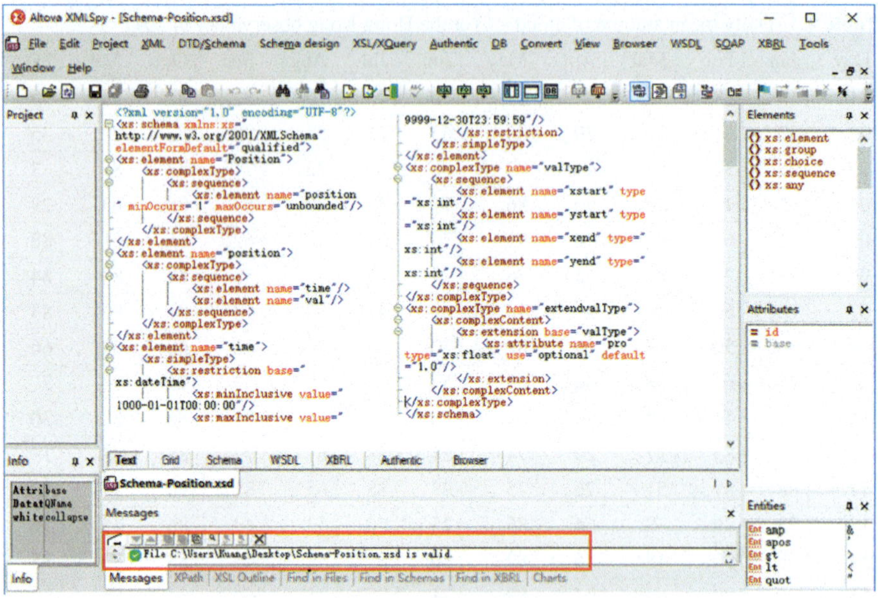

Fig. 4.14 Position schema of the fuzzy spatiotemporal data

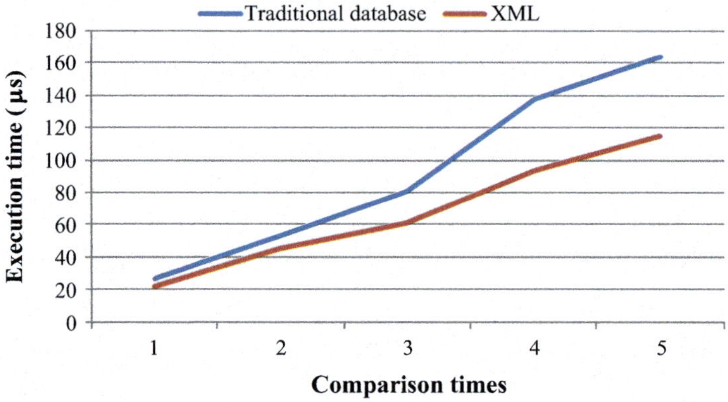

Fig. 4.15 Execution time when comparison times change (average)

In the final series of experiments, we assess the precision of our methodology. Figure 4.18 illustrates the precision of the results pertaining to fuzzy topological relations under varying *α-cut* levels. The degree of fuzziness can be adjusted in accordance with these *α-cut* levels. It is evident that the precision associated with the determination of fuzzy topological relations utilizing non-fuzzy logic (1.0) is inferior to that achieved through fuzzy logic (0, 1). This experiment further demonstrates the advantages of employing fuzzy logic for the identification of fuzzy topological

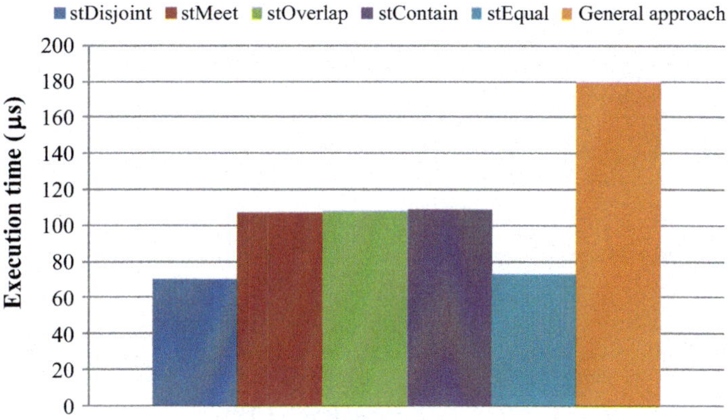

Fig. 4.16 Execution time using general approach and our proposed algorithm in XML

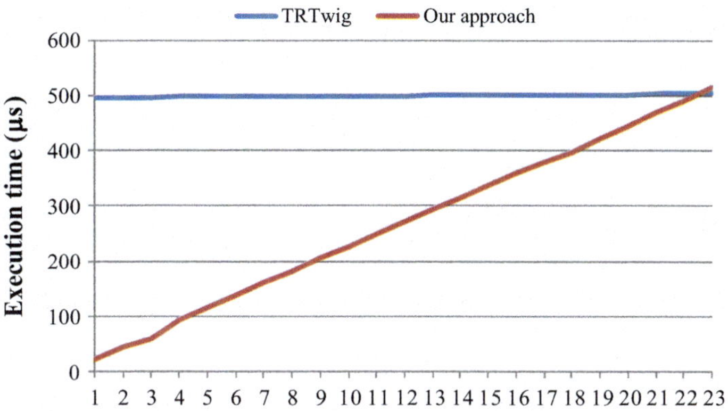

Fig. 4.17 Execution time when comparison times change between XML approaches

relations within fuzzy spatiotemporal regions over time. Additionally, it is noteworthy that the precision is maximized at medium *α-cut* levels (0.6–0.7).

4.8 Summary

In this chapter, after presenting semantics of fuzzy temporal data and fuzzy spatial data, we present a formal model for the topological relationships of fuzzy spatiotemporal regions over time, utilizing XML to encapsulate various types of spatiotemporal relations. Building upon this foundation, we propose an algorithm that leverages neighborhood relations to identify fuzzy topological relationships between two

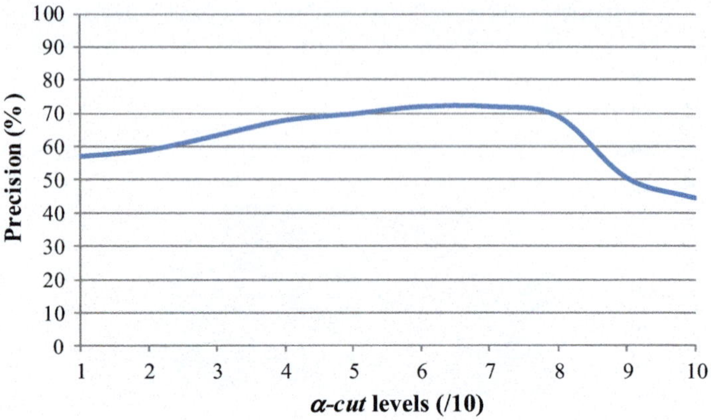

Fig. 4.18 Precisions of rainy results of fuzzy topological relations

fuzzy spatiotemporal regions over time. The results of our experiments demonstrate the effectiveness and efficiency of our method in the continuous assessment of topological relations.

References

1. X.C. Du, Research on equivalency of spatial topological relations in multiple representations. Ph. D Thesis, Wuhan University, Wuhan (2005)
2. E. Clementini, P. Di Felice, A comparison of methods for representing topological relationships. Inf. Sci. **3**(3), 149–178 (1995)
3. D. Papadias, Y. Theodoridis, T.K. Sellis, M. Egenhofer, Topological relations in the world of minimum bounding rectangles: a study with R-trees, in *Proceedings of the ACM SIGMOD International Conference on Management of Data* (1995), pp. 92–103
4. P. Guo, Study on technology and application of qualitative spatial reasoning. Ph. D Thesis, Chongqing University, Chonqing (2004)
5. E. Clementini, P. Di Felice, P. V. Oosterom, A small set of formal topological relationships suitable for end-user interaction, in *Proceedings of the 3rd International Symposium on Large Spatial Database* (1993), pp. 277–295
6. S.H. Du, Theoretics and methods research of fuzzy description and compose reasoning of spatial relations. Ph. D Thesis, Chinese Academy of Sciences, Beijing (2004)
7. J. Chen, C.M. Li, Z.L. Li, C.M. Gold, A voronoi-based 9-intersection model for spatial relations. Int. J. Geophys. Inf. Sci. **15**(3), 201–220 (2001)
8. A.I. Abdelmoty, B.A. El-Geresy, A general approach to the representation of spatial relationships. Technical Report CS-95-6. Department of Computer Studies, University of Glamorgan, Wales (1995)
9. M.J. Egenhofer, A formal definition of binary topological relationships, in *Proceedings of the 3rd International Conference of Foundations of Data Organization and Algorithms* (1989), pp. 452–472
10. M.J. Egenhofer, R. Franzosa, Point-set topological spatial relations. Int. J. Geogr. Inf. Syst. **5**(2), 161–174 (1991)

11. M.J. Egenhofer, J.R. Herring, A mathematical framework for the definition of topological relationships, in *Proceedings of the 4th International Symposium on Spatial Data Handling* (1990), pp. 803–813
12. A. Alboody, J. Inglada, F. Sèdes, Enriching the spatial reasoning system RCC8, in *Proceedings of the 16th ACM SIGSPATIAL International Conference on Advances in Geographic Information Systems* (2009), pp. 14–20
13. M. Deng, T. Cheng, X. Chen, Z. Li, Multi-level topological relations between spatial regions based upon topological invariants. GeoInformatica **11**(2), 239–267 (2007)
14. S. Badaloni, M. Falda, Classical and fuzzy neighborhood relations of the temporal qualitative algebra, in *Proceedings of the 16th International Symposium on Temporal Representation and Reasoning* (2009), pp. 147–154
15. S. Badaloni, M. Giacomin, The algebra IA^{fuz}: a framework for qualitative fuzzy temporal reasoning. Artif. Intell. **170**(10), 872–908 (2006)
16. S. Ribarić, B.D. Bašić, L. Maleš, An approach to validation of fuzzy qualitative temporal relations, in *Proceedings of the 24th International Conference on Information Technology Interfaces* (2002), pp. 223–228
17. S. Schockaert, M. De Cock, Temporal reasoning about fuzzy intervals. Artif. Intell. **172**(8–9), 1158–1193 (2008)
18. S. Schockaert, M. De Cock, E.E. Kerre, An efficient characterization of fuzzy temporal interval relations, in *Proceedings of the IEEE International Conference on Fuzzy Systems* (2006), pp. 1894–1901
19. K.C. Santosh, W. Laurent, L. Bart, Using spatial relations for graphical symbol description, in *Proceedings of the 2010 International Conference on Pattern Recognition* (2010), pp. 2041–2044
20. Y.C. Luo, H. Yan, X.Y. Cheng, The summary on description of spatial relation in image, in *Proceedings of the 2010 International Conference on Computational Intelligence and Software Engineering* (2010), pp. 1–4
21. C. Claramunt, M. Theriault, Fuzzy semantics for direction relations between composite regions. Inf. Sci. **160**(1–4), 73–90 (2004)
22. D. A. Randell, Z. Cui, A. G. Cohn, "A spatial logic based on regions and connection," *in: Proceedings of 3rd Int. Conf. on Knowledge Representation and Reasoning,* Morgan Kaufmann, 165–176, 1992.
23. S. Du, Q. Qin, Q. Wang, B. Li, Fuzzy description of topological relations I: a unified fuzzy 9-intersection model. Adv. Nat. Comput. **3612**, 1261–1273 (2005)
24. M. Egenhofer, E. Clementini, P. Felice, Topological relations between regions with holes. Int. J. Geogr. Inf. Syst. **8**(8), 129–144 (1994)
25. A. Azough, A. Delteil, M.S. Hacid, F.D. Marchi, Fuzzy conceptual graphs for handling uncertainty in semantic video retrieval, in *Proceedings of the 11th IEEE International Symposium on Multimedia* (2009), pp. 324–329
26. N.K. Verma, N.R. Pal, Prediction of satellite images using fuzzy rule based Gaussian regression, in *Proceedings of the 39th IEEE Conference on Applied Imagery Pattern Recognition Workshop* (2010), pp. 1–8
27. E. Clementini, D. Felice, Approximate topological relations. Int. J. Approx. Reason. **16**(2), 173–204 (1997)
28. A.G. Cohn, N.M. Gotts, The 'egg-yolk' representation of regions with indeterminate boundaries, in *Geographic objects with indeterminate boundaries.* ed. by P.A. Burrough, A.U. Frank (Taylor & Francis, London, 1996), pp.171–187
29. A.J. Roy, J.G. Stell, Spatial relations between indeterminate regions. Int. J. Approx. Reason. **27**(3), 205–234 (2001)
30. I. Bloch, Fuzzy spatial relationships for image processing and interpretation: a review. Image Vis. Comput. **23**(2), 89–110 (2005)
31. E. Clementini, Objects with broad boundaries, in *Book of S. Shekhar and H. Xiong: Encyclopedia of GIS* (Springer US, Boston, MA, 2008), pp. 129–144

32. E. Clementini, P.D. Felice, An algebraic model for spatial objects with indeterminate boundaries, in *Geographic Objects with Indeterminate Boundaries* (Taylor & Francis, London, 1996), pp. 155–169
33. X. Tang, Y. Fang, W. Kainz, Fuzzy topological relations between fuzzy spatial objects. Fuzzy Syst. Knowl. Discov. **4223**(S1), 324–333 (2006)
34. X. Tang, Y. Fang, W. Kainz, Topological matrices for topological relations between fuzzy regions, in *Proceedings of the 4th International Symposium on Multispectral Image Processing and Pattern Recognition* (2005), pp. 105–122
35. X. Tang, W. Kaina, Analysis of topological relations between fuzzy regions in a general fuzzy topological space, in *Proceedings of Symposim on Geospatial Theory, Processing and Applications* (2002), pp. 1–15
36. L. Bejaoui, F. Pinet, Y. Bedard, M. Schneider, Qualified topological relations between spatial objects with possible vague shape. Int. J. Geogr. Inf. Sci. **23**(7), 877–921 (2009)
37. X. Tang, Spatial object modeling in fuzzy topological spaces with applications to land cover change. Ph.D. dissertation, ITC Printing Department, Enschede, The Netherlands (2004)
38. A. Esterline, G. Dozier, A. Homaifar, Fuzzy spatial reasoning, in *Proceedings of the International Fuzzy Systems Association Conference* (1997), pp. 162–167
39. S. Schockaert, C. Cornelis, M. De Cock, E.E. Kerre, Fuzzy spatial relations between vague regions, in *Proceedings of the 3rd International IEEE Conference Intelligent Systems* (2006), pp. 221–226
40. L. Bao, X.L. Qin, J. Zhang, Q.Y. Lin, Reasoning the spatiotemporal relations between time evolving indeterminate regions, in *Proceedings of LNAI* (2006), pp. 448–458
41. P. Senellart, S. Abiteboul, On the complexity of managing probabilistic XML data, in *Proceedings of the 26th ACM SIGACT-SIGMOD-SIGART Symposium on Principles of Database Systems* (2007), pp. 283–292
42. S. Al-Khalifa, H.V. Jagadish, N. Koudas, J.M. Patel, D. Srivastava, Y. Wu, Structural joins: a primitive for efficient XML query pattern matching, in *Proceedings of the IEEE International Conference on Data Engineering* (2002), pp. 141–152
43. L. Bai, L. Yan, Z. Ma, Determining topological relationship of fuzzy spatiotemporal data integrated with XML twig pattern. Appl. Intell. **39**(1), 75–100 (2013)
44. L. Bai, L. Yan, Z. Ma, Querying fuzzy spatiotemporal data using XQuery. Integr. Comput.-Aided Eng. **21**(2), 147–162 (2014)
45. L. Bai, L. Zhu, Y. Zhao, Determining topological relations of fuzzy spatiotemporal regions over time in XML. IEEE Access **6**(1), 65620–65634 (2018)
46. H.W. Guesgen, Reasoning about distance based on fuzzy sets. Appl. Intell. **17**(3), 265–270 (2002)
47. L.A. Zadeh, Is there a need for fuzzy logic? Inf. Sci. **178**(13), 2751–2779 (2008)
48. L.A. Zadeh, Toward a generalized theory of uncertainty (GTU)-an outline. Inf. Sci. **172**(1), 1–40 (2005)
49. F. Chirigati, H. Doraiswamy, T. Damoulas, J. Freire, Data polygamy: the many-many relationships among urban spatiotemporal data sets, in *Proceedings of SIGMOD* (2016), pp. 1011–1025
50. S. Grumbach, R. Rigaux, L. Segoufin, Spatio-temporal data handling with constraints. GeoInformatic **5**(1), 95–115 (2001)
51. H.H. Le, P. Gabriel, J. Gietzel, H. Schaeben, An object-relational spatio-temporal geoscience data model. Comput. Geosci. **57**(4), 104–115 (2013)
52. M. Erwig, M. Schneider, Spatio-temporal predicates. IEEE Trans. Knowl. Data Eng. **14**(4), 881–901 (2002)
53. M. Schneider, Spatial plateau algebra for implementing fuzzy spatial objects in databases and GIS: spatial plateau data types and operations. Appl. Soft Comput. **16**(3), 148–170 (2014)
54. J. Verstraete, Deriving topological concepts for fuzzy regions: from properties to definitions. Control. Cybern. **41**(1), 113–143 (2012)
55. M. Duckham, D. Nussbaum, J.R. Sack, N. Santoro, Efficient, decentralized computation of the topology of spatial regions. IEEE Trans. Comput. **60**(8), 1100–1113 (2011)

56. E.F. Lashin, A.M. Kozae, A.A. Abo Khadra, T. Medhat, Rough set theory for topological spaces. Int. J. Approx. Reason. **40**(1), 35–43 (2005)
57. K. Qin, Z. Pei, On the topological properties of fuzzy rough sets. Fuzzy Sets Syst. **151**(3), 601–613 (2005)
58. A. Pauly, M. Schneider, VASA: an algebra for vague spatial data in databases. Inf. Syst. **35**(1), 111–138 (2010)
59. A. Dilo, R.A. de By, A. Stein, A system of types and operators for handling vague spatial objects. Int. J. Geogr. Inf. Sci. **21**(4), 397–426 (2007)
60. J.T. Bjorke, Topological relations between fuzzy regions: derivation of verbal terms. Fuzzy Sets Syst. **141**(3), 449–467 (2004)
61. S. Schockaert, M. De Codk, E.E. Kerre, Spatial reasoning in a fuzzy region connection calculus. Artif. Intell. **173**(2), 258–298 (2009)
62. L. Bai, L. Yan, Z. Ma, Fuzzy spatiotemporal data modeling and operations in XML. Appl. Artif. Intell. **29**(3), 259–282 (2015)
63. Z. Ma, L. Bai, Y. Ishikawa, L. Yan, Consistencies of fuzzy spatiotemporal data in XML documents. Fuzzy Sets Syst. **343**, 97–125 (2018)
64. Z. Ma, L. Yan, Fuzzy XML data modeling with the UML and relational data models. Data Knowl. Eng. **63**(3), 972–996 (2007)
65. T.P. Hong, C.Y. Lee, Induction of fuzzy rules and membership functions from training examples. Fuzzy Sets Syst. **84**(1), 33–47 (1996)
66. L. Bai, L. Yan, Z. Ma, Modeling topological relations between fuzzy spatiotemporal regions over time, in *Proceedings of the 21th IEEE International Conference on Fuzzy Systems* (2012), pp. 1–8
67. L. Bai, L. Yan, A general characterization of representing and determining fuzzy spatial relations. Int. Arab J. Inf. Technol. **13**(1), 70–75 (2016)
68. M.A. Cobb, F.E. Petry, Modeling spatial relationships within a fuzzy framework. J. Am. Soc. Inf. Sci. **49**(3), 253–266 (1998)
69. Summary of meteorological and tidal observations in Hong Kong in 2011. http://www.weather.gov.hk/publica/pubsmo.htm.
70. L. Bai, L. Yan, Z. Ma, C. Xu, Incorporating fuzziness in spatiotemporal XML and transforming fuzzy spatiotemporal data from XML to relational databases. Appl. Intell. **43**(4), 707–721 (2015)

Chapter 5
Algebraic Operations on Fuzzy Spatiotemporal XML Data and Querying Fuzzy Spatiotemporal Data Using XQuery

5.1 Introduction

A formal algebra is essential for applying standard database-style query optimization to XML queries. A significant number of spatiotemporal applications [1, 2] necessitate the efficient management of (fuzzy) spatiotemporal data, which is pertinent to both academic research and industrial practices [3–6]. The adaptability of a semi-structured fuzzy spatiotemporal data model [7], derived from general data models in XML [8, 9], spatiotemporal data models in XML [10], and fuzzy data models in XML [11], is fundamental to the effective management of fuzzy spatiotemporal data. Additionally, XML is inherently suited for the integration and exchange of data from diverse sources, making it a logical choice for modeling. Although fuzzy theory [3, 12] and fuzzy mathematical programming [13] have been established, researches on fuzzy spatiotemporal data have only recently emerged and warrants further investigation. In the context of spatiotemporal data represented in XML, there are ongoing efforts to amalgamate spatial and temporal attributes within a unified framework and to analyze spatiotemporal data in XML. For instance, Huang et al. [10] propose a methodology for representing and querying spatiotemporal data in XML. Liu and Wan [14] introduce a feature-based spatiotemporal data model and utilize a Native XML Database for the storage of spatiotemporal data. Furthermore, Franceschet et al. [15] present a translation algorithm that converts spatiotemporal conceptual schemas into XML schemas, along with a framework that facilitates the validation of XML documents containing spatiotemporal information against these conceptual schemas. In addition, Bai et al. [16] propose a fuzzy spatiotemporal XML data model; however, it does not accommodate algebraic operations.

The necessity of a formal algebra for the application of standard database-style query optimization to XML queries is underscored by the success of relational technologies [17, 18]. Consequently, the development of a valid formal algebra for XML queries is imperative, as it would provide a comprehensible and order-sensitive intermediate representation. Given its importance, extensive research has been conducted

L. Bai and L. Zhu, *Fuzzy Spatiotemporal XML Data Management*, Studies in Computational Intelligence 1183, https://doi.org/10.1007/978-3-031-81033-6_5

on this topic [19–22]. With the introduction of XQuery by the W3C as a standardized query language for XML, there have been concerted efforts to integrate algebraic methods into XML query processing [20, 23–25]. For instance, Fernandez et al. [24] propose an algebra for XML queries utilizing regular-expression types, which has been submitted to the W3C XML Query Working Group. Additionally, Buratti and Montesi [20] introduce a novel algebra for XML that is based on a simplified data model, wherein trees and forests correspond to relational tuples and relations. Che and Sojitrawala [23] present a new algebra termed DUMAX, specifically designed for XML and XML queries, aimed at integrating node-based and tree-based features (both of which are crucial for XML) and enhancing the execution speed of XML queries within large XML databases. Furthermore, Wang et al. [25] propose a comprehensive approach for facilitating order-sensitive XQuery-to-SQL translation, which is applicable regardless of the chosen XML-to-relational data mapping or order-encoding method. However, it is noteworthy that the aforementioned algebraic approaches do not accommodate (fuzzy) spatiotemporal data, despite the existence of alternative XML algebraic frameworks, such as probabilistic XML algebra [26] and fuzzy XML algebra [8]. Nonetheless, while these studies do not directly address fuzzy spatiotemporal data in XML, they contribute significantly to the foundational understanding of algebraic operations concerning fuzzy spatiotemporal data in XML contexts. Bai and Zhu [27] specifically investigate an algebra for the manipulation of fuzzy spatiotemporal XML data.

On the basis of algebra, to facilitate the efficient handling of such queries, two primary research domains are proposed. The first domain focuses on the integration of fuzzy logic into the querying of fuzzy spatiotemporal data. In their work, Sözer et al. [6] introduce a querying mechanism that merges a fuzzy object-oriented database model with a knowledge base. Additionally, Sözer et al. have implemented this methodology [28]. The second research domain addresses the application of fuzzy logic in querying fuzzy data within XML frameworks. Rodrigues et al. [29] propose a novel fuzzy database architecture and introduce the FSQL Server. Furthermore, Seto et al. [30] present a system designed to enable non-technical users to query extensive XML documents, incorporating both strict and fuzzy qualifiers. Lastly, Polyzotis et al. [31] embark on an investigation into approximate query answering techniques for large XML databases, utilizing an innovative and conceptually straightforward XML summarization mechanism known as TREESKETCH synopses.

Conversely, the primary research domain previously discussed is further developed through the application of fuzzy spatial SQL. While the relational language SQL is both mature and well-established, it shows inadequate for querying fuzzy spatiotemporal data. In terms of query performance, fuzzy data, or fuzzy spatiotemporal data, impose stringent constraints on the structure [9, 32] and data types [33, 34] associated with spatiotemporal data, which can lead to excessive query times due to the complexity of the underlying structure. Regarding memory efficiency, relational data is characterized as “dense” [35], indicating that each row contains a value for every column, necessitating the use of "null" values for any missing or inapplicable data. Consequently, this results in relatively high memory costs. In contrast, XML presents a fundamentally different approach to data representation, effectively addressing the aforementioned issues. For the first issue, XML's foundational data structure is a labeled ordered rooted tree [8], which facilitates the natural querying of nested data at various levels of the hierarchy. For the second issue, XML data can be "sparse," allowing for the absence of elements to be represented as "empty" or "nonexistent," thereby reducing memory costs. However, despite the integration of fuzzy logic derived from other applications (such as neural networks [36, 37] and fuzzy wavelet [3, 38, 39]) in the querying of fuzzy spatiotemporal data and fuzzy data within XML, there remains a paucity of research focused on querying fuzzy spatiotemporal data specifically utilizing XML.

XQuery, a query language developed by the World Wide Web Consortium (W3C), facilitates the selection, reorganization, and potential transformation of XML data elements, allowing users to return results in a preferred structure. Despite its expressive capabilities, XQuery is limited in its ability to accommodate query criteria based on user preferences articulated through fuzzy linguistic terms. Additionally, it lacks the capacity to differentiate query results according to specific user-defined criteria. To address this limitation, Goncalves and Tineo [40] have proposed an extension to the XQuery language that enables the declaration and utilization of fuzzy terms within query expressions. However, their approach primarily focuses on traditional data and does not extend to fuzzy spatiotemporal data, which possesses distinct characteristics that differentiate it from conventional data types. Bai et al. [4] propose an approach for querying fuzzy spatiotemporal data using XQuery. The exploration of querying fuzzy spatiotemporal data using XQuery remains insufficiently addressed in the existing literature.

This chapter introduces a comprehensive algebraic framework designed to facilitate fuzzy spatiotemporal XML queries, incorporating set operations and various additional operations. We explore multiple formal characterizations of fuzzy spatiotemporal algebraic equivalences, which provide a solid basis for the optimization of fuzzy spatiotemporal queries in XML. Furthermore, we examine the process of converting XQuery expressions into algebraic expressions, utilizing algebraic methods to encapsulate fuzzy spatiotemporal queries articulated in XQuery. After that, we initially introduce an architecture for querying fuzzy spatiotemporal data utilizing XQuery. Building upon this foundation, we implement fuzzy-set-based enhancements to XQuery, which encompass truth degrees, fuzzy spatiotemporal linguistic terms, and FLWOR expressions. Additionally, we provide illustrative example queries from the relevant application domain, examine essential aspects of their processing, and present evaluations for each query example. Ultimately, the comparative analysis highlights the benefits of our proposed methodology.

5.2 Formal Representation of Fuzzy Spatiotemporal Data in XML

The fundamental data structure utilized for fuzzy spatiotemporal XML is the data tree. To effectively manage fuzzy spatiotemporal XML data, it is essential to utilize a fuzzy spatiotemporal XML data tree. We will begin by presenting several foundational concepts [16].

Definition 5.1 Let V represent a finite set of vertices, $E \in V \times V$ denote a set of edges, and $\ell: E \to \Gamma$ be a mapping from edges to a set Γ of strings called labels. The triple $G = (V, E, \ell)$ is an edge-labeled directed data tree.

Definition 5.1 pertains to edge-labeled directed data trees and is influenced by the tree patterns found in XML; however, it does not account for several XML characteristics, including data types, ordering, and the differentiation between elements and attributes. Given that ordering is a critical component in the management of fuzzy spatiotemporal XML data, the concept of partial order is introduced to articulate the ordering within XML structures.

Definition 5.2 A relation R is a partial order (denoted as "$\leq$") on a set S if it satisfies the following properties:

- Reflexivity: $\forall \alpha \in S$, $(\alpha, \alpha) \in R$.
- Antisymmetry: if $(\alpha, \beta) \in R$ and $(\beta, \alpha) \in R$, then $\alpha = \beta$.
- Transitivity: if $(\alpha, \beta) \in R$ and $(\beta, \gamma) \in R$, then $(\alpha, \gamma) \in R$.

Building upon the aforementioned definitions, we proceed to formulate the fuzzy spatiotemporal XML data model. This involves the delineation of fuzzy spatiotemporal XML data and the corresponding fuzzy spatiotemporal XML data tree, followed by an examination of the relationships between these two constructs. Furthermore, we provide a formal definition of the structure of the fuzzy spatiotemporal XML data model. In light of the inherent characteristics of spatiotemporal data, we commence by presenting the definition of fuzzy spatiotemporal XML data as outlined below.

Definition 5.3 (*Fuzzy spatiotemporal XML data*) Fuzzy spatiotemporal data *FSP* is a 5-tuple, *FSP* = (*OID*, *ATTR*, *FP*, *FM*, *FT*), where

- *OID* is the changing history of the spatiotemporal data.
- *ATTR* is fuzzy attributes of spatiotemporal data.
- *FP* is fuzzy position of spatiotemporal data.
- *FM* is fuzzy motion of spatiotemporal data.
- *FT* is fuzzy time of spatiotemporal data.

Given the structured nature of XML data, it is inherently capable of representing fuzzy information. In the context of an XML document, membership degrees may be linked to elements, while possibility distributions can be associated with the attribute values of these elements. Consequently, this study explores the representation of fuzzy XML spatiotemporal data within an XML document that has been modified and extended as outlined in reference [8].

Definition 5.4 (*Fuzzy spatiotemporal XML data tree*) For the fuzzy XML spatiotemporal document, we have $F = (V, \psi, T, \varpi, \wp, \tau, \xi, \delta, \pi)$, including:

- $V = \{V_1, V_2, \ldots, V_n\}$ is a finite set of vertices.
- $\psi \subset \{(V_i, V_j) \mid V_i, V_j \in V\}$, (V, ψ) is a directed tree.
- $T \in \psi \subset \{(V_i, V_j) \mid V_i, V_j \in V\}$ specifies time of V_j.
- ϖ represents the nesting depth of V in the spatiotemporal data tree of the document. The nesting depth ϖ of root node is assigned a value of 1, and the ϖ of each following level adds 1.
- $\wp$ maintains the order information in the fuzzy spatiotemporal XML data tree. It is derived by enumerating the number of words from the root of the fuzzy data tree until the starting and ending of the element, respectively. Here, we employ preorder traversal.
- τ represents a set of labels.
- For the node $v \in V$ and the label $\nabla \in \tau$, $\xi(v, \nabla)$ specifies v exists with label ∇.
- $\delta \in \tau$ represents a mapping from the element node $v \in V$ to membership degree functions. It defines the potential existence of the element node.
- $\pi \in \tau$ represents a mapping from the attribute node $v \in V$ to possibility distribution functions. It defines the potential existence of the attribute node.

Definition 5.5 Suppose $F = (V, \psi, T, \varpi, \wp, \tau, \xi, \delta, \pi)$ and $f = (V', \psi', T', \varpi', \wp', \tau', \xi', \delta', \pi')$ are two fuzzy spatiotemporal XML data trees. f is a subtree of F, written $f \propto F$, when

- $V' \subseteq V$, $\psi' = \psi \cap V' \times V'$.
- if $(V_i, V_j) \in \psi$ and $V_i \in V'$, then $V_j \in V'$.
- $T' \subseteq T$.
- if $V_i \in V$, $V_i' \in V'$, and $\varpi(V_i) = \max\{\varpi_1, \varpi_2, \ldots, \varpi_n\}$, $\varpi(V_i') = \max\{\varpi_1', \varpi_2', \ldots, \varpi_n'\}$, then $\varpi(V_i') \leq \varpi(V_i)$.
- $\wp'$, ξ' and τ' denote the restriction of $\wp$, ξ and τ to the nodes in V', respectively.
- $\delta' \subseteq \delta$ and $\pi' \subseteq \pi$.

Based on the aforementioned definitions, a fuzzy spatiotemporal XML data tree may consist of several fuzzy spatiotemporal XML data subtrees. Furthermore, multiple fuzzy spatiotemporal XML data trees can collectively form a fuzzy spatiotemporal XML data forest.

Theorem 5.1 For any collection of fuzzy spatiotemporal XML data trees denoted as $F = \{f_1, f_2, \ldots, f_n\}$, it is possible to identify a corresponding fuzzy spatiotemporal XML data forest that aligns with F.

Theorem 5.2 λ denotes a fuzzy partial order relationship on a fuzzy spatiotemporal XML data tree F. It can be established that there exists a fuzzy transitivity relationship λ' that fulfills the following conditions:

- $\lambda \in \lambda'$.
- The relationship denoted by λ' is a partial order that is derived from λ, preserving both the properties of reflexivity and antisymmetry inherent in λ.

Proof Firstly, $\eta = \{\vartheta \mid \lambda \subseteq \vartheta\}$ is constructed. It is evident that $(\eta, \subseteq)$ is a partial order. Let us assume $\vartheta = \cup_{i\in I} \vartheta_i$, where $\{\vartheta_i\}_{i\in I} \subseteq \eta$ represents an ordered subtrees of F. For each $\forall i \in I$, there exists $\lambda \subseteq \vartheta_i$. By the definition of partial order, we have $\vartheta_i\,(f,f) = 1$. For each $f \in F$, it follows that $\vartheta\,(f,f) = \vee_{i\in I} \vartheta_i\,(f,f) = 1$. Therefore, ϑ exhibits reflexivity.

Secondly, for every $\vartheta_i \subseteq \vartheta_j$, where $i, j \in I$, $\forall x, y \in F$, the property of antisymmetry of F is denoted as ϖ. Consequently, it follows that $\vartheta\,(x, y)\ \varpi\ \vartheta\,(y, x) = [\vee_{i\in I} \vartheta_i\,(x, y)]\ \varpi\ [\vee_{j\in I} \vartheta_j\,(y, x)] = \vee_{i,j\in I}\,[\vartheta_i\,(x, y)\ \varpi\ \vartheta_j\,(y, x)]$. Conversely, it follows that $\vartheta_i\,(x, y)\ \varpi\ \vartheta_j\,(y, x) \le \vartheta_j\,(x, y)\ \varpi\ \vartheta_j\,(y, x) = 0$. Hence, $\vee_{i,j\in I}\,[\vartheta_i\,(x, y)\ \varpi\ \vartheta_j\,(y, x)] = 0$. Consequently, $\vartheta\,(x, y)\ \varpi\ \vartheta\,(y, x) = 0$, confirming that ϑ has the property of antisymmetry denoted by ϖ.

Finally, we will demonstrate the proof of transitivity. For every $\vartheta_i \subseteq \vartheta_j$, where i, $j \in I$, $\forall x, y, z \in F$, F exhibits transitivity $\Im$. Then, $\vartheta\,(x, y)\ \Im\ \vartheta\,(y, z) = [\vee_{i\in I} \vartheta_i\,(x, y)]\ \Im\ [\vee_{j\in I} \vartheta_j\,(y, z)] = \vee_{i,j\in I}\,[\vartheta_i\,(x, y)\ \Im \vartheta_j\,(y, z)]$. $\vartheta_i\,(x, y)\ \Im\ \vartheta_j\,(y, z) \le \vartheta_j\,(x, y)\ \Im\ \vartheta_j\,(y, z) \le \vartheta_j\,(x, z) \le \vartheta\,(x, z)$. Therefore $\vartheta\,(x, y)\ \Im\ \vartheta\,(y, z) \le \vartheta\,(x, z)$, confirming that ϑ has transitivity $\Im$.

Consequently, there exists maximal element λ' in η that satisfies $\lambda \in \lambda'$. The partial order λ' is derived from λ, preserving the properties of the reflexivity and antisymmetry of λ.

An XML document can be characterized as a labeled, ordered, rooted tree structure. In this context, fuzzy spatiotemporal data can be conceptualized as a structured tree representing metadata. Therefore, operations performed on fuzzy spatiotemporal data can be interpreted as operations conducted on tree structures. As a result, it is essential to regard fuzzy spatiotemporal XML data trees as isomorphic. □

Definition 5.6 Let fuzzy spatiotemporal XML data trees $f_1 = (V_1, \psi_1, T_1, \varpi_1, \wp_1, \tau_1, \xi_1, \delta_1, \pi_1)$ and $f_2 = (V_2, \psi_2, T_2, \varpi_2, \wp_2, \tau_2, \xi_2, \delta_2, \pi_2)$ be considered as the subtrees of $F = (V, \psi, T, \varpi, \wp, \tau, \xi, \delta, \pi)$. Then, f_1 and f_2 are isomorphic (recorded $f_1 \cong f_2$), when:

- $V_1 \cup V_2 \subseteq V$, $\psi_1 \cup \psi_2 \subseteq \psi$, $T_l \cup T_2 \subseteq T$, $\tau_1 \cup \tau_2 \subseteq \tau$.
- $\varpi_1 = \varpi_2 \le \varpi$.
- There is a one-to-one mapping $\Im\xi$: $\xi_1 \to \xi_2$, which makes $\forall \Im\xi(\xi_1) = \xi_2$.

Theorem 5.3 Fuzzy data tree F and its subtree f are isomorphic.

The proof of this theorem is based on an examination of fuzzy spatiotemporal XML data trees, along with the relevant definitions. The process is relatively straightforward.

5.3 Algebraic Operations on Fuzzy Spatiotemporal XML Data

5.3.1 Algebraic Operators

This subsection addresses the challenges associated with the formal design of fuzzy spatiotemporal XML algebra. Each operator is designed to accept collections of fuzzy spatiotemporal XML data trees as input and subsequently generate a collection of fuzzy spatiotemporal XML data trees as output.

5.3.1.1 Set Operations

In this subsection, we will delineate four fundamental operations pertaining to fuzzy sets as applied to fuzzy spatiotemporal XML data. These operations include fuzzy union ($\cup$), fuzzy intersection ($\cap$), fuzzy difference ($-$), and Cartesian product (∞). The symbols ω (a) and φ (a) are employed to signify the pattern and instance associated with the algebraic expression a.

Definition 5.7 (Fuzzy union). Suppose $f_i = (V_i, \psi_i, T_i, \varpi_i, \wp_i, \tau_i, \xi_i, \delta_i, \pi_i)$ and $f_j = (V_j, \psi_j, T_j, \varpi_j, \wp_j, \tau_j, \xi_j, \delta_j, \pi_j)$ represent fuzzy spatiotemporal XML data trees. When f_i and f_j are isomorphic, the fuzzy union ($\cup$) can be defined in the followings:

- $\omega(f_i \cup f_j) = \left(V_{i\cup j}, \psi_{i\cup j}, T_{i\cup j}, \varpi_{i\cup j}, \wp_{i\cdot\cup j}, \cdot\tau_{i\cup j}, \xi_{i\cup j}, \delta_{i\cup j}, \pi_{i\cup j}\right)$.
- $\varphi(f_i \cup f_j) \in \{e | e \in \varphi(f_i) \text{ore} \in \varphi(f_j)\}$.

Figure 5.1 illustrates a fuzzy union operation applied to two fuzzy spatiotemporal XML data trees, designated as A and B, where T_s and T_e denote the starting and ending time points, respectively. In accordance with Definition 5.7, the fuzzy spatiotemporal XML data trees A and B exhibit isomorphism. Utilizing principles from fuzzy set theory, the membership degree of the fuzzy union result, denoted as δ_r, is defined as δ_r = Max (δ_i, δ_j), where δ_i andδ_j represent the membership degrees of trees A and B, respectively. Max (δ_i, δ_j) yields the maximum value between δ_i and δ_j, with the default value of δ_r set to 1. As shown in Fig. 5.1, the time interval $[t_3, t_8]$ in the fuzzy union result $A \cup B$ is derived from the union of the time intervals $[t_3, t_6]$ and $[t_4, t_8]$. Furthermore, the "Poss" value for "thick," which is 0.85, is calculated as Max (0.85 $\cup$ 0.65) = 0.85; the "Poss" value for "moderate," also 0.85, is derived from Max (0 $\cup$ 0.85) = 0.85; and the "Poss" value for "thin," which is 0.75, results from Max (0.75 $\cup$ 0) = 0.75.

Definition 5.8 (*Fuzzy intersection*) Suppose $f_i = (V_i, \psi_i, T_i, \varpi_i, \wp_i, \tau_i, \xi_i, \delta_i, \pi_i)$ and $f_j = (V_j, \psi_j, T_j, \varpi_j, \wp_j, \tau_j, \xi_j, \delta_j, \pi_j)$ represent fuzzy spatiotemporal XML data trees. When f_i and f_j are isomorphic, the fuzzy intersection ($\cap$) can be defined in the followings:

- $\omega(f_i \cap f_j) = \left(V_{i\cap j}, \psi_{i\cap j}, T_{i\cap j}, \varpi_{i\cap j}, \wp_{i\cap j}, \cdot\tau_{i\cap j}, \xi_{i\cap j}, \delta_{i\cap j}, \pi_{i\cap j}\right)$.

<cloud density T_s = "t_3" T_e = "t_6">	<cloud density T_s = "t_4" T_e = "t_8">	<cloud density T_s = "t_3" T_e = "t_8">
<Dist type = "disjunctive">	<Dist type = "disjunctive">	<Dist type = "disjunctive">
<Val Poss = 0.85>thick</Val>	<Val Poss = 0.65>thick</Val>	<Val Poss = 0.85>thick</Val>
<Val Poss = 0.75>thin</Val>	<Val Poss = 0.85>moderate</Val>	<Val Poss = 0.85>moderate</Val>
<Dist>	<Dist>	<Val Poss = 0.75>thin</Val>
</cloud density>	</cloud density>	<Dist></cloud density>
		</cloud density>
A	B	A ∪ B

Fig. 5.1 Fuzzy union operation on two fuzzy spatiotemporal XML fragments

- $\varphi(f_i \cap f_j) \in \{e|e \in \varphi(f_i)\text{or}e \in \varphi(f_j)\}$.

Figure 5.2 illustrates a fuzzy intersection operation conducted on two fuzzy spatiotemporal XML data trees, denoted as *A* and *B*, utilizing membership degrees and possibility distributions. In accordance with fuzzy set theory, the membership degree of the fuzzy intersection result, denoted as δ_r, is determined by the equation δ_r = Min (δ_i, δ_j), where δ_i andδ_j represent the membership degrees of data trees *A* and *B*, respectively. Min (δ_i, δ_j) yields the lesser value between δ_i and δ_j. It is important to note that a fuzzy spatiotemporal object is considered non-existent and can be disregarded if δ_r equals 0. As shown in Fig. 5.2, the time interval $[t_4, t_6]$ in the fuzzy intersection result $A \cap B$ is derived from the intersection of the time intervals $[t_3, t_6]$ and $[t_4, t_8]$. Furthermore, the "Poss" value for the attribute "thick," which is 0.65, is calculated as Min (0.85 ∩ 0.65) = 0.65; the "Poss" value for "moderate" is 0, resulting from Min (0 ∩ 0.85) = 0; and the "Poss" value for "thin" is 0.75, derived from Min (0.75 ∩ 0) = 0. Consequently, the attributes "moderate" and "thin" are excluded from the fuzzy intersection result due to their membership degrees being equal to 0.

Definition 5.9 (*Fuzzy difference*) Suppose $f_i = (V_i, \psi_i, T_i, \varpi_i, \wp_i, \tau_i, \xi_i, \delta_i, \pi_i)$ and$f_j = (V_j, \psi_j, T_j, \varpi_j, \wp_j, \tau_j, \xi_j, \delta_j, \pi_j)$ represent fuzzy spatiotemporal XML data trees. Whenf_i andf_j are isomorphic, the fuzzy difference (−) can be defined in the followings:

- $\omega(f_i - f_j) = (V_{i-j}, \psi_{i-j}, T_{j-j}, \varpi_{i-j}, \xi_{i-j}, \tau_{i-j}, \xi_{i-j}, \delta_{i-j}, \pi_{i-j})$.
- $\varphi(f_i - f_j) \cdot \{e|e \in \varphi(f_i)\text{and } e \notin \varphi(f_j)\}$.

Figure 5.3 illustrates a fuzzy difference operation applied to fuzzy spatiotemporal XML data trees *A* and *B*, utilizing membership degrees and possibility distributions.

<cloud density T_s = "t_3" T_e = "t_6">	<cloud density T_s = "t_4" T_e = "t_8">	<cloud density T_s = "t_4" T_e = "t_6">
<Dist type = "disjunctive">	<Dist type = "disjunctive">	<Dist type = "disjunctive">
<Val Poss = 0.85>thick</Val>	<Val Poss = 0.65>thick</Val>	<Val Poss = 0.65>thick</Val>
<Val Poss = 0.75>thin</Val>	<Val Poss = 0.85>moderate</Val>	<Dist>
<Dist>	<Dist>	</cloud density>
</cloud density>	</cloud density>	
A	B	A ∩ B

Fig. 5.2 Fuzzy intersection operation on two fuzzy spatiotemporal XML fragments

<cloud density T_s = "t_3" T_e = "t_6"> <Dist type = "disjunctive"> <Val Poss = 0.85>thick</Val> <Val Poss = 0.75>thin</Val> <Dist> </cloud density>	<cloud density T_s = "t_4" T_e = "t_8"> <Dist type = "disjunctive"> <Val Poss = 0.65>thick</Val> <Val Poss = 0.85>moderate</Val> <Dist> </cloud density>	<cloud density T_s = "t_3" T_e = "t_4"> <Dist type = "disjunctive"> <Val Poss = 0.35>thick</Val> <Val Poss = 0.75>thin</Val> <Dist> </cloud density>
A	B	A – B

Fig. 5.3 Fuzzy difference operation on two fuzzy spatiotemporal XML fragments

In accordance with fuzzy set theory, the relationship is defined as δ_r = Min (δ_i, $\delta_j{}'$), where $\delta_j{}'$ represents the complement of δ_j, calculated as $\delta_j{}' = 1 - \delta_j$. Here, δ_i andδ_j denote the membership degrees of data trees *A* and *B*, respectively, while δ_r signifies the membership degree of the resulting fuzzy difference. As shown in Fig. 5.3, the time interval $[t_3, t_4]$ in the fuzzy difference result $A - B$ is derived from the difference between the time intervals $[t_3, t_6]$ and $[t_4, t_8]$. Furthermore, the "Poss" value for "thick," which is 0.35, is computed as Min (0.85 ∩ (1 – 0.65)) = 0.35; the "Poss" value for "moderate" is 0, derived from Min (0 ∩ (1 – 0.85)) = 0; and the "Poss" value for "thin" is 0.75, calculated as Min (0.75 ∩ (1 – 0)) = 0.75. Consequently, the "moderate" attribute is excluded from the fuzzy difference result due to its membership degree being 0.

Definition 5.10 (*Cartesian product*) Suppose $f_i = (V_i, \psi_i, T_i, \varpi_i, \wp_i, \tau_i, \xi_i, \delta_i, \pi_i)$ and $f_j = (V_j, \psi_j, T_j, \varpi_j, \wp_j, \tau_j, \xi_j, \delta_j, \pi_j)$ represent fuzzy spatiotemporal XML data trees. When f_i and f_j are isomorphic, the Cartesian product (∞) can be defined in the followings:

- $\omega\left(f_i \infty f_j\right) = \left(V_{i\infty j}, \psi_{i\infty j}, T_{i\infty j}, \varpi_{i\infty j}, \Im_{i\infty j}, \tau_{i\infty j}, \xi_{i\infty j}, \delta_{i\infty j}, \pi_{i\infty j}\right)$.
- $\varphi\left(f_i \infty f_j\right) \in \left\{(o_1, o_2) | o_l \in \varphi(f_i) \text{and} o_2 \in \varphi\left(f_j\right)\right\}$.

The Cartesian product operation involves two fuzzy spatiotemporal trees, denoted as f_i and f_j, which serve as inputs to generate output fuzzy spatiotemporal trees that represent the "Bjuxtaposition" of each pair of trees from f_i to f_j. In accordance with fuzzy set theory, the membership degree of the Cartesian product result, denoted as δ_r, is defined as δ_r = Min (δ_i, δ_j), where δ_i and δ_j represent the membership degrees of sets *A* and *B*, respectively.

Theorem 5.4 Suppose $f_i = (V_i, \psi_i, T_i, \varpi_i, \wp_i, \tau_i, \xi_i, \delta_i, \pi_i)$ and $f_j = (V_j, \psi_j, T_j, \varpi_j, \wp_j, \tau_j, \xi_j, \delta_j, \pi_j)$ represent fuzzy spatiotemporal XML data trees, then we have:

- $f_i \cup f_j$ are fuzzy spatiotemporal XML data trees.
- $f_i \cap f_j$ are fuzzy spatiotemporal XML data trees.
- $f_i - f_j$ are fuzzy spatiotemporal XML data trees.
- $f_i \infty f_j$ are fuzzy spatiotemporal XML data trees.

Definition 5.11 Suppose $f_1 = (V_1, \psi_1, T_1, \varpi_1, \wp_1, \tau_1, \xi_1, \delta_1, \pi_1), f_2 = (V_2, \psi_2, T_2, \varpi_2, \wp_2, \tau_2, \xi_2, \delta_2, \pi_2), \ldots, f_i = (V_i, \psi_i, T_i, \varpi_i, \wp_i, \tau_i, \xi_i, \delta_i, \pi_i)$ represent fuzzy spatiotemporal XML data trees. $F = (V, \psi, T, \varpi, \wp, \tau, \xi, \delta, \pi)$ is considered the reconstruction of $f_1, f_2, \ldots, f_i$, if there exists a rule $\Im$: $T \to T'$ such that F is derived from $f_1, f_2, \ldots, f_i$ based on Theorem 5.4.

5.3.1.2 Other Useful Operations

The fuzzy selection (F-Selection) operator, denoted as σ, processes fuzzy spatiotemporal data trees by employing a specific predicate that may consist of various combinations of logical operators and straightforward qualifications. This operator takes as input a collection of fuzzy spatiotemporal XML data trees, denoted as F, and applies a specified condition to each element and attribute node within these trees. The output is a comprehensive set of selected trees, which encompasses not only the content of the desired results but also the structural characteristics of the target trees.

Definition 5.12 (*F-Selection*) Let $F = (V, \psi, T, \varpi, \wp, \tau, \xi, \delta, \pi)$ represent a fuzzy spatiotemporal XML data tree, if there is a predicate ρ, then we have the definition of F-Selection as follows:

$$\sigma_\rho(F) = \{a | \mathrm{a} \Leftarrow \mathrm{F} \cap \mathrm{con}(\rho, \mathrm{a})\}.$$

Here, the predicate ρ and the variable a are associated with a collection of fuzzy spatiotemporal XML data trees. The function *con* (ρ, a) is employed to extract the relevant elements and attributes that satisfy the specified selection criteria.

Theorem 5.5 Let us consider that the fuzzy spatiotemporal XML data trees denoted as f_i $(i = 0, 1, 2, \ldots)$ are derived from the F-Selection applied to a fuzzy spatiotemporal XML data tree represented by F. Then, it can be stated that $f_i \cong F$.

Proof As demonstrated in Definition 5.12, the F-Selection operation does not generate any new fuzzy spatiotemporal objects. Consequently, if $f_i \propto F$, it follows from Theorem 5.3 that $f_i \cong F$.

The fuzzy projection (F-Projection) operator, denoted as π, can be conceptualized as the process of removing element nodes or attribute nodes that are not explicitly specified within the fuzzy spatiotemporal XML data trees. In the resultant substructure following the elimination of nodes, it is anticipated that the partial hierarchical relationships among the remaining nodes, which were present in the original input trees, will be maintained. □

Definition 5.13 (*F-Projection*) Let $F = (V, \psi, T, \varpi, \wp, \tau, \xi, \delta, \pi)$ represent a fuzzy spatiotemporal XML data tree, if there is a fuzzy projection function κ, then we have the definition of F-Projection as follows:

$$\pi_\kappa(F) = \{\kappa(a) | \mathrm{a} \in \mathrm{F} \wedge \kappa(\mathrm{a}) \in \mathrm{F}\}.$$

Theorem 5.6 Let π_κ (F) and $\pi_\kappa{}'$ (F) denote the outcomes derived from the F-Projection applied to a fuzzy spatiotemporal XML data tree represented as $F = (V, \psi, T, \varpi, \wp, \tau, \xi, \delta, \pi)$. In the event that $a \wp a'$, it follows that $\kappa (a) \wp \kappa (a')$.

Proof As demonstrated in Definition 5.13, the fuzzy projection operator is conceptualized as the removal of element nodes or attribute nodes while maintaining the existing partial hierarchical relationships within the substructure that arises from the F-Projection operation. This is due to the fact that the operator $\wp$ retains the order information present in the fuzzy spatiotemporal XML data tree. Consequently, it follows that $\kappa (a) \wp\kappa (a')$ if $a\wp a'$, where $a, a' \in F$.

The fuzzy join (F-Join) operator is utilized to identify fuzzy spatiotemporal XML data trees that satisfy a specified predicate, subsequently generating new fuzzy spatiotemporal XML data trees. This process can be understood as a fuzzy selection derived from the Cartesian product of the fuzzy spatiotemporal XML data trees. □

Definition 5.14 (*F-Join*) Let $f_i = (V_i, \psi_i, T_i, \varpi_i, \wp_i, \tau_i, \xi_i, \delta_i, \pi_i), f_j = (V_j, \psi_j, T_j, \varpi_j, \wp_j, \tau_j, \xi_j, \delta_j, \pi_j)$ represent fuzzy spatiotemporal XML data trees, and $f_1 \cong f_2$. In the presence of a predicate ρ, then we have the definition of F-Join (θ) as follows:

- $\omega\left(f_i\theta_\rho f_j\right) = \omega_\rho\left(f_i \infty f_j\right)$.
- $\varphi\left(f_i\theta_\rho f_j\right) \subseteq \varphi_\rho\left(f_i \infty f_j\right)$.

Theorem 5.7 Let $f_i = (V_i, \psi_i, T_i, \varpi_i, \wp_i, \tau_i, \xi_i, \delta_i, \pi_i), f_j = (V_j, \psi_j, T_j, \varpi_j, \wp_j, \tau_j, \xi_j, \delta_j, \pi_j)$ represent fuzzy spatiotemporal XML data trees, then we have $(f_i\ \theta f_j) \cong f_i \cong f_j$.

Proof In accordance with the characteristics of fuzzy join, it can be stated that $(f_i\ \theta f_j) = \sigma\ (f_i \infty f_j)$. Based on Theorem 5.4, it follows that $f_i \infty f_j$ represents fuzzy spatiotemporal XML data trees. Additionally, Theorem 5.5 indicates that $\sigma\ (f_i \infty f_j) \cong f_i \cong f_j$. Consequently, it can be concluded that $(f_i\ \theta f_j) \cong f_i \cong f_j$.

Fuzzy grouping (F-Grouping) operator, denoted as Γ, serves to represent a subtree within a fuzzy spatiotemporal XML data structure. This operator facilitates the aggregation or summarization of information derived from a collection of similar element nodes or attribute nodes. □

Definition 5.15 (*F-Grouping*) Let $F = (V, \psi, T, \varpi, \wp, \tau, \xi, \delta, \pi)$ represent a fuzzy spatiotemporal XML data tree, and *gro* (a, ρ) denote a grouping function applied to the variable a, which is constrained by a predicate ρ that pertains to a collection of fuzzy spatiotemporal XML data trees. Consequently, the F-Grouping Γ can be formally defined as follows:

$$\Gamma_\rho(F) = \{a|\mathrm{a} \Leftarrow \mathrm{F} \cap \mathrm{gro(a,}\ \rho)\}.$$

Fuzzy ordering (F-Ordering) operator, denoted as Θ, facilitates the reorganization of specified element nodes or attribute nodes. The ordering function, represented as *ord* (a, ρ), operates on the variable a, which is constrained to a collection of fuzzy spatiotemporal XML data trees under the predicate ρ. This predicate may take the form of a scalar function, such as *avg*. By default, the ordering is determined by the sequence of the initial element within the original set of fuzzy spatiotemporal XML data trees.

Definition 5.16 (*F-Ordering*) Let $F = (V, \psi, T, \varpi, \wp, \tau, \xi, \delta, \pi)$ represents a fuzzy spatiotemporal XML data tree. The grouping function is defined as *gro* $(a, \rho) = \{a \mid a \in F \wedge \zeta\ (a, \wp, \beta) \Leftarrow \zeta\ (a, \wp, \alpha)\}$ over the variable a under predicate ρ, where ζ: $\alpha \rightarrow \beta$. The F-Ordering Θ can be defined as follows:

$$\Theta_{\rho}(F) = \{a|\mathrm{a} \Leftarrow \mathrm{F} \cap \mathrm{ord}(\mathrm{a}, \rho)\}.$$

The Bind operator is responsible for extracting fuzzy spatiotemporal data from an input tree based on a specified filter. It generates a structure that encapsulates the variable bindings that arise from the process of pattern matching.

Definition 5.17 (*Bind*) Let $F = (V, \psi, T, \varpi, \wp, \tau, \xi, \delta, \pi)$ represents a fuzzy spatiotemporal XML data tree. The Bind function is denoted as *bind* $(a, \rho) = \{a \mid a \in F \wedge \zeta\ (a, \rho, \beta) \Leftarrow \zeta\ (a, \rho, \alpha)\}$ over the variable a bounding to a set of fuzzy spatiotemporal XML data trees according to a given filter ρ. Consequently, the Bind operator can formally be defined as follows:

$$\mathrm{Bind}(F) = \sum_{\mathrm{bind}\ (a,\rho)} \{a|a \in F \wedge \zeta(a, \rho, \beta) \Leftarrow \zeta(a, \rho, \alpha)\}.$$

The Tree operator functions as the inverse of the Bind operator, specifically applied to Tab structures. It produces a collection of trees that adhere to a specified input pattern. This operator can be utilized to create a new nested XML structure.

Definition 5.18 (*Tree*) Let $F = (V, \psi, T, \varpi, \wp, \tau, \xi, \delta, \pi)$ represent a fuzzy spatiotemporal XML data tree. Additionally, let the function *tree* (a, ρ) denote a treeing function over the variable a bounding to a set of fuzzy spatiotemporal XML data trees in accordance with a given filter ρ. Consequently, the Tree operator can be formally defined as follows:

$$\mathrm{Tree}\ (F) = \sum_{\mathrm{tree}\ (a,\rho)} \{a|a \in F \wedge\ \mathrm{tree}\ (a, \rho) \in F\}.$$

In Fig. 5.4, the Bind operator is utilized on the tree structure that represents the XML collection of positions, incorporating a filter that associates each position with its respective variables: $x, $y, $Vx, $\$f_{Vx}$, $Vy, and $\$f_{Vy}$. Subsequently, the Tree operator is applied to the output generated by the preceding Bind operation, where F [$x, $y, $Vx, $\$f_{Vx}$, $Vy, $\$f_{Vy}$] signifies the relevant filter. The positions are organized based on the variable $Vx, with each subtree containing the variable $\$f_{Vx}$ at various temporal intervals.

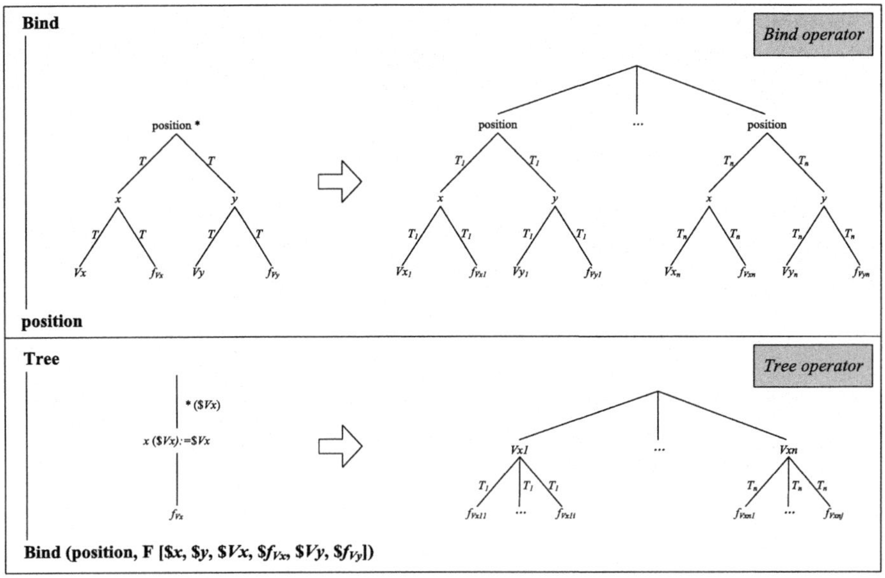

Fig. 5.4 Bind and tree operators

5.3.2 *Algebraic Equivalences and Optimization*

5.3.2.1 Equivalences

Algebraic equivalences demonstrate that substituting the corresponding relations of two expressions with identical relations yields identical results. This subsection presents a number of algebraic equivalences, followed by the introduction of strategies for a logical sequence of operations.

Proposition 5.1 *Let f_1, f_2 represent fuzzy spatiotemporal XML data trees, and let ρ_s denote the fuzzy selection predicate. Then, we can express the followings*:

$$\sigma_{\rho_s}(f_1 \cup f_2) = \sigma_{\rho_s}(f_1) \cup \sigma_{\rho_s}(f_2).$$
$$\sigma_{\rho_s}(f_1 \cap f_2) = \sigma_{\rho_s}(f_1) \cap \sigma_{\rho_s}(f_2).$$
$$\sigma_{\rho_s}(f_1 - f_2) = \sigma_{\rho_s}(f_1) - \sigma_{\rho_s}(f_2).$$

Proposition 5.2 *Let f_1, f_2 represent fuzzy spatiotemporal XML data trees, and let ρ_p denote the fuzzy projection predicate. Then, we can express the followings:*

$$\pi_{\rho_p}(f_1 \cup f_2) = \pi_{\rho_p}(f_1) \cup \pi_{\rho_p}(f_2).$$
$$\pi_{\rho_p}(f_1 \cap f_2) = \pi_{\rho_p}(f_1) \cap \pi_{\rho_p}(f_2).$$
$$\pi_{\rho_p}(f_1 - f_2) = \pi_{\rho_p}(f_1) - \pi_{\rho_p}(f_2).$$

Proposition 5.3 *Let F represent a fuzzy spatiotemporal XML data tree, and let* $\rho_{si}(i = 1, 2, \ldots, n)$ *and* ρ_{sj} $(j = 1, 2, \ldots, n)$ *denote the fuzzy selection predicates. Then, we can express the followings*:

$$\sigma_{\rho_{\mathrm{si}}}\left(\sigma_{\rho_{\mathrm{sj}}}(\mathrm{Bind}(F))\right) = \sigma_{\rho_{\mathrm{si}}}\rho_{\mathrm{sj}}(\mathrm{Bind}(F)).$$

Proposition 5.4 *Let F represent a fuzzy spatiotemporal XML data tree, and let* $\rho_{pi}(i = 1, 2, \ldots, n)$ *and* ρ_{pj} $(j = 1, 2, \ldots, n)$ *denote the fuzzy projection predicates. If* $\{\rho_{pi}\}\subseteq\{\rho_{pj}\}$, *then we can express the followings*:

$$\pi_{\rho_{pi}}\left(\pi_{\rho_{pj}}(\mathrm{Bind}(F))\right) = \pi_{\rho_{pi}\wedge\rho_{sj}}(\mathrm{Bind}(F)) = \pi_{\rho_{pi}}(\mathrm{Bind}(F)).$$

Proposition 5.5 *Let F represent a fuzzy spatiotemporal XML data tree, let* $\rho_{si}(i = 1, 2, \ldots, n)$ *denote the fuzzy selection predicate, and let* $\rho_{pj}(j = 1, 2, \ldots, n)$ *and* $\rho_{pk}(k = 1, 2, \ldots, n)$ *indicate the fuzzy projection predicates.*

- If ρ_{si} is the fuzzy selection predicate on *j*, then we can express the followings:

$$\sigma_{\rho_{si}}\left(\pi_{\rho_{pj}}(\mathrm{Bind}(F))\right) = \pi_{\rho_{Pj}}\left(\sigma_{\rho_{si}}(\mathrm{Bind}(F))\right).$$

- If ρ_{si} is the fuzzy selection predicate on *j* and *k*, then we can express the followings:

$$\sigma_{\rho_{si}}\left(\pi_{\rho_{Pj}}(\mathrm{Bind}(F))\right) = \pi_{\rho_{Pj}}\left(\sigma_{\rho_{si}}\left(\pi_{\rho_{\rho jk}}(\mathrm{Bind}(F))\right)\right).$$

Proposition 5.6 *Let* f_1, f_2 *represent fuzzy spatiotemporal XML data trees, and let* $\rho_{si}(i = 1, 2, \ldots, n)$ *and* ρ_{sj} $(j = 1, 2, \ldots, n)$ *denote the fuzzy selection predicates.*

- If $\rho_s = \rho_{si}$, then we can express the followings:

$$\sigma_{\rho_s}(\mathrm{Bind}(f_1) \infty (\mathrm{Bind}(f_2))) = \sigma_{\rho_{si}}(\mathrm{Bind}(f_1)) \infty \mathrm{Bind}(f_2).$$

- If $\rho_s = \rho_{si} \cap \rho_{sj}$, where ρ_{si} is the fuzzy selection predicate on *i*, then we can express the followings:

$$\sigma_{\rho_s}(\mathrm{Bind}(f_1) \infty (\mathrm{Bind}(f_2))) = \sigma_{\rho_{si}}(\mathrm{Bind}(f_1)) \infty \sigma_{\rho_{\mathrm{sj}}}(\mathrm{Bind}(f_2)).$$

- If ρ_{si} is the fuzzy selection predicate on *i*, where ρ_{sj} is the fuzzy selection predicate on *i* and *j*, then we can express the followings:

$$\sigma_{\rho_s}(\mathrm{Bind}(f_1) \infty (\mathrm{Bind}(f_2))) = \sigma_{\rho_{\mathrm{sj}}}\left(\sigma_{\rho_{si}}(\mathrm{Bind}(f_1))\right) \infty \mathrm{Bind}(f_2).$$

Proposition 5.7 *Let f_1, f_2 represent fuzzy spatiotemporal XML data trees, and let $\rho_{pi}(i = 1, 2, \ldots, n)$ and ρ_{pj} $(j = 1, 2, \ldots, n)$ denote the fuzzy projection predicates, then we can express the followings*:

$$\pi_{\rho_{pi}}, \rho_{pj}(\text{Bind}(f_1) \infty (\text{Bind} \cdot (f_2))) = \pi_{\rho_{pi}, \rho_{pj}}\left(Bind(f_l) \infty \pi_{\rho_{pi,}, \rho_{pj}}(\text{Bind}(f_2)\right).$$

Proposition 5.8 *Let f_1, f_2 represent fuzzy spatiotemporal XML data trees, let ρ_s denote the fuzzy selection predicate, and let ρ_j indicate the fuzzy join restriction.*

- If $\forall a \in f_1, b \in f_2, \rho_s\ (a \vee b) = \rho_s\ (a)$, then we can express the followings:

$$\sigma_{\rho_s}\left(\text{Bind}(f_1)\theta_{\rho_j}(\text{Bind}(f_2))\right) = \sigma_{\rho_s}(\text{Bind}(f_i))\theta_{\rho_j}\sigma_{\rho_{sj}}f_2.$$

- If $\forall a \in f_1, b \in f_2, \rho_s\ (a \vee b) = \rho_s\ (b)$, then we can express the followings:

$$\sigma_{\rho_s}\left(\text{Bind}(f_l)\theta_{\rho_j}(\text{Bind}(f_2))\right) = f_l\theta_{\rho_j}\sigma_{\rho_s}(\text{Bind}(f_2)).$$

5.3.2.2 Optimization

The proposed propositions facilitate the execution of fuzzy spatiotemporal XML queries through a systematic sequence of operations. Specifically, these strategies involve relocating selections and projections to the leaves of the input query tree to the greatest extent possible by employing the aforementioned propositions, shown in Algorithm 5.1. Additionally, they aim to consolidate the concatenations of selections and projections into singular selections, singular projections, or projections that follow selections. The relocation of selections can be achieved through Propositions 5.1, 5.3, 5.5, 5.6, and 5.8, while the relocation of projections can be accomplished using Propositions 5.2, 5.5, and 5.7. The integration of selections and projections can be facilitated by Propositions 5.3, 5.4, and 5.5. The optimization strategies for querying fuzzy spatiotemporal data are outlined in the subsequent sections. Notably, a single proposition, such as Proposition 5.5, may be applicable across various strategies, as the input query may necessitate a logical sequence of operations that prioritizes selections before projections or vice versa, depending on the specific requirements of practical applications.

Algorithm 5.1 Optimization strategies

Input: Initial query tree of Query Q
Output: Query tree of Query Q after optimization
Let Leaf(Q) = move Q to the bottom of the query tree
for $i = 1$ to l //l is the maximum number of optimization loops
Leaf(Q) $\leftarrow \sigma$ (Q) //moving the selections to the leaves of the input query tree according to Propositions 5.1, 5.3, 5.5, 5.6, and 5.8
Leaf(Q) $\leftarrow \pi$ (Q) //moving the projections to the leaves of the input query tree according to Propositions 5.2, 5.5, and 5.7
σ (Q) $\leftarrow \sigma(\sigma(Q))$ // integrate the concatenations of selections according to Proposition 5.3
π (Q) $\leftarrow \pi$ (π (Q)) // integrate the concatenations of projections according to Proposition 5.4
$\sigma(\pi$ (Q)) $\leftarrow \pi$ ($\sigma(Q)$) //integrate the concatenations of selections and projections according to Propositions 5.3 and 5.4
end for

5.3.3 Transformation of XQuery Expressions

Fuzzy spatiotemporal data queries can be characterized as inquiries pertaining to fuzzy spatiotemporal data, where the fuzziness may manifest in temporal, spatial, spatiotemporal, or general attributes. Regarding temporal attributes, fuzziness may be present in specific time instances or time intervals. In the context of spatial attributes, fuzziness may arise in terms of location or movement. With respect to spatiotemporal attributes, fuzziness may be evident in spatiotemporal operations. Lastly, concerning general attributes, fuzziness may pertain to the attributes themselves or the quantity of these general attributes.

The motivations for querying XML are as diverse as the applications of XML itself. Notable applications of the XQuery language include: retrieving information from relational databases for integration into web services; producing reports from database-stored data for display on the web in XHTML format; conducting searches within textual documents housed in native XML databases and presenting the findings; extracting data from databases or proprietary software and transforming it for the purpose of application integration; amalgamating content from traditionally non-XML sources to facilitate content management and delivery; and performing ad hoc queries on standalone XML documents for testing or research purposes.

XQuery, the query language designed for XML data, offers a diverse array of query expressions. To facilitate the construction of complex queries, XQuery incorporates a structural framework known as FLWOR expressions, which is an acronym derived from the initial letters of the keywords that define the various clauses within these expressions: "for", "let", "where", "order by", and "return". The fundamental framework of numerous (though not all) queries is encapsulated in the FLWOR expression. FLWOR expressions, commonly referred to as FLWORs, are employed for queries that exhibit greater complexity. In addition to facilitating more comprehensible and organized selections, these expressions enable functionalities such as the integration of data from various sources, the creation of new elements and attributes, the evaluation of functions on intermediate values, and the sorting of results. The FLWOR expression serves as the foundational structure for numerous queries. In this subsection, we will demonstrate the process of transforming XQuery expressions into algebraic expressions. Initially, we will illustrate the transformation of each clause within a FLWOR expression, followed by examples that explore the representation of fuzzy spatiotemporal XML queries. The syntax of the FLWOR expression is outlined as follows:

for clause | *let clause*
[*where* <*where_condition*>]
[*order by* <*order_expression*>]
return return_expression

5.3.3.1 The "For" Clause

A "for" clause establishes an iteration mechanism that enables the subsequent components of the FLWOR expression to be assessed multiple times, corresponding to each item within the sequence produced by the expression following the "in" keyword. This sequence, referred to as the binding sequence, may yield any sequence containing zero, one, or multiple items. In instances where the binding sequence is empty, the remaining elements of the FLWOR are not evaluated, resulting in zero iterations. A "for" clause can encompass one or more variables, which are linked to an expression that determines the binding sequence for those variables. This "for" clause is subsequently converted into an algebraic expression that generates a distinct tree for each potential binding configuration.

Definition 5.19 Consider a "for" clause that includes a single variable binding, expressed as follows: *for $i in doc* ("*F*") $\lambda_1[\gamma_1]\ldots\lambda_n[\gamma_n]$, where F represents the input fuzzy spatiotemporal XML data tree, λ_i (for $1 \leq i \leq n$) denotes a path expression, and γ_i (for $1 \leq i \leq n$) signifies a condition. This clause can be transformed into the subsequent algebraic expression:

$$\sigma_{\gamma_i}\left(\pi_{\lambda_i}\ldots\left(\sigma_{[\gamma_1]}(\pi_{\lambda_1}(\text{ Bind }(\text{``}F\text{''})))\right)\right).$$

Definition 5.20 Consider a "for" clause that incorporates multiple variable bindings, expressed as follows: *for $i, j in doc* ("*F*") $\lambda_1[\gamma_1]\ldots\lambda_n[\gamma_n]$, where F denotes the input fuzzy spatiotemporal XML data tree, λ_i and λ_j (where $1 \leq i \leq n$ and $1 \leq j \leq n$) represent path expressions, and γ_i and γ_j (where $1 \leq i \leq n$ and $1 \leq j \leq n$) denote conditions. This clause can be transformed into the following algebraic expression:

$$\sigma_{\gamma_i}\left(\pi_{\lambda_i}\cdots\left(\sigma_{[\gamma_1]}(\pi\lambda_{\lambda_l}(\text{Bind}(\text{``}F\text{''})))\right)\right)\infty\,\sigma_{[\gamma_j]}\left(\pi\lambda_{\lambda_j}\ldots\left(\sigma_{[\gamma_l]}\right](\pi\lambda_l(\text{Bind}(\text{``}F\text{''})))\right)\right).$$

5.3.3.2 The "Let" Clause

A "let" clause functions as an efficient mechanism for associating a variable with a specific value. In contrast to a "for" clause, which facilitates iteration, a "let" clause binds the entire sequence to the variable rather than sequentially binding each individual item. This construct provides a practical advantage in programming by eliminating the need to repeatedly express the same computation. Additionally, in certain implementations, it may enhance performance by ensuring that the expression is evaluated only once, rather than each time it is referenced. A "let" clause can encompass one or more variables. In contrast to a "for" clause, a "let" clause associates each variable with the complete outcome of its corresponding expression without engaging in iteration. Consequently, a "let" clause is converted into an algebraic expression that yields a singular tree.

Definition 5.21 Consider a "let" clause that involves a singular variable binding: expressed as *let* $i: = *doc* ("*F*") $\lambda_1[\gamma_1]...\lambda_n[\gamma_n]$, where F denotes the input fuzzy spatiotemporal XML data tree, λ_i ($1 \leq i \leq n$) represents a path expression and γ_i ($1 \leq i \leq n$) signifies a condition. This clause can be transformed into the subsequent algebraic expression:

$$\text{Tree F}_{\text{root}}\left(\sigma_{[\gamma_i]}\left(\pi\lambda_{\lambda_i}\ldots\left(\sigma_{[\gamma_l]}(\pi\lambda_l(\text{Bind}(\text{“}F\text{”})))\right)\right)\right)$$

Definition 5.22 Consider a "let" clause with multiple variable bindings: expressed as *let* $i: = *doc* ("*F*") $\lambda_1[\gamma_1]...\lambda_n[\gamma_n]$, *let* $j: = *doc* ("*F*") $\lambda_1[\gamma_1]...\lambda_m[\gamma_m]$, where F is the input fuzzy spatiotemporal XML data tree, λ_i, λ_j ($1 \leq i \leq n$, $1 \leq j \leq m$) represent a path expressions and γ_i, γ_j ($1 \leq i \leq n$, $1 \leq j \leq m$) represent conditions. This clause can be transformed into the subsequent algebraic expression:

$$\text{TreeF}_{\text{root}}\left(\sigma_{[\gamma_i]}\left(\pi_{\lambda_i}\ldots\left(\sigma_{[\gamma_1]}\left(\pi\lambda_{\lambda_l}(\text{ Bind (“}F\text{”))})\right)\right)\right)\right)$$
$$\infty\ \text{TreeF}_{\text{root}}\left(\sigma_{[\gamma_j]}\left(\pi\lambda_j\ldots\left(\sigma_{[\gamma_l]}\left(\pi\lambda_{\lambda_l}(\text{Bind (“}F\text{”))})\right)\right)\right)\right).$$

5.3.3.3 The "Where" Clause

The "where" clause serves to delineate criteria that refine the outcomes of the FLWOR expression. It is capable of referencing variables that have been established through a "for" or "let" clause. Beyond facilitating the articulation of intricate filtering conditions, the "where" clause is also instrumental in performing joins. Although only a single "where" clause may be incorporated within each FLWOR expression, it can consist of multiple expressions interconnected by the logical operators "and" and "or." An optional "where" clause may include one or more variables, which function as a filter for the tuples of variable bindings.

Definition 5.23 Consider a "where" clause that includes a single variable binding: expressed as *where $i* $\lambda[\gamma]$, where F_a denotes the algebraic expression that represents the input fuzzy spatiotemporal XML data trees, λ signifies a path expression, λ_i indicates a path expression that identifies the nodes associated with the variable $i, and γ represents a condition. This clause can be transformed into the subsequent algebraic expression:

$$\sigma_{\lambda_i\lambda[\gamma]}(F_a).$$

Definition 5.24 Consider a "where" clause that includes multiple variable bindings: expressed as *where* $i $\lambda_i[\gamma_i]$ *and* $j $\lambda_j[\gamma_j]$, where F_a denotes the algebraic expression that represents the input fuzzy spatiotemporal XML data trees, λ_i and λ_j signify a path expressions, λ_i' and λ_j' indicate a path expression that identifies the nodes associated with the variable $i and $j, and γ represents a condition. This clause can be transformed into the subsequent algebraic expression:

$$\sigma_{\lambda_i'\lambda_i[\gamma_i]}(F_a)\theta\sigma_{\lambda_j'\lambda_j[\gamma_j]}(F_a).$$

5.3.3.4 The "Order By" Clause

The "order by" clause comprises one or more ordering specifications, which are delineated by commas. Each specification consists of an expression accompanied by an optional modifier. The expression must yield a single value for each item subjected to sorting. In contrast to SQL, XQuery permits the ordering of items based on a value that is not explicitly returned by the expression. An "order by" clause is utilized to arrange the results of FLWOR expressions. The conversion of an "order by" clause can be accomplished through the application of the previously established ordering operator. This transformation process is relatively straightforward.

5.3.3.5 The "Return" Clause

The "return" clause is composed of the return keyword and a singular expression designated for return. This expression is evaluated once during each iteration, contingent upon the condition that the where expression evaluates to true. The resultant value of the entire FLWOR construct is a sequence of items produced by each evaluation of the return clause. In instances where multiple expressions are to be incorporated within the "return" clause, they may be aggregated into a sequence. A "return" clause delineates the outcome of the FLWOR expression, which may be modified through the application of the tree construction operator. In certain instances, a FLWOR expression may be embedded within a "return" clause.

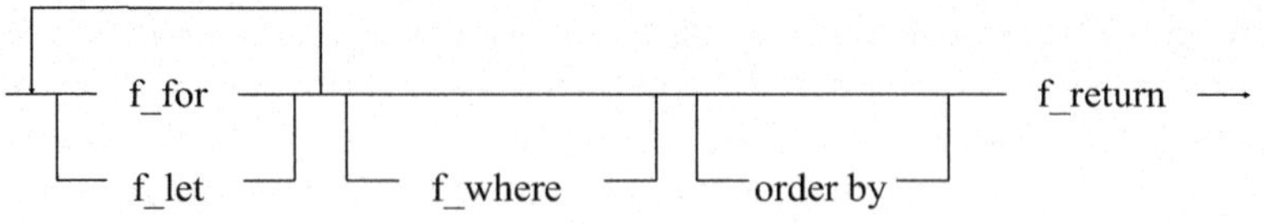

Fig. 5.5 FLWOR syntax of fuzzy spatiotemporal data

Definition 5.25 Consider a "return" clause, where F denotes the input fuzzy spatiotemporal XML data tree, E represents an expression capable of encapsulating the internal clauses such as "for", "let", "where", and "order by".

- In the absence of a nested clause in the "return" clause, the following condition is observed:

$$\text{TreeF}_{\text{root}}\left(^{*}\right)(\Sigma\lambda_i(\pi_\lambda(\text{ Bind (``}F\text{'')}))).$$

- In the absence of nested clauses in the "return" clause, the following condition is observed:

$$\text{Tree(E(result })\theta\text{F}_{\text{root}}\,(*)(\Sigma\lambda_i(\pi_\lambda(\text{ Bind(``}F\text{'')})))).$$

Figure 5.5 illustrates the FLWOR syntax applicable to fuzzy spatiotemporal data, where "f" denotes clauses that may exhibit fuzziness. It is evident from Fig. 5.5 that all clauses, with the exception of the "order by" clause, have the potential to be fuzzy. Specifically, the fuzziness in the "for" clause may pertain to the number of iterations, which can be either singular or multiple. In the "let" clause, fuzziness may arise in the quantity of binding variables, which can range from a single variable to a set of variables. The "where" clause may encompass fuzziness in its criteria, which can include fuzzy temporal criteria, fuzzy spatial criteria, fuzzy spatiotemporal criteria, or fuzzy general attribute criteria. Lastly, the "return" clause may exhibit fuzziness in the results returned, which can pertain to either the quantity or the values of the results.

5.3.3.6 Examples

In this subsection, we will provide a detailed account of the processing of fuzzy spatiotemporal XML queries, grounded in the previously established definitions of fuzzy spatiotemporal XML algebraic operations. We will utilize the spatiotemporal XML document referenced in [16] to explore the formulation of fuzzy spatiotemporal XML queries. For the sake of clarity, we will express these fuzzy XML queries using natural language. It is important to note that within fuzzy spatiotemporal XML data trees, elements and attributes may share identical names. To differentiate between elements and attributes, we will employ the symbol "$" to denote XML elements and "#" to signify XML attributes.

Query 1 Please provide the name of a cloud characterized by a "thick" density and a "Poss" value exceeding 0.7.

In accordance with Definitions 5.12, 5.17, and 5.18, it can be stated that:

Tree cname (/cloud/[$d = "thick"] (/cloud/[#p > 0.7] (/cloud(Bind(clouds)))))

Query 2 Please provide the names and densities of clouds in Shenyang that have an area exceeding 100 km^2 and are associated with Wiz Khalifa.

In accordance with Definitions 5.12, 5.13, 5.14, 5.17, and 5.18, it can be stated that:

Tree answer ((/cloud/$c), (/density/$d)) (
σ/[cloud/#CID = covered city/#CCID] (
σ/cloud/[$cc = Shenyang] (
π/cloud (Bind (clouds))) ∞ σ/area/[$a > 100] (
π/covered cities (Bind (cities)))).

Figure 5.6 presents a simplified yet significantly expressive segment of the description pertaining to fuzzy spatiotemporal algebraic operations. The algebra we have developed is rigorously defined and possesses sufficient power to encapsulate the semantics of XQuery.

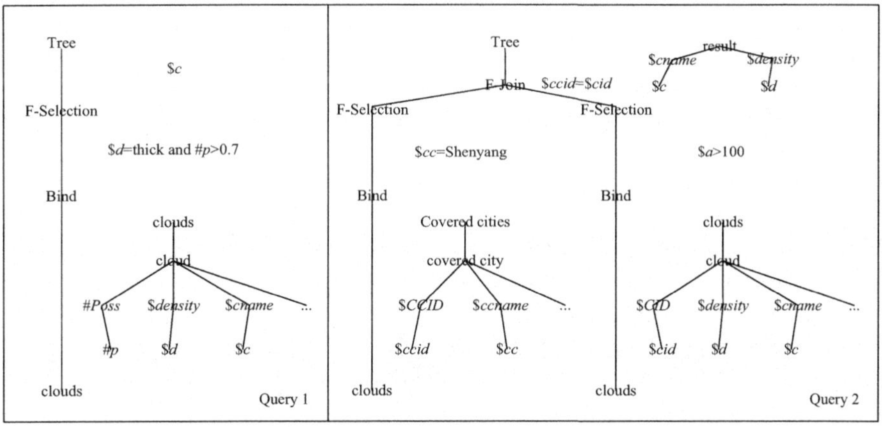

Fig. 5.6 Fuzzy spatiotemporal algebraic operations

Table 5.1 Queries used for two datasets

ID	Query types	Queries
Q1	Query with general node labels only	…//title
Q2	Query with temporal node labels	…//ATTR/ name [# t_s] | [# t_e]
Q3	Query with spatial node labels	…//position/x | y
Q4	Query with spatiotemporal node labels	…//position[# t_s] | [# t_e]
Q5	Query with fuzzy general node labels	…//title/[#p > 0.7]
Q6	Query with fuzzy temporal node labels	…//ATTR/ name [# t_s] | [# t_e] / [#p > 0.7]
Q7	Query with fuzzy spatial node labels	…//position/x | y / [#p > 0.7]
Q8	Query with fuzzy spatiotemporal node labels	…//position[# t_s] | [# t_e] / [#p > 0.7]

5.3.4 Experiments

5.3.4.1 Experimental Setting

All evaluations were conducted using C++ within the Microsoft Visual Studio 10.0 environment, executed on a Windows 7 operating system equipped with a 3.2 GHz i7 processor and 4 GB of RAM.

For the purpose of experimental evaluations, we utilized two distinct datasets. The first dataset is the FXMark dataset [41], which is derived from the widely recognized XMark dataset [42]. The second dataset is a synthetic dataset, referred to as Random, which encompasses fuzzy spatiotemporal node labels. Representative queries were selected for evaluation from each dataset. The queries employed in the experiments are presented in Table 5.1.

5.3.4.2 Experimental Results

In the initial set of experiments, we assess the accuracy of the results. The evaluations consist of queries Q1 through Q8, as detailed in Table 5.1, along with their anticipated outcomes based on XQuery semantics. Table 5.2 presents the experimental findings for both query algebra and optimized query algebra across the two datasets. Due to the absence of spatiotemporal node labels in FMark, evaluations are limited to queries Q1 and Q5 within this framework. It is evident from Table 5.2 that the precision of each query across the two datasets is consistently 100%. A notable conclusion drawn from Table 5.2 is that our algebra effectively maintains the semantics of the queries.

In the second series of experiments, we assess the effectiveness of the proposed methods. These experiments are designed to measure the comparative reduction in document size both in terms of file storage and memory utilization.

Figures 5.7 and 5.8 illustrate the average percentage reduction in size relative to the original document across both the FMark and Random datasets. A notable finding

Table 5.2 Precision of queries

		Q1	Q2	Q3	Q4	Q5	Q6	Q7	Q8
Algebra	FMark	100%	–	–	–	100%	–	–	–
	Random	100%	100%	100%	100%	100%	100%	100%	100%
Optimized algebra	FMark	100%	–	–	–	100%	–	–	–
	Random	100%	100%	100%	100%	100%	100%	100%	100%

from these figures is that both the algebraic and optimized algebraic approaches effectively decrease the document size in file format for both datasets. Furthermore, the optimized algebra demonstrates a greater reduction in document size compared to the standard algebraic method. Specifically, as depicted in Fig. 5.8, Q8 achieves the most significant reduction in document size, while Q1 results in the least reduction. This phenomenon can be attributed to the presence of fuzzy spatiotemporal node labels (including temporal, spatial, or fuzzy node labels), which correlate with a greater reduction in document size. Consequently, Q8, which incorporates fuzzy, spatial, and temporal node labels, results in the highest reduction. Q4 follows, yielding the second highest reduction due to its inclusion of spatiotemporal node labels. Q2 and Q3 exhibit nearly equivalent reductions, with Q2 containing temporal node labels and Q3 containing spatial node labels; however, Q3 slightly outperforms Q2 in terms of document size reduction, likely due to a higher frequency of spatial node labels compared to temporal ones. A similar pattern is observed with Q6 and Q7, which also achieve marginally greater reductions than Q2 and Q3, attributable to the inclusion of additional fuzzy node labels. Additionally, Q5 surpasses Q1 in document size reduction due to the presence of extra fuzzy node labels. The trends observed for Q1 and Q5 in Fig. 5.7 are consistent with this analysis.

Figures 5.9 and 5.10 illustrate the average memory utilization expressed as a percentage of the memory allocated for the original document. It is evident that the memory consumption associated with FMark and Random aligns consistently with

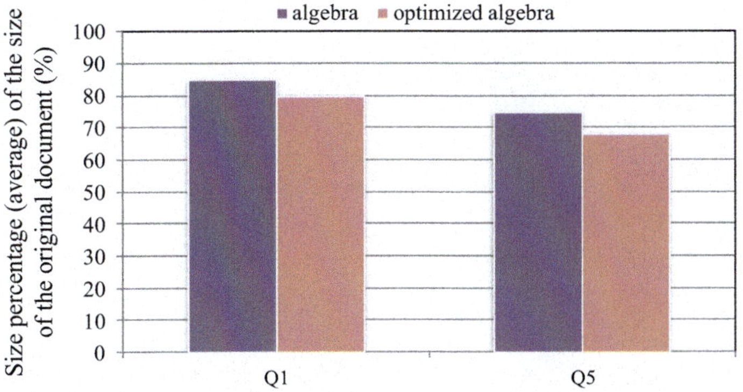

Fig. 5.7 Size percentage (average) of the size of the original document over FMark

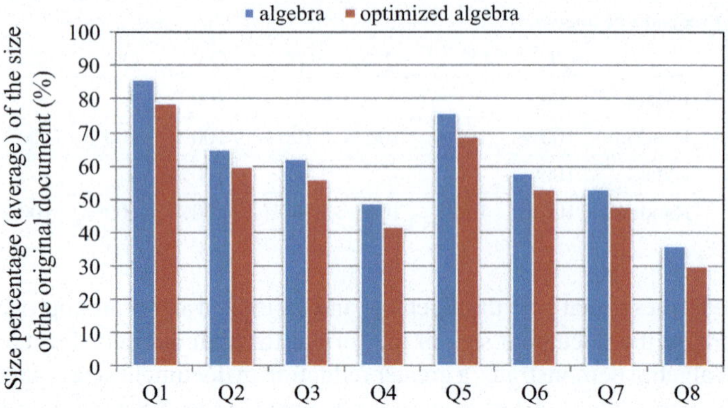

Fig. 5.8 Size percentage (average) of the size of the original document over Random

the proportional size of the original document. Notably, both algebra and optimized algebra exhibit a marginally higher memory usage relative to their size, attributable to the complexities involved in parsing and loading fuzzy spatiotemporal XML documents.

In the final series of experiments, we assess the processing time associated with various configurations. Figures 5.11 and 5.12 illustrate the execution time corresponding to different proportions of fuzzy nodes in FMark and different proportions of fuzzy spatiotemporal nodes in Random. It is evident that when the proportion of fuzzy (spatiotemporal) nodes is low, the execution time for the original query is less than that of both algebra and optimized algebra. Conversely, as the proportion of fuzzy (spatiotemporal) nodes increases, the execution time for the original query exceeds that of algebra and optimized algebra. Additionally, when the proportion of fuzzy (spatiotemporal) nodes is low, the execution time for algebra is shorter than

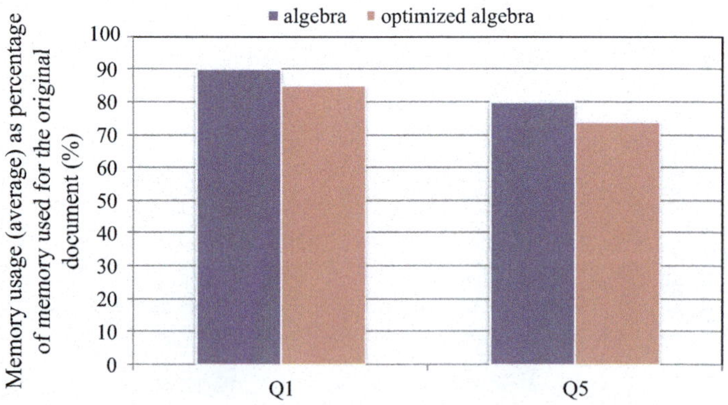

Fig. 5.9 Memory usage (average) as percentage of memory used for the original document over FMark

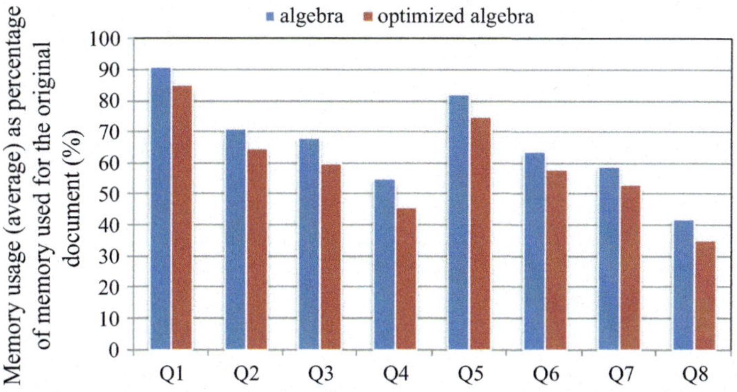

Fig. 5.10 Memory usage (average) as percentage of memory used for the original document over Random

that for optimized algebra; however, this trend reverses when the proportion of fuzzy (spatiotemporal) nodes is high. This phenomenon, while seemingly counterintuitive, can be attributed to the parsing and loading times associated with fuzzy (spatiotemporal) XML documents. Furthermore, the execution time for Random is superior to that for FMark, which can be explained by the higher density of fuzzy spatiotemporal nodes present in the Random dataset. Moreover, the execution time for both FMark and Random align with the size percentage of the original document and the memory usage as a percentage of the memory allocated for the original document.

The reported execution time represent the average durations for queries Q1 through Q8; however, it is important to acknowledge that these queries exhibit varying performance characteristics. To elucidate these differences, we conducted evaluations of their execution time on the FMark dataset, as illustrated in Fig. 5.13, and on the Random dataset, as depicted in Fig. 5.14. The analysis reveals that the execution

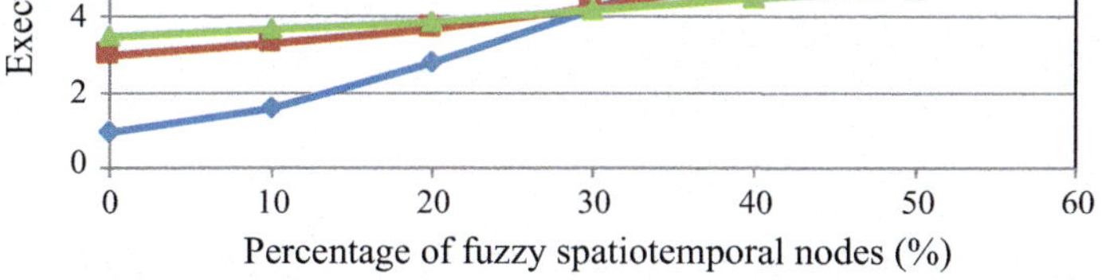

Fig. 5.11 Execution time (average) with different percentage of fuzzy (spatiotemporal) nodes over FMark

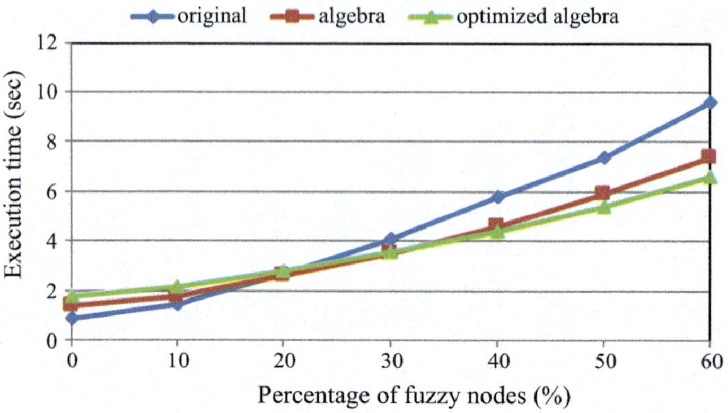

Fig. 5.12 Execution time (average) with different percentage of fuzzy (spatiotemporal) nodes over Random

time for both FMark and Random correlate with the size percentage of the original document and the memory usage as a percentage of the memory allocated for the original document. Notably, the execution time for optimized algebra are lower than those for standard algebra, indicating that the implementation of optimized algebra significantly enhances query processing efficiency.

In conclusion, the implementation of the proposed algebra, particularly the optimized version, has the potential to enhance performance outcomes. Furthermore, an increase in the fuzziness of spatiotemporal nodes correlates with a greater improvement in performance.

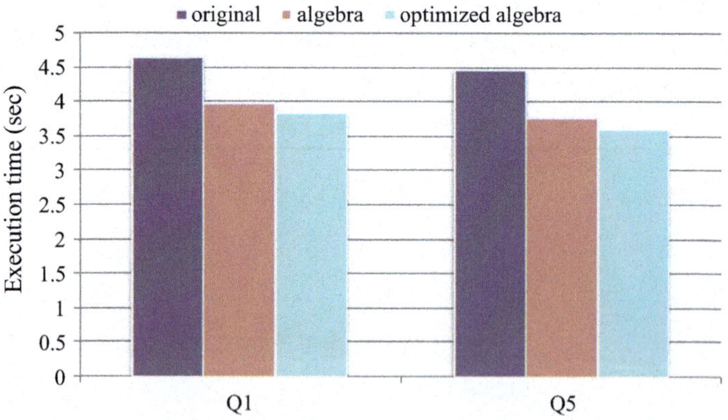

Fig. 5.13 Execution time (average) of each query over FMark

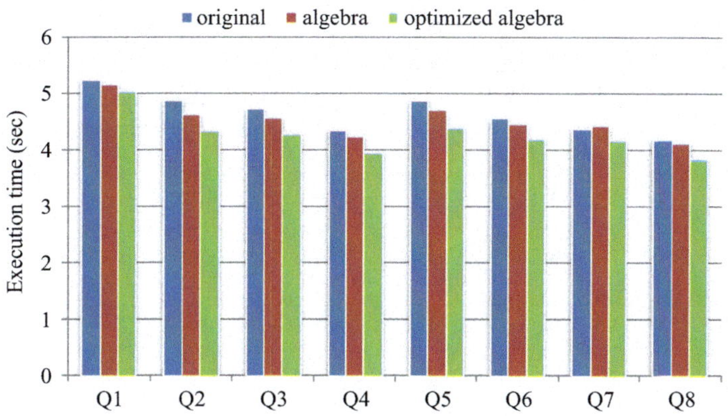

Fig. 5.14 Execution time (average) of each query over Random

5.3.4.3 Comparative Study

In this subsection, we conduct a comparative analysis of our methodology against similar approaches, focusing specifically on the elements they support. A comparison with other relational algebra and XML algebra reveals the advantages inherent in our work. Detailed comparisons between the relevant approaches and our methodology, with respect to the supported elements, are presented in Table 5.3.

In comparing the relevant methodologies with our research regarding the elements they support, we find that the approaches [3, 10, 15] provide sufficient support for measures related to spatial and temporal elements. However, their ability to accommodate all algebraic measures is constrained. Notably, the approach outlined in [3] incorporates support for fuzziness, whereas the methodologies [10, 15] do not. Conversely, the approaches [20, 21, 24, 25] lack support for measures pertaining to spatial and temporal elements, as well as fuzziness; however, they do possess capabilities for algebraic support. For instance, the approaches [20, 25] demonstrate adequate support for algebraic operators, yet their scope does not extend to algebraic equivalences or the transformation of XQuery expressions. The approach [24] offers sufficient support for both algebraic operators and algebraic equivalences, but similarly does not extend to the transformation of XQuery expressions. The methodology [21] provides adequate support for algebraic operators and the transformation of XQuery expressions, but lacks support for algebraic equivalences. Lastly, the approach [8] demonstrates comprehensive support for all algebraic measures and fuzziness, although its capabilities in supporting spatial and temporal elements are limited.

In comparison to existing research, our study provides substantial support for all metrics pertaining to both fuzzy spatiotemporal elements and algebraic components. In contrast to their findings, our model and algebra are clearly articulated and sufficiently robust to accommodate tree patterns and order-sensitive fuzzy spatiotemporal XML queries.

Table 5.3 Comparing corresponding approaches in terms of supported elements

	Spatial elements	Temporal elements	Fuzziness	Algebra		
				Algebraic operators	Algebraic equivalences	Transformation of XQuery expressions
Adeli and Jiang [3]	Y	Y	Y	N	N	N
Buratti and Montesi [20]	N	N	N	Y	N	N
Fernandez et al. [24]	N	N	N	Y	Y	N
Franceschet et al. [15]	Y	Y	N	N	N	N
Huang et al. [10]	Y	Y	N	N	N	N
Magnani and Montesi [21]	N	N	N	Y	N	Y
Ma et al. [8]	N	N	Y	Y	Y	Y
Wang et al. [25]	N	N	N	Y	N	N
Ours	Y	Y	Y	Y	Y	Y

5.4 Querying Fuzzy Spatiotemporal Data Using XQuery

5.4.1 The Architecture of Querying Fuzzy Spatiotemporal Data Using XQuery

The framework for querying fuzzy spatiotemporal data utilizing XQuery is depicted in Fig. 5.15. The spatiotemporal XML documents refers to the XML data that is subject to querying. The queried data can manifest in several formats, including: (a) text files that consist of spatiotemporal XML documents; (b) fragments of spatiotemporal XML documents obtained from the Web via a Uniform Resource Identifier (URI); (c) a collection of spatiotemporal XML documents linked to a specific URI; (d) spatiotemporal data housed within native XML databases; (e) spatiotemporal data stored in relational databases equipped with an XML interface; and (f) spatiotemporal XML documents held in memory.

A fuzzy XQuery request may be found in various formats, including text files, embedded within program code, within a query library, dynamically generated by program code, or entered by users via a command line or dialog box. Unlike conventional XQuery requests, fuzzy XQuery requests incorporate fuzzy concepts that reflect user preferences. For instance, users typically seek to obtain an approximate range of the results they desire.

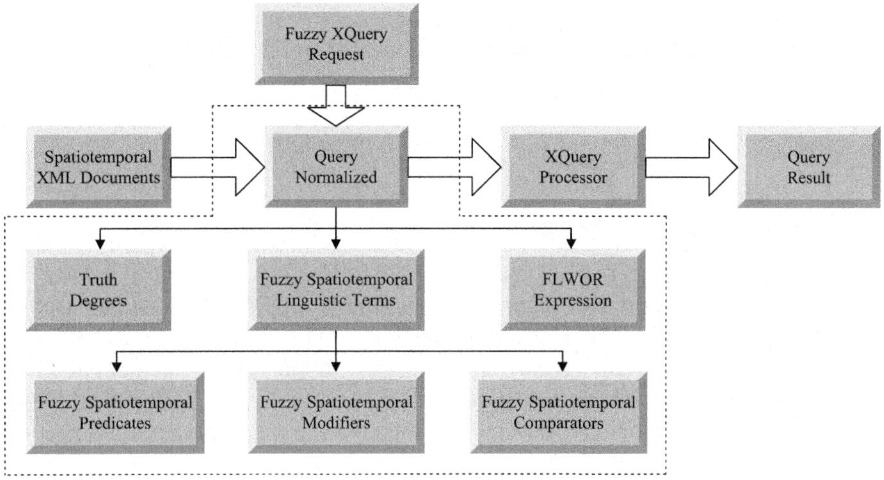

Fig. 5.15 The architecture of querying fuzzy spatiotemporal data using XQuery

Due to the inherent fuzzy concepts present in fuzzy XQuery requests, such requests are not compatible with traditional XQuery processors. Consequently, when addressing fuzzy XQuery requests pertaining to spatiotemporal XML documents, it is essential to first normalize these queries to facilitate their execution within a traditional XQuery environment. The normalization phase of the query encompasses three components: truth degrees, which are utilized to represent membership degrees or possibility distributions of the intended queries; fuzzy spatiotemporal linguistic terms, which serve to denote fuzzy concepts; and an extended FLWOR expression designed to accommodate fuzzy spatiotemporal queries. The category of fuzzy spatiotemporal linguistic terms can be further subdivided into fuzzy spatiotemporal predicates, fuzzy spatiotemporal modifiers, and fuzzy spatiotemporal comparators.

The XQuery processor is responsible for parsing, analyzing, and evaluating queries. The phases of analysis and evaluation can be likened to the processes of compiling and executing program code. During the analysis phase, the processor identifies syntax errors and other static errors that are independent of the input data. In contrast, the evaluation phase assesses the results of the query based on the provided input documents, which may lead to dynamic errors in cases such as the absence of input documents or division by zero.

Ultimately, the query processor produces a sequence of values as the results of the query. It is important to note that these results may include a degree of membership or a possibility distribution reflecting user preferences, in addition to the desired query outcomes.

In the context of querying fuzzy spatiotemporal data using XQuery, the normalized query component is pivotal, as it must be adapted to accommodate fuzzy concepts, while the other components remain largely analogous to traditional structures. Consequently, the normalized query component, delineated by the dashed line in Fig. 5.15, is the primary focus of this subsection.

5.4.2 *Fuzzy XQuery*

In this subsection, we propose fuzzy-set-based extensions to XQuery that enable users to articulate preferences regarding spatiotemporal XML documents based on their degrees of satisfaction. These extensions introduce a novel *xs:truth* data type designed to represent gradual truth values, along with the new *Val* element and *Poss* attribute, which facilitate the management of satisfaction degrees within the nodes of fuzzy XQuery expressions. Furthermore, we extend fuzzy spatiotemporal linguistic terms to define fuzzy concepts and incorporate them into query expressions. Lastly, we present various types of expressions, such as FLWOR, that have been adapted to incorporate fuzzy logic.

5.4.2.1 Truth Degrees

In order to accommodate the concept of fuzziness within XQuery, it is essential to provide a representation for membership degrees or possibility distributions derived from fuzzy conditions. Additionally, it is important to establish a standardized notation for spatiotemporal XML data that is annotated with fuzzy degrees corresponding to fuzzy queries. Consequently, we introduce a novel *xs: truth* data type, a new *Val* element, and a new *Poss* attribute in the subsequent discussion.

To enhance XQuery, it is necessary to extend the *xs: boolean* data type to accommodate gradual truth values. In this context, we propose the introduction of a new data type, *xs: truth*, which encapsulates a range of potential truth degrees, represented within the real unit interval [0.0, 1.0]. Given that *xs: truth* is a derivative of *xs: boolean*, the conversion from an *xs: boolean* value to an *xs: truth* value will yield a result of 1.0 for true and 0.0 for false. Conversely, the transformation from *xs: truth* to *xs: boolean* will result in false for a value of 0.0 and true for any other value. For instance, a source value of 0.8 (or any other non-zero value) of the *xs: truth* type will be converted to true when cast to *xs: boolean*, whereas a value of 0 will be converted to false.

The incorporation of fuzzy logic concepts into XQuery expressions is expected to yield nuanced responses to queries concerning XML documents. Consequently, it is imperative to establish a mechanism for annotating XML documents with degrees of truth. To address this need, we propose the implementation of a possibility attribute, denoted as *Poss*, in conjunction with the element *Val*, to indicate the likelihood of a specific element's presence within XML documents, with values ranging from 0.0 to 1.0. Degrees of truth may be articulated through either internal or external mechanisms. However, it is essential to ensure that all users possess a valid schema that facilitates the recognition of truth degrees within XML documents.

5.4.2.2 Fuzzy Spatiotemporal Linguistic Terms

Fuzzy spatiotemporal queries incorporate fuzzy linguistic terms that are defined by the user and expressed as fuzzy terms. The specification of these fuzzy terms may resemble standard function declarations found in XQuery. This subsection provides definitions for the formulation of fuzzy spatiotemporal terms.

Historically, topological relationships such as *disjoint*, *meet*, *overlap*, *contain*, and *equal* among points, lines, and regions have been the subject of extensive research. However, existing methodologies are not applicable to scenarios involving fuzzy spatiotemporal entities when utilizing XQuery. To illustrate this issue, consider the challenge of monitoring two fuzzy regions over time. The inquiry regarding whether these two fuzzy regions meet possesses a spatiotemporal dimension, necessitating an examination of the boundaries of these regions throughout a sequence of events and intervals in a specified temporal order. While spatial and temporal components can be represented through the incorporation of additional tags in XML, the current capabilities of XQuery do not accommodate fuzzy information. Consequently, there is a need to extend fuzzy spatiotemporal predicates.

The fundamental characteristic of spatiotemporal predicates is their ability to articulate propositions that can be classified as either true or false. In contrast, fuzzy spatiotemporal predicates not only convey the proposition but also indicate the degree of truth associated with it. A fuzzy spatiotemporal predicate is a specific type of user-defined function, which is characterized by its name, as well as the names and data types of its parameters. The returned data type is *xs: truth*, which may be omitted in certain contexts.

Definition 5.26 (*Fuzzy spatiotemporal predicate declaration*) A fuzzy spatiotemporal predicate declaration contains an expression that defines the methodology by which its outcome is derived from its parameters.

FuzzySpatiotemporalPredicateDeclaration:: =
"declare" "fuzzy" "spatiotemporal" "predicate" FPName "("Param")" ("as" "xs: truth")?
"("Expr")"
*Param:: = ("a" "," "b" "," "τ")**

The degree of a fuzzy relation is quantified by aggregating the *α-cut* levels of two fuzzy spatiotemporal regions. In Definition 5.26, *FPName* denotes the designation of the specified fuzzy spatiotemporal predicate, while *Param* refers to the parameters within *Expr*. The use of double quotation marks around *Param* indicates that further clarification is required. The variables α and β, defined within *Param*, represent the *α-cut* levels of the two fuzzy spatiotemporal regions, respectively. The variable τ, also defined in *Param*, signifies the value that characterizes the topological relationship between the two *α-cut* level fuzzy spatiotemporal regions. The number of parameters specified in Param may vary, depending on the quantity of *α-cut* levels of the two fuzzy spatiotemporal regions and the corresponding crisp topological relations. *Expr* encapsulates the general relationship between the two fuzzy spatiotemporal regions A and B, which can be established by

$$\sum_{i=1}^{n}\sum_{j=1}^{m} m(A_{\alpha i})m(B_{\alpha j})\tau(A_{\alpha i}, B_{\alpha j})?$$

The fundamental probability assignment, denoted as $m\,(A_{\alpha i})$, can be understood as the likelihood that $A_{\alpha i}$ accurately represents the fuzzy spatiotemporal region A. This assignment is articulated as follows [43]: $m\,(A_{\alpha i}) = \alpha_i - \alpha_{i+1}$ for $1 \leq i \leq n$ for some $n \in N$ with $1 = \alpha_1 > \alpha_2 > \ldots > \alpha_n > \alpha_{n+1} = 0$. $\tau\,(A_{\alpha i}, B_{\alpha j})$ represents one of the values associated with topological relationships, yielding a membership degree that characterizes the fuzzy topological relations between the two fuzzy spatiotemporal regions. For instance, the terms corresponding to the fuzzy topological relations of these regions are detailed in Table 5.4, and the definitions of valid fuzzy spatiotemporal predicate declarations are provided in the followings:

declare fuzzy spatiotemporal predicate Fcontain $(\tau_{equal}, \tau_{contain}, \tau_{overlap}, \tau_{meet}, \tau_{disjoint})$

$$\sum_{i=1}^{n}\sum_{j=1}^{m} m\left(A_{\alpha j}\right)m\left(B_{\beta j}\right)\tau\left(A_{\alpha j}, B_{\beta j}\right)?$$

Based on prior assertions, we can observe the following values: $Fcontain(A_{\alpha i}, B_{\beta j}, \tau_{equal}) = 0$, $Fcontain(A_{\alpha i}, B_{\beta j}, \tau_{contain}) = 0.35$, $Fcontain(A_{\alpha i}, B_{\beta j}, \tau_{overlap}) = 0.35$, $Fcontain(A_{\alpha i}, B_{\beta j}, \tau_{meet}) = 0.15$, $Fcontain(A_{\alpha i}, B_{\beta j}, \tau_{disjoint}) = 0.15$. It is important to note that the cumulative membership degrees of these five fuzzy topological relations total 1.0, as each fuzzy topological relation corresponds uniquely to a specific crisp topological relation.

Fuzzy queries pertaining to fuzzy spatiotemporal data can be enhanced through the application of XQuery. However, in instances where further refinement of fuzzy queries is required, it becomes necessary to adjust the existing fuzzy terms. For instance, one might query specific areas where two fuzzy regions either significantly overlap or exhibit minimal overlap, or inquire about months characterized by high

Table 5.4 Computing a fuzzy topological relation for two fuzzy spatiotemporal regions *A* and *B*

α-cut levels of fuzzy regions	$m\,(A_{\alpha i}) \times m\,(B_{\beta j})$	τ_{equal}	$\tau_{contain}$	$\tau_{overlap}$	τ_{meet}	$\tau_{disjoint}$
1.0–0.85	0.15	0	0	1	1	1
0.85–0.65	0.2	0	0	1	0	0
0.65–0.3	0.35	0	1	0	0	0
0.3–0.0	0.3	0	0	0	0	0

humidity and low cloud cover. While it is indeed feasible to execute such queries in their original form (by omitting the terms "much," "less," "high," and "low") it is essential to modify the fuzzy spatiotemporal predicates or terms in accordance with the original definitions when addressing these types of queries.

A fuzzy spatiotemporal modifier functions as a unary operator on *xs: truth* and is characterized by a fuzzy number that indicates the degree of membership. This modifier can be utilized in conjunction with fuzzy spatiotemporal predicates to generate new fuzzy spatiotemporal predicates or other fuzzy constructs. Consequently, the resultant modified fuzzy spatiotemporal terms are not explicitly defined; rather, they are derived from the membership degree of the original fuzzy spatiotemporal terms. The declaration of such a modifier includes the name of the modifier, while the returned data type and the parameter are consistently *xs: truth*, allowing for the potential omission of the latter.

Definition 5.27 (*Fuzzy spatiotemporal modifier declaration*) A fuzzy spatiotemporal modifier declaration contains an expression that defines the methodology by which its outcome is derived from its parameters.

FuzzySpatiotemporalModifierDeclaration::=
"declare" "fuzzy" "spatiotemporal" "modifier" FMName "(" Param ")" ("as" "xs: truth")?
"power" "(" Expr ")"
Param::= FPName | FName ("as" "xs: truth")?

In Definition 5.27, *FMName* refers to the designation of the fuzzy spatiotemporal modifier, whereas *FPName* denotes the title of the original fuzzy spatiotemporal predicate as outlined in Definition 5.26. Additionally, *FName* signifies the fuzzy terms, and *Param* represents the parameters manipulated by *Expr*. The term "power" pertains to the exponentiation operation applied to the original fuzzy spatiotemporal predicate. *Expr* is defined as a static positive numeric value that serves as the exponent for the original membership degree of the fuzzy spatiotemporal predicate or the fuzzy terms. For instance, the following examples illustrate valid declarations of fuzzy spatiotemporal modifiers for fuzzy spatiotemporal predicates:

declare fuzzy spatiotemporal modifier much (Fcontain) power (+2.0)
declare fuzzy spatiotemporal modifier less (Fcontain) power (+0.5)

Based on the statements and the illustration provided previously, it can be concluded that: $much\ Fcontain(A_{\alpha i}, B_{\beta j}, \tau_{equal}) = 0$, $much\ Fcontain(A_{\alpha i}, B_{\beta j}, \tau_{contain}) = 0.1225$, $much\ Fcontain(A_{\alpha i}, B_{\beta j}, \tau_{overlap}) = 0.1225$, $much\ Fcontain(A_{\alpha i}, B_{\beta j}, \tau_{meet}) = 0.0225$, $much\ Fcontain(A_{\alpha i}, B_{\beta j}, \tau_{disjoint}) = 0.0225$; $less\ Fcontain(A_{\alpha i}, B_{\beta j}, \tau_{equal}) = 0$, $less\ Fcontain(A_{\alpha i}, B_{\beta j}, \tau_{contain}) \approx 0.5916$, $less\ Fcontain(A_{\alpha i}, B_{\beta j}, \tau_{overlap}) \approx 0.5916$, $less\ Fcontain(A_{\alpha i}, B_{\beta j}, \tau_{meet}) \approx 0.3873$, $less\ Fcontain(A_{\alpha i}, B_{\beta j}, \tau_{disjoint}) \approx 0.3873$.

Utilizing fuzzy spatiotemporal predicates or their modified counterparts, an additional fuzzy modifier, specifically "not," will be incorporated. The outcome of this integration will yield a modified predicate, wherein the resultant value represents the complement of the membership degree associated with the original fuzzy spatiotemporal predicate. To illustrate this concept, consider the preceding examples: *not* $Fcontain(A_{\alpha i}, B_{\beta j}, \tau_{contain}) = 0.65$, *not* $Fcontain(A_{\alpha i}, B_{\beta j}, \tau_{disjoint}) = 0.85$, *not much* $Fcontain(A_{\alpha i}, B_{\beta j}, \tau_{overlap}) = 0.8775$, *not less* $Fcontain(A_{\alpha i}, B_{\beta j}, \tau_{meet}) \approx 0.6127$.

5.4.2.3 Fuzzy Spatiotemporal Comparators

Fuzzy spatiotemporal comparators serve the purpose of evaluating the values of two fuzzy spatiotemporal entities, as well as assessing the degree of truth associated with a given fact. Similar to fuzzy spatiotemporal modifiers, when fuzzy queries necessitate further refinement, the existing fuzzy comparative terms must be adjusted accordingly. For instance, one might query a region where the rainfall distribution exceeds 150 mm. While it is indeed feasible to conduct such queries in their original form (by omitting the term "much"), in instances where these queries are applied, it is essential to modify the fuzzy spatiotemporal comparative terms based on their original counterparts. The primary distinction between fuzzy spatiotemporal modifiers and fuzzy spatiotemporal comparators lies in whether the original fuzzy terms are comparative in nature.

A fuzzy spatiotemporal comparator represents a specialized user-defined comparison operator. It functions as a binary operand on the data type *xs:float* and is characterized by a membership degree that is determined by two parameters. This comparator can be utilized on fuzzy spatiotemporal terms that share the same data type. The declaration of a fuzzy spatiotemporal comparator includes the specification of operator names, as well as the names and data types of the two parameters involved. The output data type is consistently *xs: truth*, although this may be omitted in certain contexts.

Definition 5.28 (*Fuzzy spatiotemporal comparator declaration*) A fuzzy spatiotemporal comparator declaration contains an expression that defines the methodology by which its outcome is derived from its parameters.

FuzzySpatiotemporalComparatorDeclaration::=
"declare" "fuzzy" "spatiotemporal" "comparator" FCName "("Param", "Param")"
("as" "xs: truth")?
"(" "\$" $FName_1$ ("−" | "/")(("\$" $FName_2$) | value) ")" "function" "("Expr")"
Param::= "\$" FName ("as" SequenceType)?

In Definition 5.28, the term *FCName* refers to the designation of a fuzzy spatiotemporal comparator, while *FName*1 and *FName*2 denote the parameters being compared. The variable *value* is defined as a numeric value of the float data type. The term *function* signifies the membership degree function that fulfills the requirements of the comparison operation, and *Expr* represents the expression associated with this function. The parameters outlined in *Param* are maintained in the same sequence as their declaration, necessitating that *FName*1 and *FName*2 appear in a specific order. The output of this process is the membership degree reflecting the comparative distance between the two parameters, which is computed as the difference when the subtraction symbol ("-") is employed, and as the quotient when the division symbol ("/") is utilized. It is important to note that the first parameter in the comparison is treated as an element, while the second parameter may be either an element or a numeric value. This distinction is based on the premise that the focus of the comparison operation is primarily on the first parameter, rendering it nonsensical for the first parameter to be a numeric value. Nevertheless, it is essential that the parameters being compared are converted into numeric values during the execution of the comparison operations. Below, we provide several examples of valid declarations for fuzzy spatiotemporal comparators:

declare fuzzy spatiotemporal comparator mlt ($x as xs: float, $y as xs: float) ($x – $y) function Expr
declare fuzzy spatiotemporal comparator sim ($x as xs: float, $y as xs: float) ($x – $y) function Expr
declare fuzzy spatiotemporal comparator mol ($x as xs: float, $y as xs: float) ($x/$y) function Expr

In the aforementioned three declarations, "*mlt*" denotes "*much less than,*" "*sim*" signifies "*similar,*" and "*mol*" stands for "*more or less.*" The term "*function*" refers to the membership degree function, which may include various forms such as the trapezoidal function and similarity relations. In this particular instance, the similarity relation is employed for the second declaration, while the trapezoidal function is utilized for the other two declarations.

In the initial declaration involving the trapezium function, denoted as *T* (*x:* –*INF,* 0, 100, 300), we observe the following outcomes: the expression (100 *mlt* 20) evaluates to true, while (500 *mlt* 300) yields a result of 0.5. In the subsequent declaration, which is presented in Table 5.5, we find that ("*clear*" *sim* "*partly cloudy*") produces a value of 0.6, ("*partly cloudy*" *sim* "*cloudy*") also returns 0.6, and ("*clear*" sim "*cloudy*") is assessed as false. Lastly, in the final declaration involving another trapezium function, *T* (*x:* 2, 1, 1, 2), we observe that (600 mol 800) results in 0.5, (200 mol 400) is evaluated as false, and (350 mol 200) yields a value of 0.25.

In the aforementioned example, the initial parameter may be substituted with an element from the fuzzy spatiotemporal dataset, thereby facilitating the execution of fuzzy spatiotemporal queries. For instance, this parameter could represent attributes such as *$age*, *$motion*, or *$area*, among others. Prior to conducting any comparative

Table 5.5 The similarity relations for cloud attribute

Cloud	Clear	Partly cloudy	Cloudy	Closed
Clear	1.0	0.6	0	0
Partly cloudy	0.6	1.0	0.6	0.4
Cloudy	0	0.6	1.0	0.8
Closed	0	0.4	0.8	1.0

operations, it is essential to convert these parameters into numeric values, as the data types of the parameters utilized in the comparison must be consistent.

5.4.3 FLWOR Expression

XQuery, the query language designed for XML data, offers a diverse array of query expressions. To facilitate the construction of complex queries, XQuery incorporates a structural framework known as FLWOR expressions, which is an acronym derived from the keywords that define the clauses of these expressions: "for," "let," "where," "order by," and "return." The FLWOR expression serves as the foundational structure for many, though not all, queries, and its fuzzy extension does not necessitate alterations to the syntactic rules governing FLWOR expressions. Consequently, fuzzy spatiotemporal queries can be articulated within XQuery without the need for the development of a new query language. This section will elucidate the methods for extending these expressions to accommodate fuzzy spatiotemporal queries.

The "for" clause facilitates the iteration over sequences that contain spatiotemporal data. In XQuery, variable names are denoted by the use of dollar signs. Each variable within the "for" clause is linked to a sequence derived from another expression, allowing for the exploration of all potential combinations of values for the variables based on their respective sequences. When the expression associated with a variable yields a sequence of nodes, it may also provide the membership degree for that variable. Each node within this sequence possesses a membership degree corresponding to a *Val* element and a *Poss* attribute, which are computed during the iteration process. Consequently, a global membership degree is calculated, analogous to the relative possibilities for each combination of variable values.

The "let" clause serves to assign a value to a variable; however, unlike the "for" clause, it does not initiate an iteration. Instead, it assigns the result of another expression to a variable. The membership degree of a variable defined elsewhere in the XQuery statement is equivalent to the membership degree derived from the combined expression within the "let" clause.

A filtering criterion is established through the "where" clause, which, in our extension, can accommodate any expression, including fuzzy expressions. This may encompass fuzzy attributes with possibility distributions, fuzzy elements with membership degrees, fuzzy temporal data, and so forth. The optional "where" clause functions as a filter for the variables generated by both the "for" and "let" clauses.

The "order by" clause is responsible for sorting the desired results. The fuzzy extension discussed herein does not directly influence the "order by" clause; however, it could, if necessary, sort both fuzzy element nodes and attribute nodes alongside crisp ones. For instance, sorting could be performed based on the possibility distribution of a specific element node.

Finally, the "return" clause delineates the outcome of the FLWOR expression. The ultimate result is generated through iteration. In scenarios involving fuzziness, the membership degree of each iterated node is computed in accordance with the "for," "let," "where," and potentially the "order by" clauses. If the result necessitates the creation of a new node within the XML document, the computed membership degree should be incorporated as an attribute of the newly created node.

To exemplify an extended FLWOR expression, consider a variable denoted as *$clouds*, which contains a document detailing the presence of clouds over a specified temporal range. A fuzzy spatiotemporal query is necessitated, wherein the density of clouds is to be examined within the time interval from t_2 to t_8 (it is assumed that the time points t_1 to t_8 are organized in ascending order). The conditions for the query stipulate that the possibility distribution of the *α-level* cloud position must exceed 0.75; x_{min} of the cloud position is set to 2, while y_{max} must be greater than 6, as defined within a Rectangular Plane Coordinate System. Additionally, the possibility distribution of the cloud's motion should surpass 0.65, with the *xaxis* of the cloud motion represented by " → ", indicating that the cloud is moving eastward. Furthermore, the possibility distribution of the cloud that covers the city of "Shenyang" must be greater than 0.8, and the possibility distribution of clouds characterized by the color "dark" should exceed 0.7. Consequently, the query can be articulated as follows:

```
for $e in doc ("clouds.xml")/clouds//cloud density
let $a: = $e../position
let $b: = $e../motion
let $c: = $e../covered cities/Dist/Val = "Shenyang"
let $d: = $e../cloud color/Dist/Val = "dark"
where $a/Val/@Poss > 0.75 and
        $a//xmin = 2 and $a//ymax > 6 and
        $b/Val/@Poss > 0.65 and $b//xaxis = " → " and
        $c/@Poss > 0.8 and $d/@Poss > 0.7 and
        $e/@Ts > t2 and $e/@Te < t8
order by $e//Val/@Poss descending
return $e//Val/@Poss, $e//Val/*
```

The "for" clause iterates over all potential pairs of values for the variable *$e*, derived from sequences of labeled nodes representing cloud density. Each of these nodes possesses a membership degree associated with a *Val* element and a *Poss* attribute, which are computed during the iteration process.

For each instance of *$e*, the "let" clause assigns various components of the cloud to specific variables: the positional aspect of the cloud is assigned to variable *$a*, the motion aspect to variable *$b* (which includes a *Val* element and a *Poss* attribute), the covered cities aspect to variable *$c* (also with a *Val* element and a *Poss* attribute), and the color aspect of the cloud to variable *$d* (again with a *Val* element and a *Poss* attribute). Consequently, the variables *$a*, *$b*, *$c*, and *$d* referenced elsewhere in the XQuery statement correspond to the meanings ascribed by these two "let" clauses.

The "where" clause establishes a filtering criterion for the variable *$e*. It filters the *α-level* cloud based on several conditions: the possibility distribution of the cloud's position must exceed 0.75; x_{min} of the cloud's position must equal 2; and y_{max} of the cloud's position must exceed 6. Additionally, it filters the cloud's motion, requiring that the possibility distribution of the cloud's motion be greater than 0.65, with the *xaxis* of the cloud's motion designated as " → ", indicating that the cloud is moving eastward. The filtering also applies to the covered cities, necessitating that the possibility distribution for the city of "Shenyang" exceeds 0.8. Furthermore, the cloud's color is filtered such that the possibility distribution for the color "dark" must be greater than 0.7. Lastly, the time interval for this cloud density is constrained to the range $[t_2, t_8]$. It is important to note that the *Dist* construct for the covered cities is conjunctive, as the cloud may simultaneously cover multiple cities, whereas the *Dist* construct for cloud color is disjunctive, given that the cloud can only exhibit one color at any given moment.

The "order by" clause organizes the results according to the possibility value of the cloud density, which corresponds to the *Poss* attributes of the possibility values.

Ultimately, the "return" clause outputs the cloud density along with its associated possibility value. In this instance, it may yield results such as "thick" with a possibility value of 0.85 and "thin" with a possibility value of 0.75, contingent upon the specific applications involved.

5.4.4 Symbolic Notations for Fuzzy Spatiotemporal Data Using XQuery

In this subsection, we will introduce symbolic notations for fuzzy spatiotemporal data utilizing XQuery. For the sake of simplicity, our focus will be restricted to two-dimensional spatial entities. Furthermore, real-time can be converted into a user-defined temporal framework.

Spatiotemporal query := "for" qualifier ("let" qualifier)* "where" spatial qualifier "and" temporal qualifier "and" attribute qualifier "return" qualifier "^" qualifier
Spatial qualifier: $= x$ qualifier "^" y qualifier
x qualifier: = qualifier$_x$ y qualifier: = qualifier$_y$
temporal qualifier: = valid time "^" transaction time
valid time: = qualifier$_v$ transaction time: = qualifier$_t$
attribute qualifier: = qualifier$_a$
qualifier: = C | F | S | R | E | * OP

The <qualifier> encompasses seven distinct types: certainty value (C), fuzzy value (F), single value (S), range value (R), range value set (E), arbitrary value (*), and predicate operation (OP). To differentiate the qualified components within XQuery, a subscript is appended to <qualifier>. The subscripts x and y denote the x-axis and y-axis values, respectively, within the spatial qualified component. In the context of the temporal qualified component, the subscripts v and t represent valid time and transaction time, respectively. Additionally, the subscript a signifies the attribute associated with the attribute qualified component.

Conversely, fuzzy spatiotemporal queries encompass both explicit qualified queries and implicit qualified queries. The former are derived directly from the explicit attributes of the qualifiers, while the latter are obtained through the application of temporal, spatial, or spatiotemporal predicates associated with the respective qualifiers. Fuzzy spatiotemporal queries can be categorized based on the values of the qualified components, including but not limited to time instant queries, time period queries, spatial position queries, spatial motion queries, spatiotemporal state queries, and general attribute queries. It is important to note that the operational predicates (OP) include temporal, spatial, and spatiotemporal predicates, which may function as either unary or binary operations. For a comprehensive examination of the specific temporal, spatial, and spatiotemporal predicates, please refer to [44].

An explicit temporal query concerning fuzzy spatiotemporal data can be derived from the explicit characteristics of a temporally qualified clause, which can be defined in the following manner:

Spatiotemporal query: = "for" qualifier ("let" qualifier)* "where" qualifier_x ^ qualifier_y "and" $\textbf{qualifier}_v$ ^ $\textbf{qualifier}_t$ "and" qualifier_a "return" qualifier "^" qualifier, where qualifier_v and qualifier_t are the explicit attributes of temporal the qualified clause.

Next, we will give several query examples to illustrate the explicit temporal query.

Q1: Retrieve the possible current position of target 1 when it's color is black.

Symbolic notation: for * let *S* let *S* where S_v ^ S_t and S_a return *F* ^ $*_x$ and $*_y$.

This query is an explicit query, and the query object is the target 1, which is a single-value query, so it is represented as *S* after for; The location of the query includes horizontal and vertical coordinates, so it can be bound by two lets; "current position" means that both the valid time and the transaction time are current time, so it is denoted as *S*; The color is black, which limits the attribute and is single-valued data; The returned result is the possible position, which is a fuzzy value, so it is *F* after return; The returned result is not limited to space, so the corresponding space qualifiers are $*_x$ and $*_y$.

Q2: Retrieve the possible position of all the targets when it's color is black.

Symbolic notation: for *S* let *S* let *S* where E_v ^ S_t and S_a return *F* ^ $*_x$ and $*_y$.

In Q2, "all the targets" means that there is no restriction on the query object and can be any value, so it is expressed as *.

Q3: Retrieve the history position of target 1 when it's color is black.

Symbolic notation: for S let S let S where $*_v$ ^ S_t and S_a return F ^ $*_x$ and $*_y$.

The "historical position" only limits the transaction time, but does not impose any restrictions on the valid time, so the valid time is expressed as *.

Q4: Retrieve the possible position in $[t_2, t_4]$ and $[t_7, t_8]$ when it's color is grey.

Symbolic notation: for S let S let S where E_v ^ S_t and S_a return F ^ $*_x$ and $*_y$.

The limit of effective time is a set of two time periods and is therefore denoted *E*.

Q5: Retrieve the recorded position of target 1 when it's color is black.

Symbolic notation: for S let S let S where S_v ^ $*_t$ and S_a return C ^ $*_x$ and $*_y$.

In Q5, "recorded position" refers to all transaction times, so the transaction time is expressed as *; In addition, the recorded position is a definite value, so the return of the returned result is *C*.

Explicit spatial query of fuzzy spatiotemporal data is a query that can be obtained directly through the explicit attributes of spatial qualifiers. Corresponding to the symbolic notation, qualifier_x and qualifier_y are spatial qualifiers, and other conditional qualifiers may be implemented by the temporal qualifiers qualifier_v, qualifier_t, and the attribute qualifier qualifier_a. Such a query can be expressed as:

Spatiotemporal query: = "for" qualifier ("let" qualifier)* "where" $\mathbf{qualifier_x}$ ^ $\mathbf{qualifier_y}$ "and" qualifier_v ^ qualifier_t "and" qualifier_a "return" qualifier "^" qualifier, where qualifier_x and qualifier_y are the explicit attributes of the spatial qualified clause.

Next, we will give several query examples to illustrate the explicit temporal query.

Q6: Retrieve all the possible black targets at Shenyang at present.

Symbolic notation: for * let S where S_x ^ S_y and S_v ^ S_t and S_a return F ^ *

The center of Shenyang (41°47′44″N, 123°26′53″E) indicates that the position information is limited to a fixed point, so the horizontal and vertical coordinates are single values; The return result is a possible target, so *F* after return; The returned result is all targets, and the query object is not limited, so it is denoted as *.

Q7: Retrieve all the possible black targets once at Shenyang.

Symbolic notation: for * let S where S_x ^ S_y and $*_v$ ^ S_t and S_a return F ^ *

In Q7, "all the possible black targets once at Shenyang" has no limit on the validity time, so it is expressed as *.

Q8: Retrieve the possible position of target 1 within Shenyang at present.

Symbolic notation: for S let S let S where R_x ^ R_y and S_v ^ S_t and S_a return F ^ $*_x$ and $*_y$.

In Q8, "within Shenyang" is a spatial range value, so the horizontal and vertical coordinates are limited to *R*.

Q9: Retrieve the possible position of all the targets at Shenyang at present.

Symbolic notation: for * let S let S where R_x ^ R_y and S_v ^ S_t and S_a return F ^ $*_x$ and $*_y$.

In Q9, all the targets" does not limit the query object, so it is denoted as *.

Q10: Retrieve all the targets, which may cross Shenyang at $[t_3, t_5]$.

Symbolic notation: for * let S where R_x ^ R_y and R_v ^ S_t and S_a return F ^ *

The limit of the valid time is a time range and is therefore denoted as *R*.

Q11: Retrieve all the targets, which may cross Shenyang.

Symbolic notation: for * let S where R_x ^ R_y and $*_v$ ^ S_t and S_a return F ^ *

This query has no limit on the valid time, so the valid time is expressed as *.

Explicit qualified query of fuzzy spatiotemporal data attributes is a query that can be obtained directly through the explicit attributes of attribute qualifiers. Corresponding to the symbolic notation, qualifier_a is an attribute qualifier, and other conditional qualifiers may be implemented by the spatial qualifiers qualifier_x, qualifier_y and the temporal qualifiers qualifier_v, qualifier_t. Such a query can be expressed as:

Spatiotemporal query: = "for" qualifier ("let" qualifier)* "where" qualifier$_x$ ^ qualifier$_y$ "and" qualifier$_v$ ^ qualifier$_t$ "and" **qualifier$_a$** "return" qualifier "^" qualifier, where qualifier$_a$ is the explicit attributes of the general attribute qualified clause.

Q12: Retrieve the possible position of the target 1 at present.

Symbolic notation: for S let S let S where S$_v$ ^ S$_t$ and *$_a$ return F ^ *$_x$ and *$_y$.

This query is a fuzzy attribute explicitly qualified query. Since there is no qualification on the attribute, the attribute is expressed as *.

Q13: Retrieve all the targets, whose color is the same as that of target 1.

Symbolic notation: for S let S where *$_x$ ^ *$_y$ and *$_v$ ^ *$_t$ and OP$_a$ return F ^ *

Since there is no limit on location and time, it is expressed as *; "whose color is the same as that of target 1" implies information that an operation *Equal* needs to be performed to compare the color attributes of each object with the color attributes of the target 1, where *Equal* is an attribute operation.

Spatiotemporal implicit limited query of fuzzy spatiotemporal data is a query that can only be obtained by operating the query object through temporal predicate, spatial predicate or spatiotemporal predicate. Corresponding to the symbolic notation proposed in the previous subsection, such a query can be expressed as:

Spatiotemporal query: = "for" qualifier ("let" qualifier)* "where" **OP$_x$** ^ **OP$_y$** "and" **OP$_v$** ^ **OP$_t$** "and" qualifier$_a$ "return" qualifier "^" qualifier, where OP$_x$, OP$_y$, OP$_v$, and OP$_t$ can be obtained by temporal predicate, spatial predicate, or spatiotemporal predicate. They can be single or multiple.

Q14: Retrieve all the black targets, whose time staying at Shenyang is the same as target 1.

Symbolic notation: for * let S where R$_x$ ^ R$_y$ and OP$_v$ ^ S$_t$ and S$_a$ return F ^ *

This query is an implicit query, and the information implied in "whose time staying at Shenyang is the same as target 1" is that *tEqual* operation needs to be performed on the valid time of each object and the valid time of target 1, where *tEqual* is a temporal predicate.

Q15: Retrieve all the black targets, which may threaten the target 1 at present.

Symbolic notation: for * let S where OP$_x$ ^ OP$_y$ and S$_v$ ^ S$_t$ and S$_a$ return F ^ *

This query is an implicit query, and the information contained in "which may threaten the target 1 at present" is the predicate operation *sMeet* of the motion trajectory and the motion trajectory of the target 1, where *sMeet* is a spatial predicate.

Q16: Retrieve all the black targets, which may be threatened by the target 1 at Shenyang.

Symbolic notation: for * let S where OP$_x$ ^ OP$_y$ and OP$_v$ ^ S$_t$ and S$_a$ return F ^ *

This query is an implicit query, and the information contained is that the spatiotemporal operation *stOverlap* needs to be performed on the motion trajectory of each object and the motion trajectory of target 1, where *stOverlap* is a spatiotemporal predicate.

The existing researches on fuzzy spatiotemporal data queries primarily encompasses temporal database query languages, traditional spatiotemporal database query languages, and moving object database query languages. The symbolic notations proposed for fuzzy spatiotemporal data are capable of representing queries formulated in both traditional spatiotemporal database query language (STQL) and moving object database query language (FTL). Furthermore, our methodology allows for the classification of queries based on the specified criteria. For instance:

(a) STQL query:

select *e*, *o*
from each targets *e*, *o*
where *o*. targeted = "1" and *e*. color = "black" and distance (at (*e*. route, epos), at (*o*. route, opos)) < 5000 and PR_inside (opos, Region (Shenyang))

The above query can be represented using our proposed symbolic notation as follows:

for * let S where OP_x ^ OP_y and OP_v ^ $*_t$ and S_a return C ^ *.

(b) FTL query:

retrieve *e*
from Targets
where end_time (Intersect (*e*. threatenarea, object (1). position)) < now and begin_time (Intersect (*e*. threatenarea, object (1). position)) and begin_time (INSIDE (*e*, Region (Shenyang)) and INSIDE (object (1), Region (Shenyang)) And *e*. color = "black"

The above query can be represented using our proposed symbolic notation as follows:

for * let S where OP_x ^ OP_y and OP_v ^ $*_t$ and S_a return C ^ *.

(c) FTL query:

retrieve *e*
from Targets
where end_time (Intersect (*e*. threatenarea, object (1). position) and INSIDE (*e*, Region (Shenyang))) < 1200 and begin_time (Intersect (*e*. threatenarea, object (1). position) and INSIDE (*e*, Region (Shenyang))) > 1000 and *e*. color = "black"

The above query can be represented using our proposed symbolic notation as follows:

for * let S where OP_x ^ OP_y and R_v ^ $*_t$ and S_a return C ^ *

While the qualified clauses of query(a) and query(b) differ, they are categorized under the same classification. Conversely, despite the formal similarities between query(b) and query(c), they are classified differently due to the nature of their valid time; specifically, the valid time for query(b) must be derived through operations, whereas the valid time for query(c) is represented as a range value.

5.4.5 *Query Result Processing Using XQuery*

XQuery serves as a query language specifically designed for XML data. A significant benefit of our methodology is that it enables the formulation of complex spatiotemporal queries utilizing XQuery, thereby eliminating the need for the development of a new query language. The architecture of the querying process is illustrated in Fig. 5.16.

Upon receiving a query request, we execute the query in accordance with the spatiotemporal data model and relevant spatiotemporal functions. Following the processing of the results, we ultimately derive the desired outcome. The necessity for result processing arises from the fact that the outcomes generated based on the model may be implicit in nature. For instance, when a query is made regarding velocity, the results returned include *xaxix, yaxis, xval, yval, xcp, ycp*, as dictated by the model. Although these results convey velocity information, such information remains implicit. Consequently, it is essential to process the results further. In the case of velocity, we can ascertain the velocity value by $v = \sqrt{\text{xval}^2 + \text{yval}^2}$ and determine its direction by analyzing the orientations of the *xaxis* and *yaxis*. Another illustrative example pertains to the area of spatiotemporal objects, where the returned results consist of *xstart*, *ystart*, *xend*, and *yend*. Given that this result is also implicit, we can compute the area of the spatiotemporal object using the formula $area = (xend - xstart) \times (yend - ystart)$.

Next, we will demonstrate the methodology for querying spatiotemporal data utilizing XQuery. XQuery, a query language designed for XML, is fundamentally

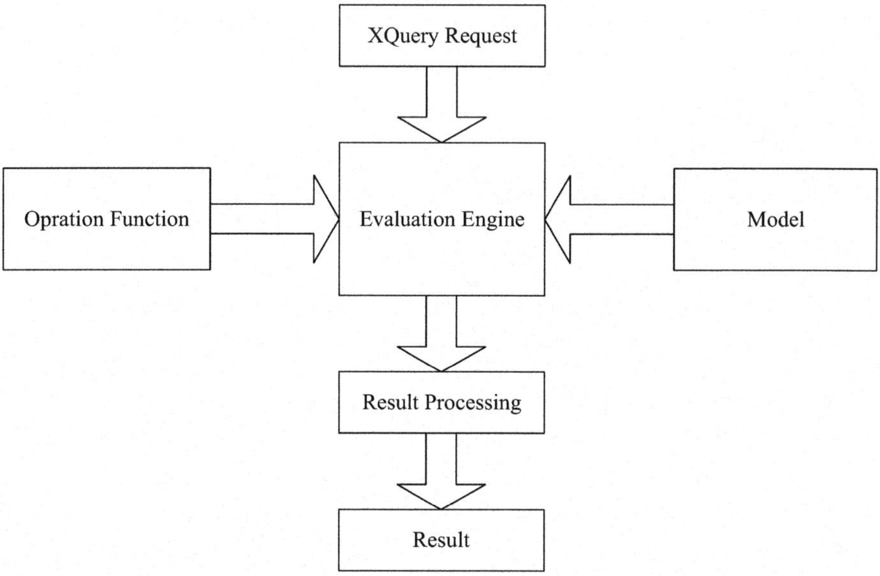

Fig. 5.16 Architecture of querying process

based on the structure of XML documents. A significant advantage of our approach is that it allows for the formulation of complex spatiotemporal queries within XQuery without necessitating the development of a new query language; instead, we introduce a set of new functions. The FLWOR expression, recognized as the most effective and widely utilized construct, can be employed directly, with its syntax remaining unchanged. We present eight spatiotemporal operations, and the results of queries are derived from the invocation of these operations. Consequently, we can associate spatiotemporal operations following the return statement, which can be expressed as follows:

return Funtion.

Funtion indicates spatiotemporal functions, which contain *stDisjoint* (*$a, $b*), *stMeet* (*$a, $b*), *stOverlap* (*$a, $b*), *stIntersects* (*$a, $b*), *stContain* (*$a, $b*), *stEqual* (*$a, $b*), *stWhen* (*$n, $t*), and *stHistory* (*$n*, $\$T_s$, $\$T_e$). Consequently, when query using XQuery, the former part FLWOR remain unchanged, and bind spatiotemporal functions after *return*.

When querying spatiotemporal data utilizing XQuery, it is essential to first extract the attributes of the spatiotemporal object in accordance with the established model, subsequently returning the relevant attributes through FLWOR expressions. For instance, to ascertain the position of a spatiotemporal object, one must initially retrieve the values of *xstart*, *ystart*, *xend*, and *yend*. Following this, appropriate functions can be invoked to obtain the desired results.

To illustrate, if the objective is to determine whether the topological relationship between two spatiotemporal objects is characterized by *stContain*, one would first employ FLWOR expressions to retrieve MBR of the two objects. Subsequently, the function *stContain*(*$a, $b*) can be called. The relationship is classified as *stContain* if the function returns true, whereas it is not classified as *stContain* if the result is false.

Because there is similarity between topological operations, they can be generally represented as:

for $a in Object [OID = "$o_1$"] and
$b in Object [OID = "$o_2$"]
return $\perp$

$\perp$ contains *stDisjoint*(*$a, $b*), *stMeet*(*$a, $b*), *stIntersects*(*$a, $b*), *stContain*(*$a, $b*), *stOverlap*(*$a, $b*), and *stEqual*(*$a, $b*), which retrieve whether spatiotemporal object o_1 and o_2 are disjoint, meet, intersect, contain, overlap, or equal, respectively.

The aforementioned queries extract the six topological operations pertaining to two spatiotemporal objects. Subsequently, we will present two queries that will retrieve geometric operations as detailed below:

for $e in Object [OID = "$o_1$"]
return $\prod$

$\prod$ contains *stWhen*(*$e*, *t*) and *stHistory*(*$e*, T_s, T_e), which retrieve the state of the spatiotemporal object at a given time instant *t*, or retrieve changing history of the spatiotemporal object at a time interval. *stWhen($e)* is similar as the snapshot of a spatiotemporal object, and is a special form of *stHistory($e*, T_s, T_e) since $T_s = t = T_e$.

Utilizing our representation framework, complex queries can be articulated with ease. Below, we delineate the primary categories of spatiotemporal queries as informed by XQuery.

QUERY 1: Topological relationship queries. Find the topological relationship of spatiotemporal object o_1 and o_2 after object o splits into object o_1 and o_2:

```
for $a in Object [OID = "o1"] and
$b in Object [OID = "o2"]
let $a: = o1/predecessor [.="o"] and o1/ts > 0
let $b: = o2/predecessor [.="o"] and o2/ts > 0
return stDisjoint ($a, $b)
return stMeet ($a, $b)
return stIntersect ($a, $b)
return stContain ($a, $b)
return stOverlap ($a, $b)
return stEqual ($a, $b)
```

Here, *stDisjoint*, *stMeet*, *stIntersect*, *stContain*, *stOverlap*, and *stEqual* are user-defined functions utilized to assess the topological relationships between two spatiotemporal objects. The initial "let" clause yields o_1, which is the successor of object o following the time instant of 0. Similarly, the subsequent "let" clause produces o_2, which is also a successor of o after the same time instant. These two clauses collectively indicate that object o has bifurcated into o_1 and o_2 subsequent to time instant 0. In reality, the spatiotemporal object o underwent this division into o_1 and o_2 at time instant 3. As previously noted, there exist six distinct spatiotemporal topological relationships between pairs of spatiotemporal objects; thus, we employ these six operations to ascertain the topological relationship, as it is imperative that a specific spatiotemporal topological relationship is identifiable at any given time instant.

From the positional data of the spatiotemporal object o_1, it is straightforward to derive the MBR for o_1 following the time instant 0, given that o_1 is a static spatiotemporal object. In contrast, the spatiotemporal object o_2 exhibits dynamic behavior, with a change in its attributes occurring at time 16. For the sake of simplicity, this discussion will focus on the states at time points 0, 1, 16, and 17, while similar analyses can be applied to other time points in an analogous manner.

The MBR for the spatiotemporal objects o_1 and o_2 at the time instants 0, 1, 16, and 17 are illustrated in Fig. 5.17. The positions of spatiotemporal object o_2 at time instants 1 and 17 are derived from its motion characteristics at time instants 0 and 16, respectively. Consequently, the topological relationships between spatiotemporal objects o_1 and o_2 at the specified time instants can be readily determined as *stDisjoint*, *stOverlap*, *stMeet*, and *stContain*, respectively. For the sake of clarity, this discussion

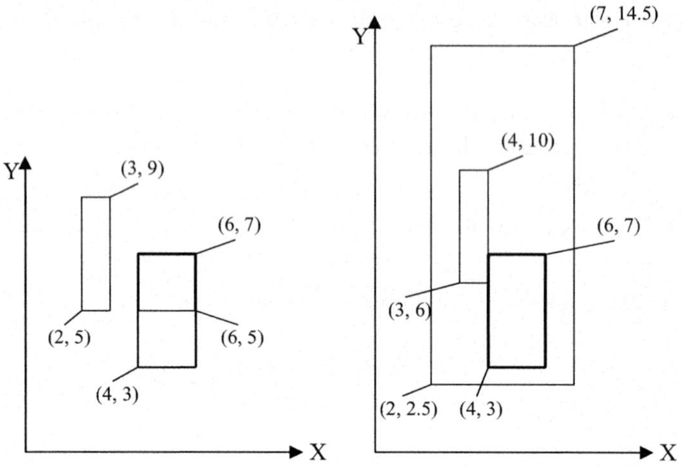

Fig. 5.17 MBR of spatiotemporal object o_1 and o_2 at time instant 0, 1, 16, and 17

will focus solely on the topological relationships of o_1 and o_2 at time instants 0, 1, 16, and 17; however, analogous relationships can be established for other time instants.

QUERY 2: Snapshot queries. Find velocity of spatiotemporal object o_2 at the time instant 16.

for $e in Object [OID = "$o_2$"]
return stWhen ($e/../κ, "16")
return stWhen ($e/../value, "16")

The two components of the "return" motion associated with the spatiotemporal object o_2 correspond to its velocity. The time instant denoted as "16" is equivalent to T_s = "16" = T_e. For illustrative purposes, the motion component of the spatiotemporal object o_2 at the specified time instant of 16 is depicted. From this analysis, we can derive that *xaxis* is → , *yaxis* is ↑, *xval* is 5, *yval* is 3, *xcp* is 9/7, and *ycp* is 17/16.

QUERY 3: Evolutionary queries. Find changing history of spatiotemporal object o_2's motion part at time interval $T_e = [3, 30]$ since o_1 comes into being.

for $e in Object [OID = "$o_1$"]
let T_s: = max{0, o_2/type[. = "1"]}
return stHistory($e, T_s, T_e)

The let clause yields T_e = [3, 30] due to the fact that the time at which o_1 is instantiated occurs prior to 3. The type value of "1" indicates that this spatiotemporal object has been created. Consequently, stHistory is capable of producing the expected outcome.

The state of motion can be determined at time instants 0 and 16. The states of motion for the intervals between time instants 1 to 15 and 17 to the present are consistent with those observed at time instants 0 and 16. Above, at time interval T_e = [3, 15], we can get *xaxis* is →, *yaxis* is ↓, *xval* is 2, *yval* is 0.5, *xcp* is 2, *ycp* is 4/7; at time interval T_e = [16, 30], we can get *xaxis* is →, *yaxis* is ↑, *xval* is 5, *yval* is 3, *xcp* is 9/7, *ycp* is 17/16.

The above queries present three straightforward examples that demonstrate the ease with which the eight spatiotemporal operations can be accessed through XQuery. Furthermore, more complex queries can be developed from these foundational examples, as they encompass the eight fundamental and robust spatiotemporal queries that effectively encapsulate the semantics of XQuery.

5.4.6 Application Study

This subsection presents three examples of each fuzzy spatiotemporal linguistic term, along with an assessment of their processing methodologies.

Example 5.1 (Fuzzy spatiotemporal predicate query). Figure 5.18 illustrates the global observation of cloud cover and precipitation at 03:00 UTC on May 3, 2012, as reported in [45]. This observation is derived from cloud quantity data, and each term associated with the fuzzy topological relationship between the two fuzzy spatiotemporal regions, G_1 and G_2, is detailed in Table 5.6.

Query: Retrieve the overcast (lawn green denoted by G_1) and cloudy overcast (sky blue denoted by G_2) areas at 03 UTC 3 May 2012.

This query is formulated using XQuery as follows:

for $a, $b in doc (“weather.xml”)/spatiotemporal data
let $a: = “G₁”
let $b: = “G₂”
where $a//../position/@Ts = 03 UTC 03 May 2012 and $b//../position/@Ts = 03 UTC 03 May 2012 and Foverlap ($a, $b, τ) > 0
return $a/../position ∩ $b/../position, $\tau_{overlap}$

The evaluation of the query is conducted in the followings:

(i) G_1 and G_2 represent fuzzy regions, and the objective of the query is to obtain the intersection, if it exists, of these two fuzzy regions. Consequently, it is essential to first establish the appropriate declaration of the fuzzy spatiotemporal predicate that is relevant to the specific application in question. It is important to note that the declaration of the fuzzy spatiotemporal predicate is contingent upon the particular application context.

declare *fuzzy spatiotemporal predicate Foverlap* (α, β, (τ_{equal}, $\tau_{contain}$, $\tau_{overlap}$, τ_{meet}, $\tau_{disjoint}$)?)

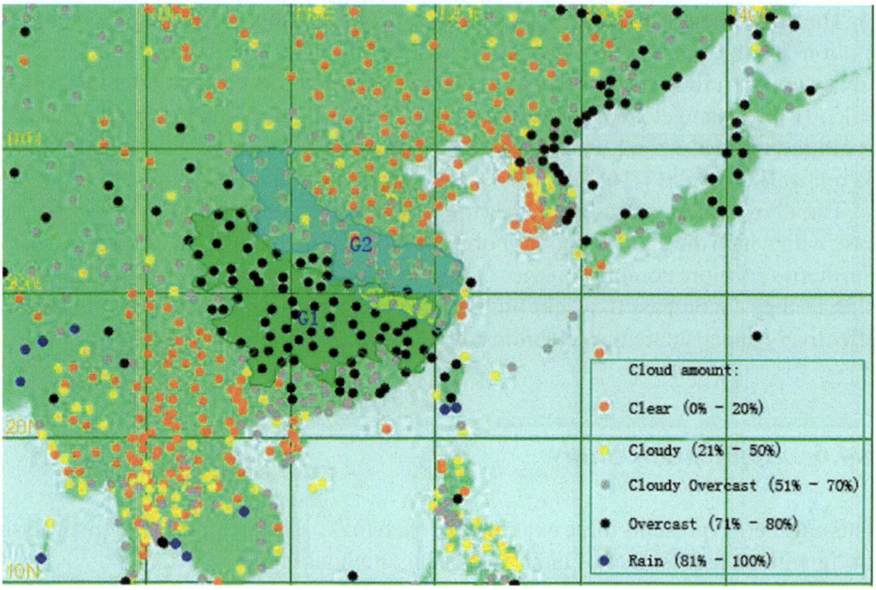

Fig. 5.18 Global observing of clouds and rain at 03 UTC 3 May 2012

Table 5.6 Computing a fuzzy topological relation for two fuzzy spatiotemporal regions G_1 and G_2

α-cut levels of fuzzy regions	$m\ (G_{1\alpha i}) \times m\ (G_{2\beta j})$	τ_{equal}	$\tau_{contain}$	$\tau_{overlap}$	τ_{meet}	$\tau_{disjoint}$
1.0–0.8	0.2	0	0	0	0	1
0.8–0.6	0.2	0	0	0	0	1
0.6–0.25	0.35	0	0	0	1	0
0.25–0.0	0.25	0	0	1	0	0

$$\sum_{i=1}^{n}\sum_{j=1}^{m} m\left(G_{1\alpha j}\right)m\left(G_{2\beta j}\right)\tau\left(G_{1\alpha j},\ G_{2\beta j}\right)?$$

(ii) The initial two conditions within the “where” clause establish filtering criteria for the variables *$a* and *$b*, respectively. These criteria are employed to refine the temporal parameters of the specified query.

(iii) The final condition in the “where” clause introduces a filtering criterion for the variables *$a* and *$b*. This criterion is employed to refine the membership degree of the fuzzy spatiotemporal relation *Foverlatp*. It is calculated that $Fcontain(A_{\alpha i}, B_{\beta j}, \tau_{equal}) = 0$, $Fcontain(A_{\alpha i}, B_{\beta j}, \tau_{contain}) = 0$, $Fcontain(A_{\alpha i}, B_{\beta j}, \tau_{overlap}) = 0.25$, $Fcontain(A_{\alpha i}, B_{\beta j}, \tau_{meet}) = 0.35$, and $Fcontain(A_{\alpha i}, B_{\beta j}, \tau_{disjoint})$

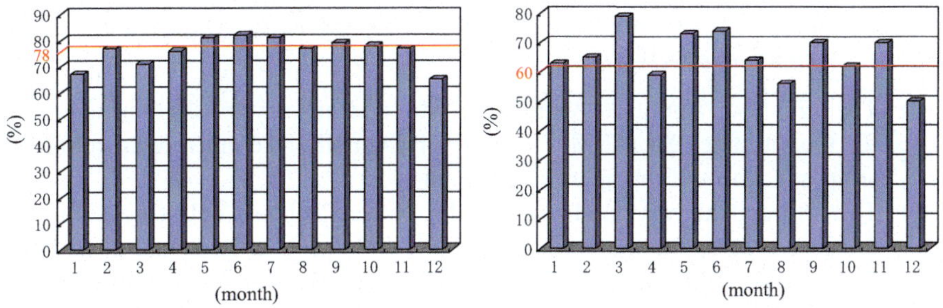

(a) Average relative humidity (b) Mean cloud amount

Fig. 5.19 Meteorological observations of Hong Kong in 2011

= 0.4. Accordingly, $Fcontain(A_{\alpha i}, B_{\beta j}, \tau_{overlap})$, $Fcontain(A_{\alpha i}, B_{\beta j}, \tau_{meet})$, and $Fcontain(A_{\alpha i}, B_{\beta j}, \tau_{disjoint})$ satisfy the filtering criterion.

(iv) Consequently, the "return" clause would yield the yellow region depicted in Fig. 5.18, characterized by a membership degree of 0.25.

Example 5.2 (*Fuzzy spatiotemporal modifier query*) Figure 5.19 presents meteorological data pertaining to Hong Kong for the year 2011 [46]. Specifically, Fig. 5.19a illustrates the average relative humidity recorded in Hong Kong during that year, while Fig. 5.19b depicts the mean cloud cover observed in the same period.

Query: Retrieve the month of high humidity and the month of low cloud amount of Hong Kong in 2011.

This query is formulated using XQuery as follows:

```
for $a, $b in doc ("meteorological.xml")/spatiotemporal data
let $a: = "average relative humidity"
let $b: = "mean cloud amount"
where $a//../humidity/@Ts = 2011 and $b//../cloud amount/@Ts = 2011 and
$a//../humidity > high (humidity) and $b//../cloud amount < low (cloud amount)
return $a//../month, $b//../month
```

The evaluation of the query is conducted in the followings:

(i) Data pertaining to the meteorological observations in Hong Kong for the year 2011 indicates a prevalence of high humidity levels and a low amount of cloud cover. To facilitate this analysis, it is necessary to establish definitions for the fuzzy spatiotemporal modifiers "high" and "low," with the degree of power being contingent upon the specific requirements of the application. It is important to acknowledge that the definitions and values associated with the fuzzy spatiotemporal modifiers are inherently dependent on the context of their application.

declare fuzzy spatiotemporal modifier high (humidity) power (+lg(78 − humidity)) *declare fuzzy spatiotemporal modifier low (cloud amount) power (+lg(*60 *– cloud amount))*

(ii) The initial two conditions within the "where" clause establish filtering criteria for the variables *$a* and *$b*, respectively, which are employed to refine the temporal aspect of the query in question.

(iii) The subsequent two conditions in the "where" clause delineate additional filtering criteria for the variables *$a* and *$b*, specifically aimed at isolating the desired months. In accordance with the definitions of fuzzy spatiotemporal modifiers, a month exhibiting humidity levels exceeding 78% and a month characterized by cloud cover below 60% fulfill the requirements of these fuzzy spatiotemporal modifiers. As illustrated in Fig. 5.19, the months of May (81%), June (82%), July (81%), and September (79%) meet the criteria for the fuzzy spatiotemporal modifier classified as high, while the months of April (59%), August (56%), and December (50%) correspond to the fuzzy spatiotemporal modifier classified as low.

(iv) Consequently, the "return" clause will yield the months of May, June, July, and September for the first retrieval (indicated by values exceeding 78%, as represented by the red line in Fig. 5.19), and the months of April, August, and December for the second retrieval (indicated by values below 60%, also represented by the red line in Fig. 5.19).

Example 5.3 (*Fuzzy spatiotemporal comparator query*) There are two episodes of heavy rain in Hong Kong on 22 May which necessitated the issuance of the red rainstorm warning. Figure 5.20 shows rainfall distribution of Hong Kong on 22 May 2011 [46]. Regions characterized by varying rainfall distributions are distinguished by distinct colors, which are represented by a numerical range.

Query: Retrieve the areas of Hong Kong on 22 May 2011, among which the rainfall distribution is much greater than 150 mm.

This query is formulated using XQuery as follows:

```
for $a in doc ("meteorological.xml")/spatiotemporal data
let $a: = "rainfall distribution"
where $a//../position/@Ts = 22 May 2011 and $a//../position/rainfall distribution
= mgt (rainfall distribution)
return $a//../position, $a//../position/Val/@Poss
```

The evaluation of the query is conducted in the followings:

(i) The present query involves a fuzzy spatiotemporal query referred to as "much greater than" (*mgt*) for the sake of simplicity. Consequently, it is necessary to establish a definition for the fuzzy spatiotemporal comparator *mgt*. In this particular instance, the variable *y* in the declaration is assigned a value of 150. Furthermore, a dividing point of 400 is selected for the membership function. It is important to acknowledge that the declaration of the fuzzy spatiotemporal comparator and the values utilized within the functions are contingent upon the specific application.

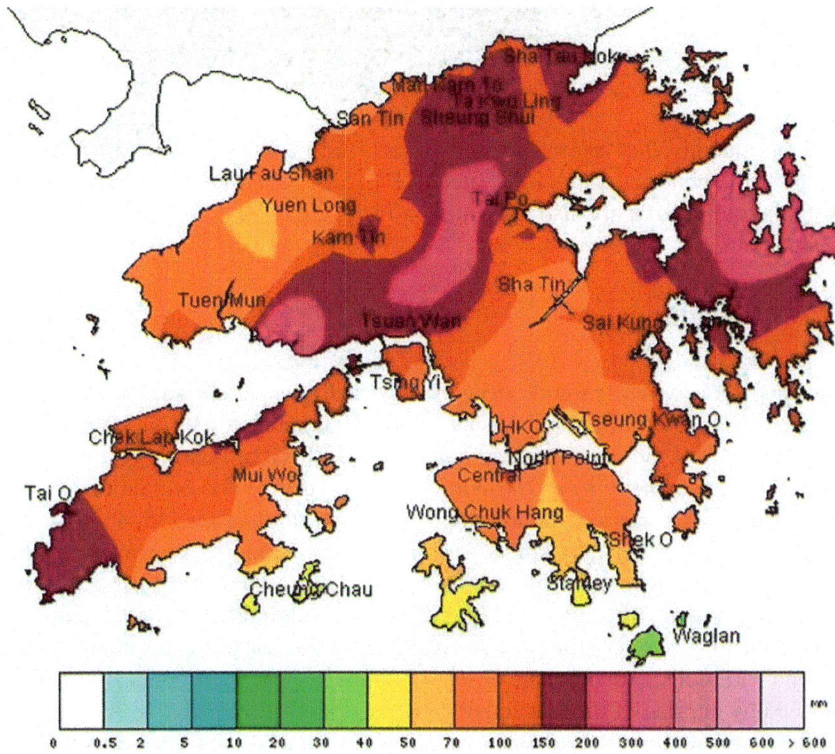

Fig. 5.20 Rainfall distribution of Hong Kong on 22 May 2011

declare fuzzy spatiotemporal comparator mgt ($x as xs: float, $y as xs: float)

$$(\$\,x - \$\,y)\ f\,(x) = \begin{cases} 0, & x \le y \\ (x - y)/(400 - y), & y < x < 400 \\ 1, & x \ge 400 \end{cases}$$

(ii) The initial condition within the "where" clause establishes a filtering criterion for the variable *$a*, which is employed to refine the temporal aspect of the query.

(iii) The subsequent condition in the "where" clause similarly delineates a filtering criterion for the variable *$a*, aimed at specifying the desired geographical regions. According to the declaration of the fuzzy spatiotemporal comparator, the membership degree for areas where the rainfall amount ranges from 150 to 200 mm is between 0 and 0.2; for areas with rainfall amounts between 200 and 300 mm, the membership degree ranges from 0.2 to 0.6; for areas with rainfall amounts between 300 and 400 mm, the membership degree ranges from 0.6 to 1.0; and for areas where the rainfall exceeds 400 mm, the membership degree is equal to 1.0.

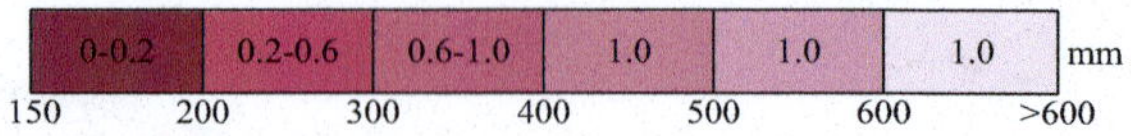

Fig. 5.21 Query result of Example 5.3

(iv) Consequently, the "return" clause will yield the areas depicted in Fig. 5.20, with the corresponding colors illustrated in Fig. 5.21. The membership degrees for each colored area are also presented in Fig. 5.21.

5.4.7 Evaluations

In this subsection, we will provide an analysis of each query example. The evaluations were conducted using Microsoft Visual Studio 10.0 and Intel VTune Performance Analyzer v9.0.009, on a system equipped with a 1.8 GHz Pentium IV processor, 1 GB of RAM, and operating on the Windows XP platform. The evaluations for each query example were carried out utilizing two real-world datasets [45, 46].

In order to assess the performance of each query example, we conducted three distinct sets of experiments. The initial set focused on measuring the query time for each example, as illustrated in Fig. 5.22. The results indicate that Example 5.1 exhibited the longest query time, while Example 5.3 demonstrated the shortest query time. The variations in query time observed in this set of experiments are primarily influenced by the size of the respective datasets.

In the second set of experiments, we assessed the memory costs associated with each query example, which are illustrated in Figs. 5.23 and 5.24. Figure 5.23 indicates that the total memory cost for querying Example 5.2 is the highest, while Example 5.3 exhibits the lowest total memory cost. The total memory costs are influenced not only by the size of each dataset but also by the costs incurred in querying the relevant nodes within the fuzzy spatiotemporal XML data tree. Consequently, we further analyzed the memory costs specifically related to node querying. As depicted in Fig. 5.24, the memory cost for node querying in Example 5.1 surpasses that of the other two examples, while the memory costs for node querying in Examples 5.2 and 5.3 are nearly identical. This discrepancy can be attributed to the nature of the

Fig. 5.22 Evaluation of query time

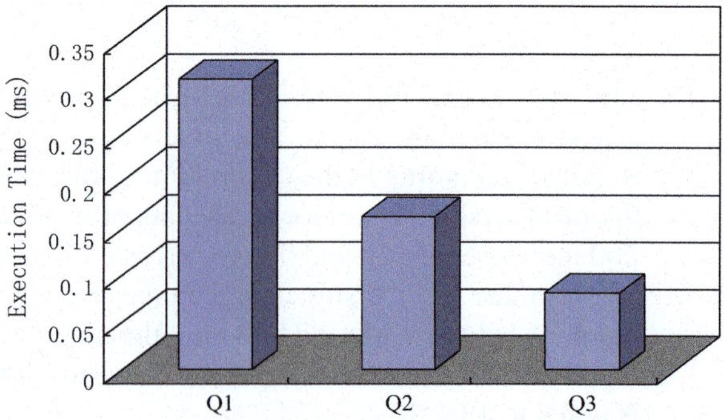

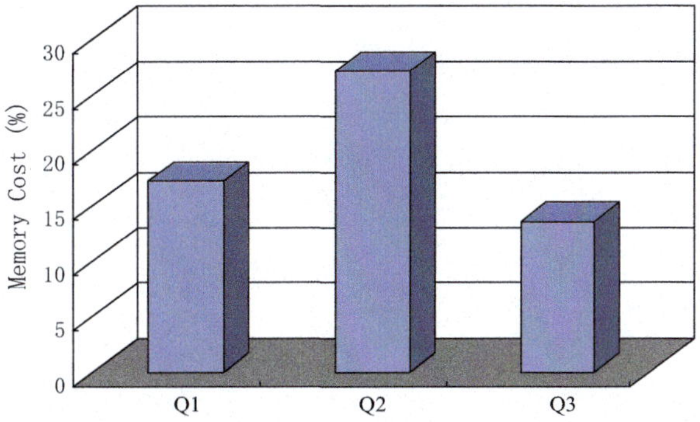

Fig. 5.23 Evaluation of memory costs on total querying

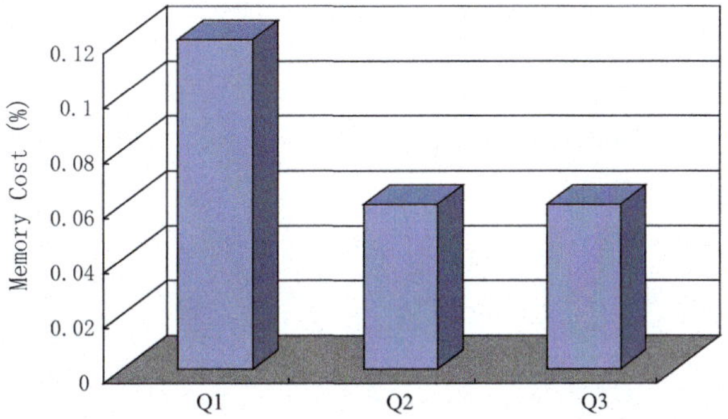

Fig. 5.24 Evaluation of memory costs on node querying

desired nodes; both Examples 5.2 and 5.3 target attribute nodes, whereas Example 5.1 focuses on position nodes. Additionally, it is noteworthy that querying position nodes incurs a higher memory cost compared to querying attribute nodes.

In the final group of experiments, we conducted a temporal complexity analysis for each query example, as presented in Figs. 5.25, 5.26, and 5.27. It is important to note that the temporal intervals for the complexity analysis differ (33, 60, and 26). This variation can be attributed to the evaluations being conducted on the individual datasets corresponding to each query example.

5.5 Summary

In this chapter, we investigate a fuzzy spatiotemporal XML data model and delineate an algebraic framework for the manipulation of fuzzy spatiotemporal XML data. Despite the intricate structure of the associated trees, the proposed algebra incorporates several operators that utilize a consistent foundational structure for their parameters. Moreover, the algebra is designed to accommodate the presence of order within a tree. We also present various formal characterizations of fuzzy spatiotemporal algebraic equivalences and examine the transformation of XQuery

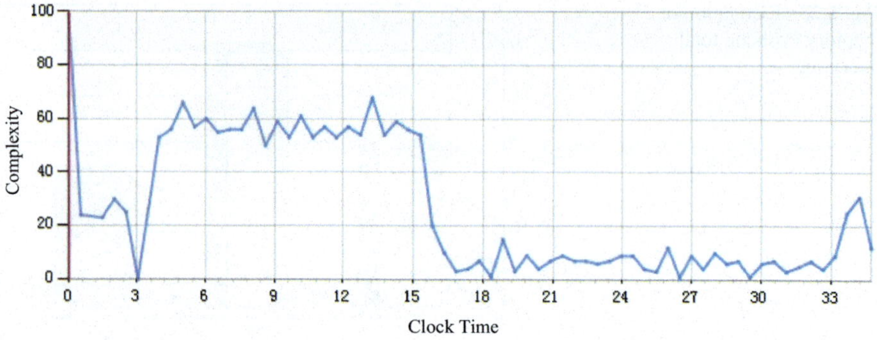

Fig. 5.25 Temporal complexity analysis of Example 5.1

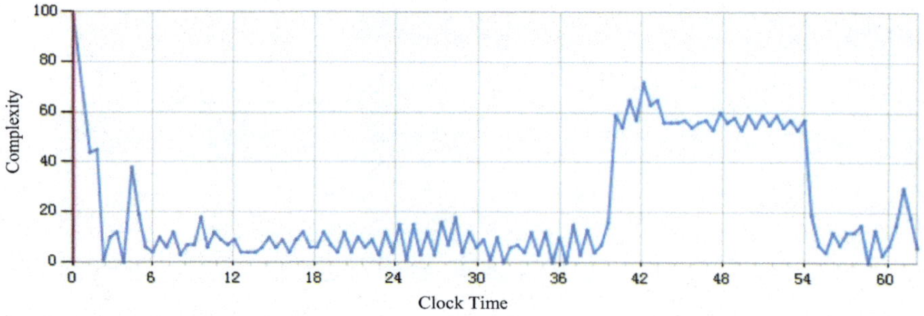

Fig. 5.26 Temporal complexity analysis of Example 5.2

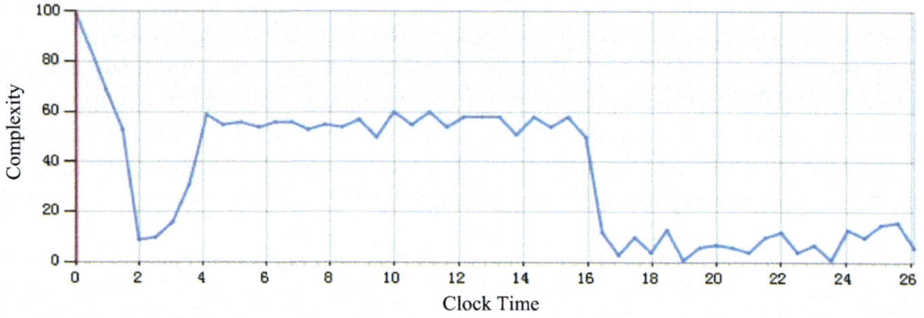

Fig. 5.27 Temporal complexity analysis of Example 5.3

expressions into algebraic expressions. The algebraic operations applied to fuzzy spatiotemporal data in spatiotemporal applications establish essential underpinnings for the management of spatiotemporal XML data. To effectively query the relevant information, it is imperative to ascertain the appropriate input and output formats.

Our primary objective in this definition is to serve as a foundation for query evaluation and optimization. After that, we first propose the architecture of querying fuzzy spatiotemporal data using XQuery. On this basis, we make fuzzy-set-based extensions to XQuery including truth degrees, fuzzy spatiotemporal linguistic terms, and FLWOR expressions. Furthermore, we present example queries from the application domain, discuss crucial details of their processing, and present evaluations of each query example. Finally, the comparative study demonstrates the advantages of our approach.

References

1. S. Benferhat, J. Ben-Naim, O. Papini, E. Würbel, An answer set programming encoding of prioritized removed sets revision: application to GIS. Appl. Intell. **32**(1), 60–87 (2010)
2. R. Mehrotra, A. Sharma, Evaluating spatio-temporal representations in daily rainfall sequences from three stochastic multisite weather generation approaches. Adv. Water Resour. **32**(6), 948–962 (2009)
3. H. Adeli, X. Jiang, Neuro-fuzzy logic model for freeway work zone capacity estimation. J. Transp. Eng. **129**(5), 484–493 (2003)
4. L. Bai, L. Yan, Z. Ma, Querying fuzzy spatiotemporal data using XQuery. Integr. Comput. Aided Eng. **21**(2), 147–162 (2014)
5. M. Koyuncu, A. Yazici, IFOOD: an intelligent fuzzy object-oriented database architecture. IEEE Trans. Knowl. Data Eng. **15**(5), 1137–1154 (2003)
6. A. Sözer, A. Yazici, H. Oğuztüzün, O. Tas, Modeling and querying fuzzy spatiotemporal databases. Inf. Sci. **178**(19), 3665–3682 (2008)
7. P. Senellart, S. Abiteboul, On the complexity of managing probabilistic XML data, in *Proceedings of the 26th ACM SIGACT-SIGMOD-SIGART Symposium on Principles of Database Systems* (2007), pp. 283–292
8. Z. Ma, J. Liu, L. Yan, Fuzzy data modeling and algebraic operations in XML. Int. J. Intell. Syst. **25**(9), 925–947 (2010)
9. F. Rizzolo, A.A. Vaisman, Temporal XML: modeling, indexing, and query processing. VLDB J. **17**(5), 1179–1212 (2008)
10. B. Huang, S. Yi, W.T. Chan, Spatio-temporal information integration in XML. Futur. Gener. Comput. Syst. **20**(7), 1157–1170 (2004)
11. Z. Ma, L. Yan, Fuzzy XML data modeling with the UML and relational data models. Data Knowl. Eng. **63**(3), 972–996 (2007)
12. C. Kahraman, Fuzzy theory and technology with applications. J. Intell. Fuzzy Syst. **19**(4.5), 231–233 (2008)
13. A. Baykasoğlu, T. Göçken, A review and classification of fuzzy mathematical programs. J. Intell. Fuzzy Syst. **19**(3), 205–229 (2008)
14. X.H. Liu, Y.C. Wan, Storing spatio-temporal data in XML native database, in *Proceedings of the 2nd International Workshop on Database Technology and Applications* (2010), pp. 1–4
15. M. Franceschet, A. Montanari, D. Gubiani, Modeling and validating spatio-temporal conceptual schemas in XML schema, in *Proceedings of the 18th International Conference Database and Expert Systems Applications* (2007), pp. 25–29
16. L. Bai, L. Yan, Z. Ma, Fuzzy spatiotemporal data modeling and operations in XML. Appl. Artif. Intell. **29**(3), 259–282 (2015)
17. H. Prade, C. Testemale, Generalizing database relational algebra for the treatment of incomplete or uncertain information and vague queries. Inf. Sci. **34**(2), 115–143 (1984)

18. S. Paparizos, Y. Wu, L.V.S. Lakshmanan, H.V. Jagadish, Tree logical classes for efficient evaluation of XQuery, in *Proceedings of the 2004 ACM SIGMOD International Conference Management of Data* (2004), pp. 71–82
19. C. Beeri, Y. Tzaban, SAL: an algebra for semistructured data and *XML*, in *Proceedings of the International Workshop on the Web and Databases* (1999), pp. 37–42
20. G. Buratti, D. Montesi, A data model and an algebra for querying XML documents, in *Proceedings of the 17th International Conference Database and Expert Systems Applications* (2006), pp. 482–486
21. H.V. Jagadish, L.V.S. Lakshmanan, D. Srivastava, K. Thompson, TAX: a tree algebra for XML, in *Proceedings of the 8th International Workshop on Database Programming Languages* (2001), pp. 149–164
22. M. Magnani, D. Montesi, XML and relational data: towards a common model and algebra, in *Proceedings of the 9th International Database Engineering and Application Symposiums* (2005), pp. 96–101
23. D. Che, R.M. Sojitrawala, DUMAX: a dual mode algebra for XML queries, in *Proceedings of the 2nd International Conference on Scalable Information Systems* (2007), pp. 1–4
24. M. Fernandez, J. Simeon, P. Wadler, An algebra for XML query, in *Proceedings of the 20th International Conference Foundations Software Technology and Theoretical Computer Science* (2000), pp. 11–45
25. L. Wang, S. Wang, B. Murphy, E.A. Rundersteiner, Order-sensitive XML query processing over relational sources: an algebraic approach, in *Proceedings of the 9th International Database Engineering and Application Symposium* (2005), pp. 175–184
26. E. Hung, L. Getoor, V.S. Subrahmanian, PXML: a probabilistic semi-structured data model and algebra, in *Proceedings of the 19th International Conference Data Engineering* (2003), pp. 467–478
27. L. Bai, L. Zhu, An algebra for fuzzy spatiotemporal data in XML. IEEE Access **7**(1), 22914–22926 (2019)
28. A. Sözer, A. Yazici, H. Oğuztüzün, P.E. Petry, Querying fuzzy spatiotemporal databases: implementation issues. Uncertain. Approach Spatial Data Model. Process. **271**(10), 97–116 (2010)
29. R.D. Rodrigues, A.J.O. Cruz, R.T. Cavalcante, Aliança: a proposal for a fuzzy database architecture incorporating XML. Fuzzy Sets Syst. **160**(2), 269–279 (2009)
30. J. Seto, S. Clement, D. Duong, K. Kianmehr, R. Alhajj, Fuzzy query model for XML documents, in *Proceedings of the 10th International Conference on Intelligent Data Engineering and Automated Learning* (2009), pp. 333–340
31. N. Polyzotis, M. Garofalakis, Y. Loannidis, Approximate XML query answers, in *Proceedings of the ACM SIGMOD International Conference on Management of Data* (2004), pp. 263–274
32. B. Bennett, A.G. Cohn, Multi-dimensional modal logic as a framework for spatio-temporal reasoning. Appl. Intell. **17**(3), 239–251 (2002)
33. C. Claramunt, M. Theriault, Managing time in GIS: an event-oriented approach, in *Proceedings of the International Workshop on Temporal Databases: Recent Advances in Temporal Databases* (1995), pp. 142–161
34. Z. Du, Y.S. Jeong, M.K. Jeong, S.G. Kong, Multidimensional local spatial autocorrelation measure for integrating spatial and spectral information in hyperspectral image band selection. Appl. Intell. **36**(2), 542–552 (2012)
35. G. Bordogna, M. Pagani, G. Pasi, R. Villa, A flexible news filtering model exploiting a hierarchical fuzzy categorization, in *Proceedings of the 7th International Conference on Flexible Query Answering Systems, Lecture Notes in Computer Science* (2006), vol. 4027, pp. 170–184
36. X. Jiang, H. Adeli, Dynamic fuzzy wavelet neuroemulator for non-linear control of irregular building structures. Int. J. Numer. Meth. Eng. **74**(7), 1045–1066 (2008)
37. A. Karim, H. Adeli, Comparison of the fuzzy-wavelet RBFNN freeway incident detection model with the California algorithm. J. Transp. Eng. **128**(1), 21–30 (2002)
38. H. Adeli, X. Jiang, Dynamic fuzzy wavelet neural network model for structural system identification. J. Struct. Eng. **132**(1), 102–111 (2006)

39. S. Freitag, W. Graf, M. Kaliske, Recurrent neural networks for fuzzy data. Integr. Comput. Aided Eng. **18**(3), 265–280 (2011)
40. M. Goncalves, L. Tineo, Fuzzy XQuery, in Z.M. Ma & L. Yan (Eds.), Soft Computing in XML Data Management, STUDFUZZ, vol. 255, pp. 133–163 (2010)
41. J. Liu, Z. Ma, Efficient processing of twig query with compound predicates in fuzzy XML. Fuzzy Sets Syst. **229**, 33–53 (2013)
42. XMARK the XML-benchmark project. http://monetdb.cwi.nl/xml/index.html
43. M. Schneider, A design of topological predicates for complex crisp and fuzzy regions, in *Proceedings of the 20th International Conference on Conceptual Modeling* (2001), pp. 27–30
44. M. Erwig, M. Schneider, Spatiotemporal predicates. IEEE Trans. Knowl. Data Eng. **14**(4), 881–901 (2002)
45. Worldweather. http://www.worldweather.cn/cloud/
46. Meteorological Observations of Hong Kong in 2011. http://www.weather.gov.hk/wxinfo/pastwx/ywx2011.htm

Chapter 6
Querying Fuzzy Spatiotemporal XML Data Integrated with Holistic Twig Pattern

6.1 Introduction

Among the various operations performed on fuzzy spatiotemporal XML data, querying is recognized as a critical operation. Currently, the querying process in XML primarily pertains to general data [1, 2], spatial data [3], temporal data [4], spatiotemporal data [5, 6], fuzzy data [7, 8], and fuzzy spatiotemporal data [9, 10]. However, the effective and precise querying of fuzzy spatiotemporal XML data remains an unresolved challenge in the field.

The hierarchical and ordered nature of XML complicates the process of querying XML data. Tekli et al. [1] investigate a semantic-aware methodology for conducting keyword searches within XML. Lukás et al. [2] propose a model that incorporates specifications for both output and non-output query nodes, aligning with the semantics of XQuery and XPath. Spatiotemporal data encompasses both spatial and temporal characteristics, prompting numerous studies focused on querying spatial and temporal XML data. In the context of spatial XML data, Córcoles and González [11] present a method for querying both spatial and non-spatial information encoded in the Geographical Markup Language (GML), utilizing the Resource Description Framework (RDF) to facilitate the integration of spatial XML documents. Additionally, Chang and Park [12] introduce a novel development model that leverages XML Web Services for the dynamic integration of distributed Geographic Information System (GIS) components, thereby enhancing the interoperability of Internet GIS applications. Regarding temporal XML data, Nørvåg [13] provides an overview of the anticipated features of a temporal XML query language and outlines the necessary query operators for executing such queries. Furthermore, a query language that extends XPath to accommodate temporal queries is proposed in [14]. The exclusive consideration of spatial and temporal features is insufficient for effectively supporting spatiotemporal applications. For instance, when monitoring a vehicle's movement along a roadway, its position is subject to change over time. To accurately query the vehicle's location at a specific moment, it is imperative to integrate both spatial

L. Bai and L. Zhu, *Fuzzy Spatiotemporal XML Data Management*, Studies in Computational Intelligence 1183, https://doi.org/10.1007/978-3-031-81033-6_6

and temporal features. Consequently, the querying of spatiotemporal XML data has garnered significant scholarly attention. Chen and Revesz [6] introduce a Layer Algebra, which serves as an innovative extension of relational algebra tailored for XML, thereby establishing a foundation for the evaluation and optimization of XML queries. They also present DataFox (Datalog For XML), a rule-based XML query language that merges the straightforwardness of Datalog with the capability to handle spatiotemporal data within constraint databases. Bai et al. [5] propose a methodology for XML keyword queries, introducing the concept of query time series (QTS) based on the degree of data revision. They investigate the logical relationships among keywords in QTS by separately analyzing intra-coupling and inter-coupling logic relationships. Subsequently, they develop a keyword similarity calculation method, optimizing its parameters through experimental validation. Recognizing that fuzziness is a fundamental characteristic of many spatiotemporal applications, Bao et al. [10] present a framework named XML2HBase, designed to enhance query performance through a novel encoding scheme. Additionally, the work referenced as [9] proposes an integration model for fuzzy spatiotemporal XML data, utilizing feature expression and a field mapping table to decompose queries and exclude local data sources that do not yield results, thereby improving query efficiency. This approach also involves transforming the XQuery query language into a format compatible with local data sources, followed by the resolution of semantic heterogeneity through the application of the field mapping table, ultimately presenting the query results as unified data source views via XML Schema. In [15], Bai et al. extend XQuery with fuzzy-set-based enhancements, incorporating truth degrees, fuzzy spatiotemporal linguistic terms, and FLWOR expressions, following their proposal of an architecture for querying fuzzy spatiotemporal data using XQuery.

Currently, quantitative query operations in XML predominantly utilize twig patterns [16–18]. In the work of Bruno et al. [16], a holistic path join algorithm known as PathStack is introduced, which operates without branching to match root-to-leaf paths in XML queries. PathStack is characterized as optimal in terms of I/O and CPU usage among all sequential algorithms that process the entire input, exhibiting worst-case complexities that are linear in relation to the sum of input and output sizes, while remaining independent of the sizes of intermediate results. However, when query paths yield numerous solutions that do not contribute to the final results, the use of PathStack becomes suboptimal, as the overall computational cost for a twig pattern is influenced not only by the sizes of the input and final output but also by the sizes of intermediate results. To address this inefficiency, they propose a novel algorithm named TwigStack, which employs a series of linked stacks to efficiently represent partial results of individual query root-to-leaf paths. This method is also optimal in terms of I/O and CPU usage among all sequential algorithms that process the entire input for twigs consisting solely of ancestor–descendant edges. Nevertheless, TwigStack is unable to manage the size of intermediate results for queries that include parent–child edges. In a subsequent study, Lu et al. [18] introduce a new holistic twig join algorithm called TwigStackList, which maintains performance comparable to TwigStack for query patterns limited to ancestor–descendant edges, yet demonstrates significantly greater efficiency than TwigStack for queries that incorporate parent–child edges. TwigStackList necessitates the reading of additional elements from all non-leaf node streams, and its processing becomes increasingly complex when query nodes possess multiple ancestors with identical labels. To mitigate this issue, an algorithm named TJFast is proposed [17], which is based on an extended Dewey labeling scheme. TJFast exclusively scans query leaf nodes, resulting in the access of considerably fewer elements compared to algorithms that utilize region encoding, while also effectively handling queries with wildcards in internal nodes. Although existing research on querying XML data primarily concentrates on general XML structures, these investigations can be expanded to include fuzzy and spatiotemporal characteristics, thereby enhancing their applicability to fuzzy spatiotemporal XML data queries. For instance, Bai et al. [19] present an algorithm called FSPTwigFast, designed to match fuzzy spatiotemporal XML twig patterns.

In this chapter, we investigate the challenges associated with querying fuzzy spatiotemporal XML data that is integrated with holistic twig structures. Initially, we enhance the Dewey code to enable the annotation of fuzzy spatiotemporal data. Following this, we propose a methodological approach for matching fuzzy spatial and fuzzy temporal attributes within fuzzy spatiotemporal XML documents. Additionally, we present a probabilistic calculation that quantifies the degree of fuzziness associated with each node. The experimental results indicate that our querying method demonstrates superior efficiency in handling fuzzy spatiotemporal XML data. Furthermore, our algorithm is capable of supporting queries related to fuzzy temporal XML, fuzzy spatial XML, and fuzzy spatiotemporal XML.

6.2 Basic Knowledge

6.2.1 Fuzzy Spatiotemporal XML Data Model

Spatiotemporal data possesses unique characteristics that differentiate it from the more conventional lists and tables of alphanumeric data typically found in traditional datasets. This distinction results in a limitation of the data model to a tree structure within XML documents. In their study, Bai et al. [15] articulate fuzzy spatiotemporal XML data through five dimensions: *OID*, *ATTR*, *FP*, *FT* and *FM*. In our approach, we represent this data using a 3-tupple (*ATTR*, *FS*, *FT*), which is informed by the characteristics of fuzzy spatiotemporal data and tailored to the requirements of querying, based on the aforementioned five dimensions. The dimensions pertinent to spatiotemporal data are outlined as follows:

- *ATTR*: It is employed to characterize the static characteristics of spatiotemporal data, such as the name of the landowner, land area, population, and other relevant factors. The number of attributes present in spatiotemporal data can vary, often contingent upon the specific application domain.

- *FS*: It delineates the fuzzy spatial characteristics inherent in spatiotemporal data, which encompasses fuzzy points, fuzzy lines, and fuzzy regions.
- *FT*: It delineates the temporal characteristics of spatiotemporal data, encompassing both fuzzy time points and fuzzy time intervals.

The three dimensions are capable of representing the attributes of fuzzy spatiotemporal XML data. This data model is applicable to a variety of XML document types, including general XML documents, fuzzy spatial XML documents, fuzzy temporal XML documents, and fuzzy spatiotemporal XML documents.

6.2.2 XML Twig Pattern

In the context of a query twig pattern *Q* and an XML document *D*, a match of *Q* within *D* is determined by establishing a mapping from the nodes in *Q* to the elements in *D*. This mapping must satisfy two conditions: (i) the predicates associated with the query nodes must be fulfilled by the corresponding elements in the database, with the wildcard symbol "*" being capable of matching any single tag; and (ii) the hierarchical relationships, specifically parent–child and ancestor–descendant relationships, between the nodes in the query must be preserved in the corresponding elements of the database. The results of query *Q*, which contains *n* nodes, can be represented as a collection of *n-array* tuples. Each tuple $(q_1, ..., q_n)$ comprises the database elements that correspond to a unique match of *Q* within *D*.

6.2.3 Dewey ID Labeling Scheme

Tatarinov et al. [20] introduce the *Dewey ID* labeling scheme, which is designed to indicate the position of an element within an XML document. In this scheme, each element is represented by a vector: (i) the root element is denoted by an empty string ε; (ii) for any non-root element *u*, the label is defined as $label(u) = label(s) \cdot x$, where *u* is the *x*-th child of *s*. The *Dewey ID* facilitates the efficient evaluation of structural relationships among elements, such that element *u* is considered an ancestor of element *s* if and only if *label*(*u*) is a prefix of *label*(*s*). A notable characteristic of the *Dewey ID* is that the ancestors of an element can be determined solely from its label. For instance, if an element *u* is labeled "1.2.3.4", its parent can be identified as "1.2.3" and its grandparent as "1.2", and so forth. Given this property, it can be further posited that if the names of all ancestors of *u* can be derived exclusively from *label*(*u*), then XML path pattern matching can be effectively reduced to string matching. For example, if the label "1.2.3.4" corresponds to the path "a/b/c/d", it becomes relatively straightforward to ascertain whether the element conforms to a specified path pattern, such as "//c/d".

6.3 Fuzzy Spatiotemporal Attributes Processing

6.3.1 *Extended Dewey Code for Fuzzy Spatiotemporal XML Data*

Dewey code demonstrates effective performance in delineating structural relationships, specifically parent–child and ancestor–descendant relationships, among nodes within fuzzy spatiotemporal XML documents. It is important to note that the structural relationships present in fuzzy spatiotemporal XML documents differ significantly from those found in conventional XML documents. This distinction arises from the existence of certain fuzzy spatial and fuzzy temporal nodes that are defined by specific time periods and locations; these relationships may dissipate as time and position change. Unfortunately, the current Dewey code lacks the capability to retrieve a node's name from the root to a specific node, necessitating an extension of its functionality. Furthermore, the Dewey code is primarily applicable to general XML data, rendering it inadequate for fuzzy spatiotemporal XML data, which is characterized by its unique spatial and temporal attributes. Consequently, fuzzy spatiotemporal nodes within these documents require individual marking and querying. The previously proposed extended Dewey code does not adequately address the complexities of fuzzy spatiotemporal data, indicating a need for further enhancements.

Definition 6.1 In the context of XML, a label denoted as t is associated with the set $CT\ (t) = \{t_0, t_1, \ldots, t_{n-1}\}$, which represents all child nodes of t derived from the DTD structure of the XML document. The notation $label(u)$ ($= label\ (s){\cdot}x$) is employed to indicate the code corresponding to node u, where s is identified as the parent node of u. The value of x can be determined through the following computation:

$$\text{If } u \text{ is a fuzzy spatiotemporal XML data:}$$

$$x = \begin{cases} 01;\ \textit{fuzzy spatiotemporal point} \\ 10;\ \textit{fuzzy spatiotemporal line} \\ 11;\ \textit{fuzzy spatiotemporal region} \end{cases}.$$

Let us consider the case where u represents the *k-th* tag of a node t_s, with k taking values from 0 to n – 1. In this context, the Dewey code associated with the left non-spatiotemporal sibling of node u is denoted as $label(s){\cdot}y$.

$$x = \begin{cases} -1; & \textit{text node} \\ k; & \textit{left most child} \\ \lfloor\left(\frac{y}{n}\right)\rfloor \times n + k\ ; & (y \bmod n) < k \\ \lfloor\left(\frac{y}{n}\right)\rfloor \times n + k\ ; & (y \bmod n > k) \end{cases}$$

The representation of the spatiotemporal node requires the utilization of two integer values:

- 01 denotes fuzzy spatiotemporal point.
- 10 denotes fuzzy spatiotemporal line.
- 11 denotes fuzzy spatiotemporal region.

Utilizing a finite state transducer (FST), it is possible to convert the code into sequences of element names that extend from the root to a specified node, following the provision of the extended Dewey code for that particular node.

Definition 6.2 FST is formally represented by a 5-tuple denoted as $(I, S, \vartheta, \delta, \Psi)$, where:

- $I = Z \cup \{-1\}$, Z is non-negative integer set;
- $S = \Sigma \cup \{\text{PCDATA}\}$, PCDATA represent the state of text;
- Initial state of ϑ corresponds to the name of the root element in XML document;
- The function of state translating is defined as follows: for $\forall\, t \in \Sigma$, if $x = 01$, then $\delta\ (t,x) = \text{fstp}$, if $x = 10$, then $\delta\ (t,x) = \text{fstl}$, if $x = 11$, then $\delta\ (t,x) = \text{fstr}$, else, if $x = -1$, then $\delta\ (t,x) = \text{PCDATA}$; else $\delta\ (t,x) = t_k$, $k = x \bmod n$;
- Ψ illustrates the name of current state.

For instance, let us examine the fuzzy spatiotemporal XML fragment depicted in Fig. 6.1. In accordance with the DTD, we utilize the extended Dewey scheme 0 to denote the element labeled *val*. The label for the element *c*, which is a child of *val*, can be determined through the following process: *val* contains four child nodes, with *c* being the third child. Here, we define $k = 2$ and $n = 4$. If the most recent left non-fuzzy spatiotemporal sibling of *c* is assigned the label 0.0, we find that $y = 0$ and $y \bmod 4 = 0$, which is less than *k*. Consequently, $x = \lfloor 0/4 \rfloor \times 4 + 2 = 2$. Given that the element *fstp* represents a fuzzy spatiotemporal point, we conclude that $x = 01$.

Definition 6.3 In a fuzzy spatiotemporal XML document, nodes *p* and *c* represent two distinct fuzzy spatiotemporal entities. The notation ℑ(p) is used to indicate the spatiotemporal range associated with node *p*, while ℑ(c) represents the spatiotemporal range of node *c*. When the spatiotemporal range of node *c* is encompassed within the spatiotemporal range of node *p*, this relationship is expressed as ℑ(c) IO ℑ(p).

```
<!ELEMENT root (val*)>
<!ELEMENT val (a*, fspp*, b*, c*)>
<!ELEMENT a (#PCDATA)>
<!ELEMENT fstp (#PCDATA)>
<!ELEMENT b (#PCDATA)>
<!ELEMENT c (#PCDATA)>
```

(a) DTD for fuzzy spatiotemporal XML

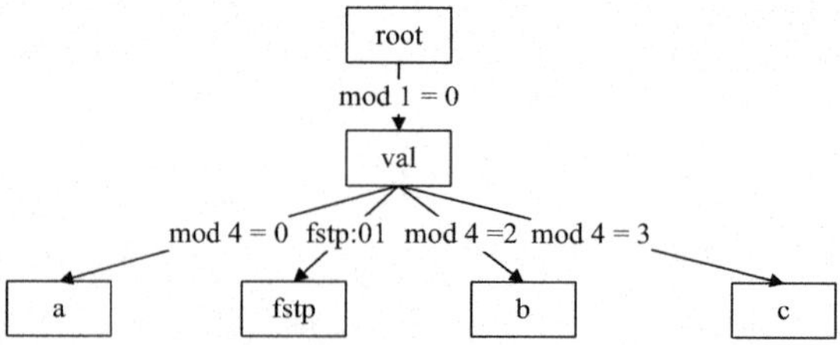

(b) A sample FST for DTD

Fig. 6.1 DTD and FST of a fuzzy spatiotemporal XML fragment

Lemma 6.1 *In a fuzzy spatiotemporal XML document, nodes p and c represent two distinct fuzzy spatiotemporal entities. If node p is identified as the ancestor of node c, it follows that the spatiotemporal range relationship between these two nodes can be expressed as* $\Im(c)$ *IO* $\Im(p)$.

Proof The structural relationship between two fuzzy spatiotemporal nodes, distinct from the relationship observed between crisp nodes, is contingent upon the spatiotemporal characteristics inherent to the fuzzy spatiotemporal nodes. These nodes are defined within specific temporal and spatial contexts, ceasing to exist when the time falls outside the designated temporal attributes or when the spatial location is outside the defined spatial attributes. Consequently, the structural relationship between the two nodes is rendered non-existent upon the disappearance of the spatiotemporal nodes.

Based on the established initial condition, it is posited that node *p* serves as the ancestor of node *c*. We assume that the spatiotemporal range of node *c* is not encompassed within the spatiotemporal range of node *p*, denoted as $\Im$(c) not IO $\Im$(p). Consequently, there exists a specific time period or spatial area within $\Im$(c) that is not included in $\Im$(p), due to the aforementioned relationship. Under these circumstances, node *c* is excluded from the fuzzy spatiotemporal XML document as a result of the absence of node *p*. The removal of node *c* consequently results in inconsistencies within the data of the fuzzy spatiotemporal XML document.

Based on the preceding analysis, if node *p* is identified as the ancestor of node *c*, then the spatiotemporal relationship between node *p* and node *c* can be expressed as $\Im$(c) IO $\Im$(p). □

Definition 6.4 In a fuzzy spatiotemporal XML document, nodes *i* and *n* are identified as two distinct entities. The notation $\Re(root, n)$ represents the path from the root node to node *n*, while the Dewey codes for nodes *i* and *n* are denoted as D_i and D_n, respectively. $D_i = D_n.\ t^*$ indicates that $\Re\ (root, i)$ subsume $\Re\ (root, n)$, which is formally represented as $\Re\ (root, n) \precsim \Re\ (root, i)$.

Lemma 6.2 *Let* D_i *and* D_n *represent the Dewey code associated with nodes i and n. If* $\Re\ (root, n) \precsim \Re\ (root, i)$, *then* $\Re\ (root, n)$ *is the sub graph of* $\Re\ (root, i)$.

Proof In the context of fuzzy spatiotemporal XML, it is feasible to obtain the corresponding positional information for any node utilizing the extended Dewey scheme. The relationship between D_i and D_n can be expressed as $\Re\ (root, n) \precsim \Re\ (root, i)$, where the relationship is denoted by the symbol $D_i = D_n.\ t^*.\ t^*$, which may represent either a character or a string.

Let $t^* = k$, indicating that t^* represents a character. Consequently, it follows that $Node_n$ serves as the parent of $Node_i$. It is evident that $\Re\ (Node_{root}, Node_n) < \Re\ (Node_{root}, Node_i)$; therefore, $\Re\ (root, n)$ constitutes a subgraph of $\Re\ (root, i)$.

Let $t^* = (k_1, k_2, \ldots, k_m)$, indicating that t^* represents a string. Consequently, it follows that $Node_n$ serves as the ancestor of $Node_i$.

The notation $\Re$ (root, n) < $\Re$ (root, i) indicates that VS(root, n) $\subseteq$ VS(root, i) and ES(root, n) $\subseteq$ ES (root, i). In this context, VS $(root,\alpha)$ and ES $(root,\alpha)$ represent the collections of nodes and edges, respectively, that connect the root to the node. Moreover, it is evident that $\Re$ $(root, n)$ constitutes a subgraph of $\Re$ $(root, i)$, signifying that $Node_i$ is the descendant of $Node_n$.

As a result of the aforementioned analysis, if $D_i = D_n.t^*$, then $\Re$ $(root, n)$ constitutes a sub graph of $\Re$ $(root, i)$. □

Lemma 1 and Lemma 2 elucidate significant characteristics of the labeling scheme, which can be effectively employed to expedite the assessment of structural relationships between two fuzzy spatiotemporal nodes, including parent–child and ancestor–descendant relationships.

6.3.2 Membership Evaluation

In fuzzy spatiotemporal XML documents, the fuzziness associated with child nodes contribute to the fuzziness of parent nodes. We employ probability to represent the fuzziness of a fuzzy node, while the overall fuzziness of a parent node can be assessed from a conditional probability perspective.

Definition 6.5 Consider a fuzzy twig fragment F comprising i nodes $(v_0, v_1, \ldots, v_{i-1})$. The fuzziness of this fragment is represented through probabilities. The overall fuzziness of F, denoted as $p_{whole}(F)$, is determined by the product of the individual probabilities associated with each node present in F. This relationship can be expressed mathematically as follows:

$$\begin{aligned} p_{\text{whole}}\,(\mathrm{F}) = \mathrm{p}(v_0, v_{12}, \ldots, v_{i-1}) \\ \mathrm{P}(v_0) \times \mathrm{p}(v_1, v_2, \ldots, v_{i-1}|\mathrm{v}_0) \\ \mathrm{P}(v_0) \times \mathrm{p}(v_1|\mathrm{v}_0) \times \mathrm{p}(v_2, \ldots, v_{i-1}|\mathrm{v}_0\mathrm{v}_1) \\ \mathrm{P}(v_0) \times \mathrm{p}(v_1|\mathrm{v}_0) \times \mathrm{p} \cdot (v_2, \ldots, v_{i-1}|\mathrm{v}_0\mathrm{v}_1) \\ \times \ldots \times \mathrm{p}(v_{i-1}|\mathrm{v}_0\mathrm{v}_1\mathrm{v}_2, \ldots, \mathrm{v}_{i-2}) \end{aligned}$$

In a fuzzy spatiotemporal XML tree, four distinct types of fuzzy structures are illustrated in Fig. 6.2.

- Type-1: fuzzy nodes that are exclusively present within a singular path.
- Type-2: fuzzy nodes emerge within the branches, and the nodes exhibit independence from one another.
- Type-3: fuzzy nodes are present within the branches, and these nodes convey information pertaining to spatiotemporal points.
- Type-4: fuzzy nodes emerge in a complex structure.

In the context of fuzzy spatiotemporal nodes, the computation of probability diverges from that of crisp nodes. Specifically, when a node is identified as a fuzzy

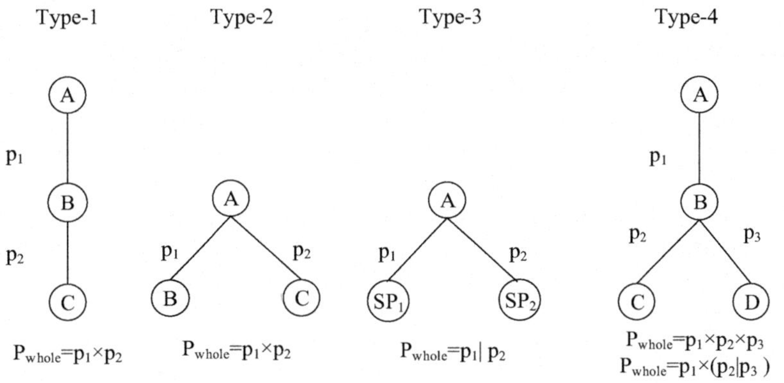

Fig. 6.2 Whole probability in different twig structures

spatiotemporal point, it encompasses multiple potential nodes, each characterized by varying degrees of accuracy that reflect their likelihood of being an actual node. Should the query node correspond to any node within this set, the probability associated with the fuzzy spatiotemporal point is equivalent to the probability of the matched point. In scenarios where events p_1 and p_2 are mutually exclusive, only one can occur at any given time, leading to the conclusion that the overall probability p_{whole} corresponds to the probability of the event that has transpired. For fuzzy spatiotemporal lines, the probability is determined using a methodology analogous to that employed for spatiotemporal points. Conversely, when dealing with fuzzy spatiotemporal regions, the probability fluctuates in accordance with the variations in each concentric ring within the entire region. The aggregate probability can be approximated by considering the probability at the center of the circular region. Furthermore, any intricate fuzzy structures can be understood as combinations of these fundamental elements.

6.3.3 Fuzzy Spatiotemporal Attributes Matching

Fuzzy spatiotemporal data can be categorized into three distinct types: fuzzy spatiotemporal points, fuzzy spatiotemporal lines, and fuzzy spatiotemporal regions. However, the conceptual frameworks for these categories are not adequately compatible with XML representation. In response to this limitation, we propose a methodology for encoding fuzzy spatiotemporal points, lines, and regions within an XML framework as outlined below:

- A fuzzy spatiotemporal point is characterized by a 3-tuple denoted as $\{(x, y), (t_0,t_1), \varepsilon\}$. In this representation, (x, y) indicates the spatial coordinates at a specific time denoted by (t_0,t_1). The notation (t_0,t_1) signifies the temporal point at which

$t_0 = t_1$ and the corresponding time interval during which $t_0 < t_1$. The variable ε quantifies the likelihood that the point corresponds to the anticipated location.

- In the context of fuzzy spatiotemporal lines, the inherent curvature of the fuzzy line necessitates its division into multiple linear segments. For the purpose of simplification, we will focus on the analysis of a single straight segment. This line can be represented as a 3-tuple $\{(x_1,y_1), (x_2,y_2),\varepsilon\}$, where (x_1,y_1) and (x_2,y_2) denote the coordinates of the starting and ending points of the line, respectively, and ε represents the probability that this line corresponds to the anticipated line.
- A fuzzy spatiotemporal region can be conceptualized as a circular area composed of multiple concentric rings that encircle a central core. The probability associated with this region diminishes as the radius of the rings increases. This region can be mathematically represented by an $(n + 1)$-tuples denoted as $\{(x, y), \{\varepsilon, r\}_n\}$, where (x, y) signifies the core of the region. The notation $\{\varepsilon, r\}_n$ refers to the n-th pair of tuples $\{\varepsilon, r\}$, in which ε denotes the degree of fuzziness of the ring region, while r represents the radius of the respective ring.

The querying of fuzzy spatiotemporal XML data necessitates a nuanced approach to matching fuzzy spatiotemporal attributes, which is both critical and intricate. Various categories of fuzzy spatiotemporal attributes, such as fuzzy spatiotemporal points, fuzzy spatiotemporal lines, and fuzzy spatiotemporal regions, require distinct methodologies for effective processing.

When fuzzy spatiotemporal attributes are utilized to characterize a fuzzy spatiotemporal point, a set of points is generated, each associated with a probability that indicates the potential location of the anticipated node. In the context of a query twig that incorporates fuzzy spatiotemporal points with associated probabilities, the initial step involves comparing these probabilities to eliminate nodes that do not meet the specified probability criteria. Subsequently, the positional information of the query node is compared with that of the remaining nodes. If a positional match is identified, the corresponding node may be considered a potential query result, with the probability of this result being equivalent to the probability associated with the match between the identified node and the expected node.

For instance, consider the point set characterized by the following probabilities: $\{\{(3, 4), 0.8\}, \{(2, 7), 0.4\}, \{(4, 2), 0.5\}, \{(1, 4), 0.7\}\}$, as illustrated in Fig. 6.3. In this scenario, a spatiotemporal query node is defined with a position of (3, 4) and an associated probability of 0.6. In the first step, two nodes are eliminated based on a comparison of their probabilities. The information pertaining to the remaining nodes is depicted in Fig. 6.3b. Although the probabilities of these nodes meet the required threshold, their positions do not all conform to the specified criteria. In the second step, nodes that fail to meet the positional requirements are filtered out following a comparison between the query node and the remaining nodes. The outcome of this filtering process is presented in Fig. 6.3c. Ultimately, the search yields the anticipated fuzzy spatiotemporal node, with the actual probability of the identified node being 0.8.

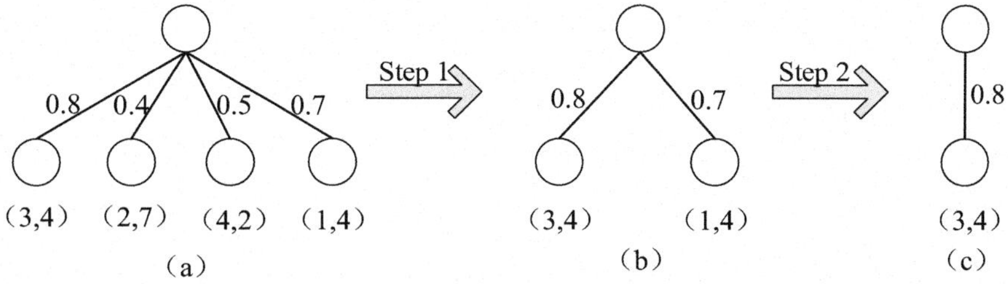

Fig. 6.3 The process of spatiotemporal point query

When fuzzy spatiotemporal attributes are utilized to represent a spatiotemporal line, it is possible for the line to exhibit curvature and irregularity. Such a curved line can be decomposed into multiple shorter straight segments. To ascertain the equivalence of two curved lines, one can evaluate the congruence of their respective short segments. For the sake of simplification, we can treat the spatiotemporal line as a straight line defined by its starting and ending points. The determination of whether two lines are identical can then be made by comparing the respective starting and ending points for congruence.

For the spatiotemporal region, each region is characterized by its core, boundary, and associated probability. Should a region meet the specified query criteria, it may be classified within the fuzzy spatiotemporal region.

Lemma 6.3 *Consider a spatiotemporal region of XML data characterized by its core o_1, boundary r_s, and similarity, denoted as follows: α-from $r = 0$ to $r = r_1$, β-from $r = r_1$ to $r = r_2$, γ-from $r = r_2$ to $r = r_s$. A query region R_b is defined by its core o_2, boundary r_q, and a probability η. If this query region meets the specified criteria for the region query, then the conditions $r_s > r_q$ and $|r_s - r_q| \geq o_1o_2$ hold true.*

Proof Initially, we assume that $r_s < r_q$. It is evident that the query region R_b is not contained within the specified region R_a. Consequently, the hypothesis cannot be validated due to the failure to meet the established query criteria.

We proceed under the assumption that $|r_s - r_q| < o_1o_2$. It is evident that the core of R_b must reside within R_a when the specified query criteria are met. Let $|r_s - r_q| = o_1o_2 + d$, where d represents the smallest distance between the boundaries of R_a and R_b. If the initial assumption holds true, then the distance $d = |r_s - r_q| - o_1o_2 < 0$. Based on fundamental principles, a distance cannot be negative. Consequently, this leads to the conclusion that the initial assumption is not valid.

Overall, it is understood that if the region R_a meets the criteria for the region query R_b, then the conditions $r_s > r_q$ and $|r_s - r_q| \geq o_1o_2$ must be satisfied. □

6.4 Holistic Matching Algorithms

6.4.1 Notations

In this chapter, Q represents a twig query pattern, while P_f signifies the path pattern extending from the root of Q to the node f ($f \in Q$). The algorithm FSPTwigFast employs several operations on the query node q, including: leafNodes(q), which retrieves the set of leaf query nodes within the sub-twig query pattern rooted at q; isFSP(q), which assesses whether node q qualifies as a fuzzy spatiotemporal node; parent(q), which identifies the parent query nodes of q; and isLeafNode(q), which determines if node q is a leaf node. The FSPTwigFast algorithm establishes an association between each leaf query node f within the twig query pattern and a corresponding stream T_f, while also linking each query node s to a list L_s. The elements within the list L_s are represented as a 2-tuple (element e, an array of pointers). The operations applicable to this list T_f include: eof (T_f), which checks if the stream pointer T_f has reached its end; advance (T_f), which progresses the stream pointer to the subsequent node; delete (T_f), which removes the current element indicated by the head pointer; and getElement (T_f), which retrieves the current element referenced by the head pointer. Furthermore, the operations on the list L_s include: addtoList (L_s, element a, pointers), which appends the 2-tuple (element a, pointers) to the list L_s; and addPointer (L_p, tuple_p, pointer to tuple_f), which adds a pointer to tuple_f to the pointer array of the 2-tuple tuple_p within the list L_p.

6.4.2 FSPTwigFast

The FSPTwigFast algorithm is designed to address fuzzy temporal and fuzzy spatial phenomena in the context of querying fuzzy spatiotemporal XML data. It utilizes the DBL function to ascertain the processing order of leaf nodes and constructs a result list that progresses from the leaf nodes to the root, minimizing the number of intermediate results. This algorithm is capable of computing responses to query path patterns within fuzzy spatiotemporal XML data and operates as a single-phase algorithm. It invokes the locateMatchedLabel function to filter the stream T_f of leaf node f (as indicated in Lines 1–2). The locateMatchedLabel function is responsible for eliminating spatiotemporal nodes that do not conform to the query criteria and for filtering out elements that do not align with the specified path pattern P_f. Subsequently, the FSPTwigFast algorithm calls the DBL function (TopBranchingNode) to obtain an array of sorted leaf nodes corresponding to the TopBranchingNode (as shown in Line 3). Line 6 processes the leaf query node f and establishes output lists for the elements in T_f at Lines 7–9.

The DBL sort plays a crucial role in the FSPTwigFast algorithm, as it establishes the sequence in which leaf query nodes are processed. The function DBL(n) retrieves all branch nodes b and leaf nodes f within the sub-twig query tree that is rooted at node n. The nodes b and f are characterized by the absence of any intervening branch nodes along the path from the root node n to both b and f.

Algorithm 6.1 FSPTwigFast

```
for each f ∈ leafNodes(root)
   locateMatchedLabel(f)
SortNodeArray[] = DBLSort(TopBranchingNode)
i = 0
while (¬end(root) and i < SortNodeArray.length)
   f = SortNodeArray[i]
while (¬eof(T_f))
   buildList(f, getElement(T_f), parent(f))
   advance(T_f)
i++
outputSolutions()
```

Function locateMatchedLabel(f)

/* Assume that the path from the root to element getElement(T_f) is $n_1/n_2/.../n_k$ and P_f denotes the path pattern from the root to leaf node f */

```
while (¬eof(T_f))
if(¬((n_1/n_2/.../n_k matches pattern P_f) ∧ fitness(n_k, f)))
   delete(T_f)
else
   advance(T_f)
reset the pointer of T_f to the first element of T_f
```

Function fitness(e,f)

```
if(isFSP(f))
   if(!isFSP(e))
      return false // α is the similarity permitted by query pattern
   else
      if (f is a fuzzy point node)
       if (∃f_k satisfies f_k · x_k = e · x_0  && f_k · y_k = e · y_0 ∧ ε > α)
        return true
        else return false
      if (f is a fuzzy line node)
      if (∃f_k satisfies (f_k · x_0, f_k · y_0) = (e · x_0, e · y_0) && (f_k · x_1, f_k · y_1) = (e · x_1, e · y_1) ∧ ε > α)
        return true
       else return false
      if (f is a fuzzy region node) //r_max is the maximum radius of f which satisfies ε > α
      if ((r_max − r_e ) > d((f · x_0, f · y_0), (e · x_0, e · y_0)))
          return true
         else return false
else if(e == f)
   return true
else return false
```

Function DBLSort(n)

```
/* At the beginning, the value nodeCount(n) for each query node n is zero */
01  if (isLeafNode(n))
02      locateMatchedLabel(n)
03      nodeCount[n] = T_n.length
04  else
05      s = L_DBL(n)
06      for each node a in list s
07          DBLSort(a)
08      sort the list s by the value nodeCount[a] ascending for each a in s
09      nodeCount[n] is the minimal node count in list s
```

Consider a twig query Q and an XML database D. For a given query node $n \in Q$, we assign a variable nodeCount to quantify the number of XML nodes that correspond to n. In the case where n is a leaf node (as indicated in Line 1), nodeCount is determined by the total number of elements in T_f (as detailed in Lines 2–3). Conversely, if n is not a leaf node, nodeCount is defined as the minimum value of nodeCount across all query nodes within DBL (n) (as outlined in Lines 5–9).

Example 6.1 In Fig. 6.4, the outcomes of DBL (a) are represented by nodes b and c, serving as the branching node. The results of DBL(c) yield nodes f and g. It is assumed that nodeCount(b) is 100, nodeCount(f) is 30, and nodeCount(g) is 400. The order of processing the query nodes for DBL (a) is c followed by b. The nodeCount(c) for the branching node c is 30, which corresponds to the minimum value between $\{f, g\}$. For DBL(c), the processing sequence is f and g subsequent to the DBL sorting. Consequently, the DBL sequence for the leaf query nodes is f, g, *and* b, following the DBL sorting for the twig query pattern Q. This represents the order of processing the leaf nodes within the FSPTwigFast algorithm.

In the FSPTwigFast algorithm, Line 8 invokes the buildList function to generate the result list for each query node. The buildList function serves as the fundamental component of the entire algorithm, as detailed below.

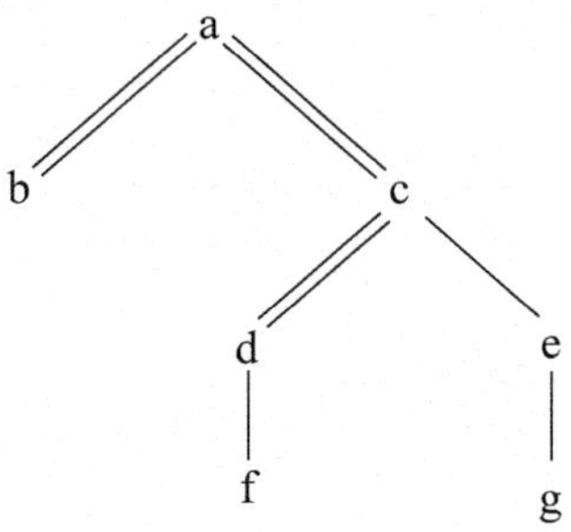

Fig. 6.4 Query twig pattern Q

Function buildList(*f*, *e*, *p*)

```
01 if (f == SortNodeArray[0])
02    if (tuple(e, null) ∉ L_f)
03       addtoList(L_f, e, null)
04 else // assume the index of f in SortNodeArray[] is i
05    a = closestCommonAncestor (SortNodeArray[i],SortNodeArray[i − 1])
06    if (P_a matches P_f)
07       if (tuple(e, null) ∉ L_f)
08          addtoList(L_f, e, null)
09 buildInternalList(f, e, p, pointer to the tuple(e, null))
```

In the algorithm FSPTwigFast, Lines 1–18 are dedicated to the creation of a result list for leaf query nodes. Line 9 invokes the function buildInternalList to generate the result list for internal query nodes. If the node f is the first element in the SortNodeArray and the 2-tuple (e, null) is not already present in the list, the algorithm appends the 2-tuple (e, null) as indicated in Lines 1–3. Conversely, if f is not the first node in the SortNodeArray (as noted in Line 4), the algorithm will add the tuple (e, null) to the list in Line 8, provided that certain conditions are met.

- The list L_a of the closest common ancestor nodes of f and the preceding query node of f within the SortNodeArray has included the 2-tuple (element e_2, pointers) as indicated in Line 5.
- The 2-tuple (element e_2, pointers), along with its ancestor nodes, aligns with the specified path query pattern P_f corresponding to e (refer to Line 6).
- There is an absence of the tuple (e, null) in L_f (Line 7).

The function buildInternalList is invoked within the function buildList to generate result lists for inner nodes subsequent to the creation of result lists for leaf nodes, as illustrated below:

Function buildInternalList(*f*, *e*, *p*, tuple_*f* (*e*, pointers))

```
/* let s be a set of element b that is an ancestor of e such that b can be match node p in the path
solution of e to path pattern P_f */
01 if (p != null)
02    for each b ∈ s
03       if (tuple_p(b, pointers array_p) ∈ L_p ∧ P_p matches P_f)
04          addPointer(L_p, tuple_p, pointer to tuple_f)
05       else
06          addtoList(L_p, b, pointer to tuple_f)
07          buildInternalList (p, b, parent(p), tuple_new)
```

In the function buildInternalList, the variable s represents the set of elements b that serve as ancestor nodes of e. The element b may correspond to node p within the path solution of e in the path pattern P_f. For each element b in the set s (as indicated in Line 2), if a tuple tuple_p (b, pointers array_p) exists, then this tuple, along with its ancestor nodes, aligns with the path query pattern associated with e (as noted in Line 3). Subsequently, a pointer referencing tuple_f is incorporated into the pointer array of tuple_p (as shown in Line 4). If no tuple_p (b, pointers array_p) is found in the list, a new tuple (b, pointer to tuple_f) is appended to the list L_p (as described in Line 6). Finally, the function buildInternalList is invoked recursively (as indicated in Line 7).

Example 6.2 In the context of the fuzzy spatiotemporal XML twig query pattern illustrated in Fig. 6.5, the leaf query nodes are designated as C and D. The associated streams for these nodes are represented as $T_c = \{c_1, c_4\}$ and $T_d = \{d_1, d_2, d_3, d_4\}$, respectively. The labels corresponding to node C are identified as $\{c_1, c_2, c_3, c_4\}$. c_2 has been excluded following a comparison of its similarity against a predetermined threshold. Additionally, c_3 has been filtered out due to its lack of alignment with the position of the query node C. The TopBranchingNode within the twig query pattern is identified as B. As depicted in Fig. 6.5, the parent–child relationships are established as parent(C) = B, parent (D) = B, and parent (B) = A. Consequently, the set DBL (B) is defined as $\{C, D\}$. In the algorithm FSPTwigFast, Line 3 invokes the function DBLSort (TopBranchingNode). This results in SortNodeArray[] = $\{C, D\}$, as nodeCount(C) = 2 and nodeCount(D) = 4.

The subsequent following presents a comprehensive description of the procedure for constructing the output list, as illustrated in Fig. 6.6.

Initially, the first leaf query node, denoted as f, is identified as C in Line 6. For each element within the stream T_c, the function buildList is invoked. During the initial execution of buildList, the function getElement (T_c) returns c_1, as c_1 represents the current element of T_c. A 2-tuple (c_1, null) is appended to the list L_c, as C represents the first node in the SortNodeArray and no prior 2-tuple (c_1, null) exists in the list L_c. Subsequently, Function buildInternalList(C, c_1, B, tuple_f(c_1, null)) is called to construct the internal query node list. It is evident from Line 2 in buildInternalList that $S = \{b_1\}$. Given that L_b is null, a new 2-tuple tuple_new (b_1, there is only one

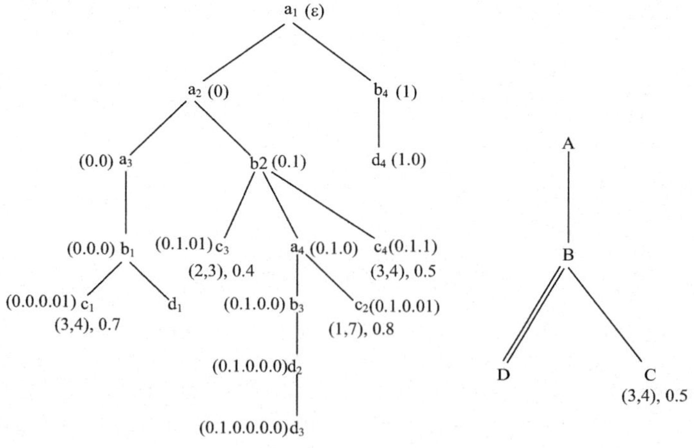

Fig. 6.5 The example of fuzzy spatiotemporal XML twig query

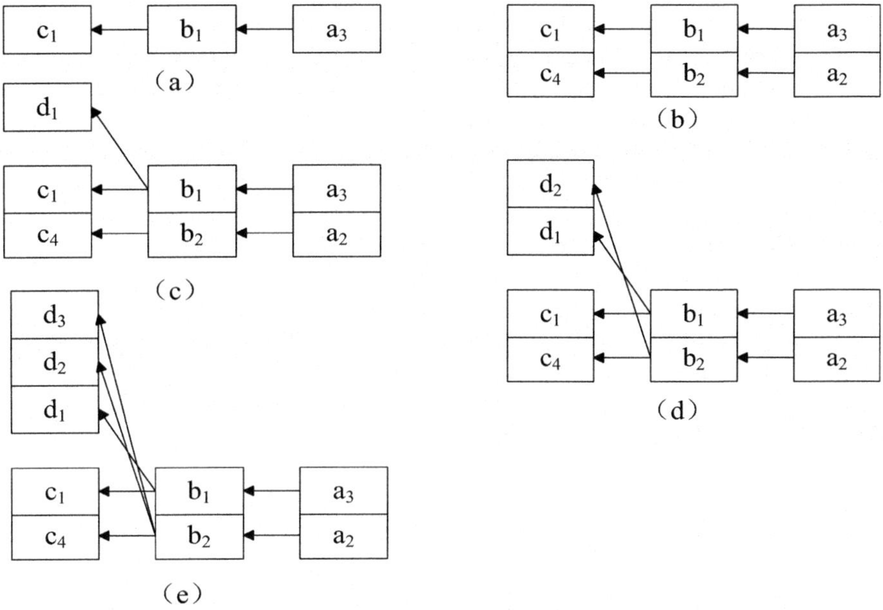

Fig. 6.6 List structure in the process of twig query pattern

pointer that points to tuple_f) is added. Following the completion of the result list for the leaf node f, the function buildInternalList(B, b_1, A, tuple_new) is recursively invoked to generate the result list for the inner nodes. A 2-tuple (a_3, there is only one pointer that points to tuple_new in pointer array) is incorporated into the list L_a, as $S = \{a_3\}$ and L_a remains empty and is null. The structural representation of the list is illustrated in Fig. 6.6a.

Secondly, the current element is designated as c_4 within the stream T_c during the invocation of the buildList function. This process mirrors that of c_1; wherein a 2-tuple (c_4, null) is incorporated into L_c. Following this, we execute buildInternalList(C, c_4, B, tuple_f(c_4, null)) function to introduce a new 2-tuple tuple_new(b_2, there is only one pointer that points to tuple_f in pointer array), given that $S = \{b_2\}$. Ultimately, we append another 2-tuple (a_2, there is only one pointer that points to tuple_f in pointer array). The structural representation of the list is illustrated in Fig. 6.6b.

Thirdly, we designate node D as f in Line 6 of FSPTwigFast, following the processing of all nodes within the stream T_f of leaf node C. It is important to note that node D does not occupy the initial position in the SortNodeArray. The preceding node to D in the SortNodeArray is node C, which shares the most recent common ancestor, node B. The 2-tuple denoted as tuple_p (b_1, pointer array), along with its ancestor, corresponds to the path pattern P_d associated with d_1. Consequently, we introduce a 2-tuple, tuple_f (d_1, null), into L_d and incorporate a pointer that directs to tuple_f within the pointer array of tuple_p, given that L_d is currently null. The structural representation of this list is illustrated in Fig. 6.6c.

Fourthly, in the fourth iteration of the buildList function, the current element in T_d is identified as d_2. A 2-tuple, denoted as tuple_f (d_2, null), is incorporated into L_d, mirroring the procedure utilized for d_1. The preceding node of node D is Node C within the SortNodeArray, which shares the most recent common ancestor with node B. The 2-tuple tuple_p(b_2, pointer array), along with its corresponding ancestor node, aligns with the path pattern associated with d_2. Consequently, a pointer directed towards tuple_f is integrated into the 2-tuple (b_2, pointer array) of L_b. The structural representation of the list is illustrated in Fig. 6.6d.

Fifthly, a 2-tuple denoted as tuple_f (d_3, null) is incorporated into L_d during the fifth invocation of the buildList function. A pointer referencing tuple_*f* is subsequently included in the 2-tuple (b_2, pointer array) of L_b. The structural representation of the list is illustrated in Fig. 6.6e.

Finally, we invoke outputSolutions function to obtain the results of the query: (a_3, b_1, d_1, c_1), (a_2, b_2, d_2, c_4), (a_2, b_2, d_3, c_4).

6.4.3 Analysis of FSPTwigFast

Theorem 6.1 FSPTwigFast is capable of accurately retrieving all potential matching results from a fuzzy spatiotemporal XML document D in response to a given twig query Q.

Proof Consider a twig query Q represented as ($A_1, A_2, \ldots, A_n$) with a root node A_1, where $\{A_k, A_{k+1}, \ldots, A_n | 1 \le k \le n\}$ denotes the leaf nodes and $A_j(1 \le j \le n)$ serves as the top branching node. When ($a_1, a_2, \ldots, a_n$) matches ($A_1, A_2, \ldots, A_n$), $a_i(i = k, k + 1, \ldots, n)$ should be included following the filtering process outlined in Lines 1–2 of FSPTwigFast. Subsequently, in Lines 5–10, the algorithm identifies the next leaf query node f based on the results obtained from the DBL sorting procedure. The function buildList is then invoked to generate lists for all elements within T_f. It is important to note that $\{a_1, a_2, \ldots, a_n\}$ is entirely contained within L_{A_i} as a result of the buildList process. For nodes characterized as fuzzy spatiotemporal, those that fail to meet the probability criteria or do not align with the spatiotemporal attributes are eliminated following the filtering of probability and fuzzy spatiotemporal characteristics. Consequently, all nodes within T_f are confirmed to correspond with the leaf query nodes in the twig query pattern. Ultimately, the algorithm invokes the function outputSolutions to present the query result ($a_1, a_2, \ldots, a_n$) that aligns with ($A_1, A_2, \ldots, A_n$). When the FSPTwigFast algorithm produces the output ($a_1, a_2, \ldots, a_n$), it can be inferred that $a_i(i = 1, 2, \ldots, n)$ is included in the list L_{A_i}. It is evident that a_j must reside along the path from A_1 to A_j, as established during the list construction process. Line 11 of FSPTwigFast subsequently outputs the query result ($a_1, a_2, \ldots, a_n$) that corresponds with the twig query pattern ($A_1, A_2, \ldots, A_n$) thereby confirming that ($a_1, a_2, \ldots, a_n$) indeed matches ($A_1, A_2, \ldots, A_n$). □

Theorem 6.2 Given a twig query Q and a fuzzy spatiotemporal XML document D, the worst-case temporal complexity is characterized as O ($n_1 \times d$). Here, n_1 denotes the total number of leaf nodes in T_f, while n_2 signifies the total number of leaf nodes in T_f subsequent to the execution of MatchedLabel. d represents the maximum between depth of Q and depth of D.

Proof In the FSPTwigFast algorithm, the initial steps, specifically Lines 1–2, are dedicated to filtering all leaf query nodes within the set T_f, resulting in a time complexity of O ($n_1 \times d$). Subsequently, Lines 5–10 involve the construction of a list by invoking the buildList function for each leaf query node in T_f, which also incurs a temporal complexity of O $(n_2 \times d) n_1 > n_2$. Consequently, the overall time complexity of the algorithm is determined to be $(n_1 \times d)$. □

6.5 Experiments

In this subsection, we delineate the experimental findings regarding the evaluations of our proposed querying algorithm across various data sets, including general data sets, temporal data sets, fuzzy spatial data sets, and fuzzy spatiotemporal data sets. Following this, we conduct a comparative analysis of the querying performance across different categories of fuzzy spatiotemporal XML data. Ultimately, we assess the performance of FSPTwigFast in relation to TwigStack [16] and AND/OR-Twig [21].

6.5.1 Experimental Setup

The algorithms utilized in the experiments were implemented using JDK 1.7. All experimental procedures were conducted on a system equipped with a 3.2 GHz Pentium IV processor and 4 GB of RAM, operating on the Windows 7 platform.

For the purposes of this study, we utilize a real-world XML dataset pertaining to hurricane forecasts in the Atlantic [22]. This dataset was selected due to its capacity to represent various forms of XML data, including general XML data (average wind speed in the Atlantic), spatial XML data (hurricane intensity in a specific region), temporal XML data (the duration of windstorms), and spatiotemporal XML data (hurricane intensity in a specific region at a given time). The dataset possesses certain characteristics that facilitate the representation of the inherent fuzziness associated with spatiotemporal XML data. To enhance the dataset's suitability for experiments involving the querying of fuzzy spatiotemporal XML data, we implement a fuzzy processing technique as outlined in [23].

The experimental outcomes vary according to the different types of twigs utilized. Consequently, four query twigs related to general XML data, spatial XML data, temporal XML data, and spatiotemporal XML data are presented in Table 6.1 and illustrated in Fig. 6.7. Furthermore, an assessment of the processing details

Table 6.1 Query twigs in the experiments

Queries	
Q_1	Querying general XML data
Q_2	Querying temporal XML data
Q_3	Querying fuzzy spatial XML data
Q_4	Querying fuzzy spatiotemporal XML data

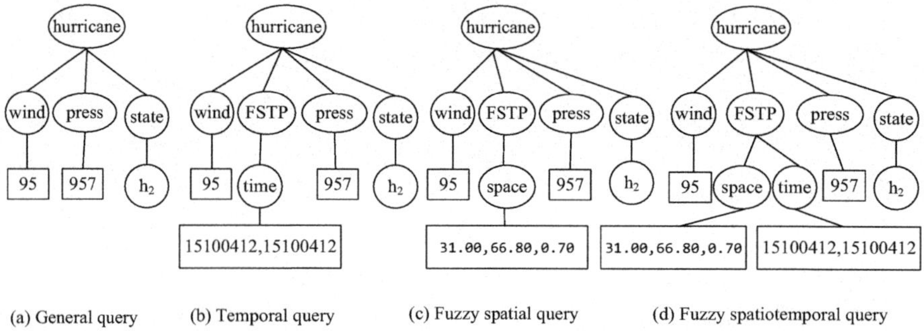

Fig. 6.7 Examples of query twigs

is conducted, along with an analysis of the comparative execution times among the various twigs.

***Q*₁** (twig query of general XML data)

Q_1 represents a twig query applied to general XML data, as illustrated in Fig. 6.7a. This query specifically targets the hurricane identified by the state designation h_2. The hurricane is characterized by static attributes, including a wind speed of 95 and a pressure measurement of 957.

The XML datasets contain numerous wind and pressure nodes. Initially, we identify the corresponding nodes and proceed to sort the leaf nodes. This sorting process is instrumental in minimizing both execution time and memory usage. Nodes that lack a common ancestor with preceding nodes are excluded from the result list. Following this, adjacent leaf nodes are merged in relation to their common ancestor. Ultimately, the ancestor node is traced back to the root node, culminating in the formation of the twig query result through the construction of an inner nodes list.

Q_2 (twig query of temporal XML data)

Q_2 represents a twig query pertaining to temporal XML data, as illustrated in Fig. 6.7b. This query specifically targets the hurricane identified by the state h_2. The hurricane is characterized by static attributes, including a wind speed of 95 and a pressure reading of 957, recorded at 12 o'clock on October 4, 2015.

The methodology for querying general XML data remains consistent with that of querying general XML data. However, the querying of temporal data introduces the inclusion of temporal attributes. These temporal attributes can be categorized into two types: temporal points, where the starting time is equivalent to the ending time, and temporal intervals, where the starting time is less than the ending time. The determination of whether a temporal XML node t (t_{start}, t_{end}) corresponds with a temporal query node q (t_0, t_1) can be established through the following criteria:

(1) If $t_0 = t_1$, this signifies that the temporal attribute denotes a specific temporal point.

1. If $t_{start} = t_{end}$, this indicates that the XML node represents a specific temporal point. Conversely, if $t_0 = t_{start}$, this signifies that the node has been successfully matched.
2. Otherwise, the node is not matched.

(2) If $t_0 < t_1$, this signifies that the temporal attribute denotes a temporal period.

1. If $t_{start} < t_{end}$, this indicate that the XML node represents a temporal period. Conversely, if $t_{start} < t_0$ && $t_1 < t_{end}$, this signifies that the node has been successfully matched.
2. Otherwise, the node is not matched.

(3) Otherwise, the expression of temporal attributes is deemed illegal.

***Q*₃** (twig query of fuzzy spatial XML data)

Q_3 represents a twig query concerning temporal XML data, as illustrated in Fig. 6.7c. This query specifically targets a hurricane identified by the state designation h_2. The hurricane is characterized by static attributes, including a wind speed of 95 and a pressure of 957. The likelihood of occurrence at a longitude of 31 and a latitude of 66.8 is quantified at 0.7.

The twig structure represents a fuzzy spatial query. The methodology for querying general XML data parallels that of querying Q_1. In the context of fuzzy spatial attributes, it is essential to align the fuzzy spatial characteristics and assess their probabilities. Consider the query fuzzy spatial node QFS ($Long_0$, Lat_0, T_0) alongside the fuzzy spatial XML nodes FS_1 ($Long_0$, Lat_0, T_1), FS_2 ($Long_0$, Lat_0, T_2) and FS_3 ($Long_1$, Lat_1, T_3). Both FS_1 and FS_2 share identical spatial coordinates with QFS; however, the temporal conditions indicate that $T_0 < T_1$ and $T_0 > T_2$. This implies that only FS_1 qualifies as a matched node. If we assume $T_1 > T_0$ and $T_3 > T_0$, node FS_3 is excluded from consideration due to the mismatch in spatial information relative to QFS. Consequently, it can be concluded that FS_1 is the sole matched node.

***Q*₄** (twig query of fuzzy spatiotemporal XML data)

Q_4 represents a twig query pertaining to fuzzy spatiotemporal XML data, as illustrated in Fig. 6.7d. This query twig integrates fuzzy spatial and temporal information, specifically targeting a hurricane identified by the state h_2. The hurricane is characterized by static attributes, including a wind speed of 95 and a pressure of 957, occurring at 12 o'clock on October 4, 2015. Additionally, it possesses a probability of occurrence at a longitude of 31 and a latitude of 66.8, with an associated probability value of 0.7. The querying process for Q_4 is a composite of the methodologies employed in Q_1, Q_2 and Q_3.

6.5.2 Experimental Results

In the initial set of experiments, we evaluate the performance of queries Q_1, Q_2, Q_3, and Q_4 in relation to the variation in the number of twig results, as illustrated in Fig. 6.8. As the number of twig results increases, the *nodeCount* for each leaf node of the twig queries also rises, leading to a corresponding increase in the number of nodes that satisfy the conditions of the twig query nodes. The functions *buildList* and *buildInternalList* are invoked more frequently with the escalation in the number of twig results, which consequently results in an increase in the time expenditure associated with the four twig queries. Among these queries, the fuzzy spatiotemporal twig exhibits a greater number of sub-twigs compared to either the fuzzy spatial twig or the temporal twig. Additionally, the number of sub-twigs for both the fuzzy spatial twig and the temporal twig exceeds that of the crisp twig when the queries are characterized by an equivalent number of twig results. Each sub-twig necessitates the invocation of the *fitness* function and the construction of a result list. Furthermore, since Q_4 is classified as a fuzzy twig query, it may require additional processing time to address the inherent fuzziness of the nodes. Figure 6.8 demonstrates that execution time escalates with the increase in the number of twig results; as more nodes are matched to generate the final results, the time consumed correspondingly rises. A notable observation from Fig. 6.8a is that TwigStack and AND/OR-Twig outperform FSPTwigFast. Conversely, in Fig. 6.8b–d, FSPTwigFast exhibits superior performance compared to TwigStack and AND/OR-Twig. This discrepancy can be attributed to the fact that Q_1 exclusively comprises crisp nodes, whereas Q_2 to Q_4 incorporate temporal, fuzzy spatial, or fuzzy spatiotemporal nodes. Moreover, our algorithm demonstrates enhanced performance in terms of execution time due to reduced time costs associated with an equivalent number of twig results. The primary time expenditure for FSPTwigFast is associated with the *buildList* function. In the context of FSPTwigFast, there is no additional time cost incurred for processing nodes that do not contribute to the final result, as every node processed to construct the result list is guaranteed to be matched to form the final outcome. Consequently, FSPTwigFast is more efficient as the number of twig results increases. The experimental findings indicate that FSPTwigFast exhibits a comparative advantage when dealing with temporal, fuzzy spatial, or fuzzy spatiotemporal nodes.

In the second set of experiments, we analyze the performance of queries Q_1, Q_2, Q_3, and Q_4 in relation to varying data sizes, as illustrated in Fig. 6.9. The FSPTwigFast algorithm necessitates a complete traversal of the XML tree during the querying process. As the data size increases, the number of XML nodes accessed also rises, resulting in increased time consumption. The fuzzy spatiotemporal twigs require more intricate matching criteria; specifically, Q_4 encompasses more conditions than

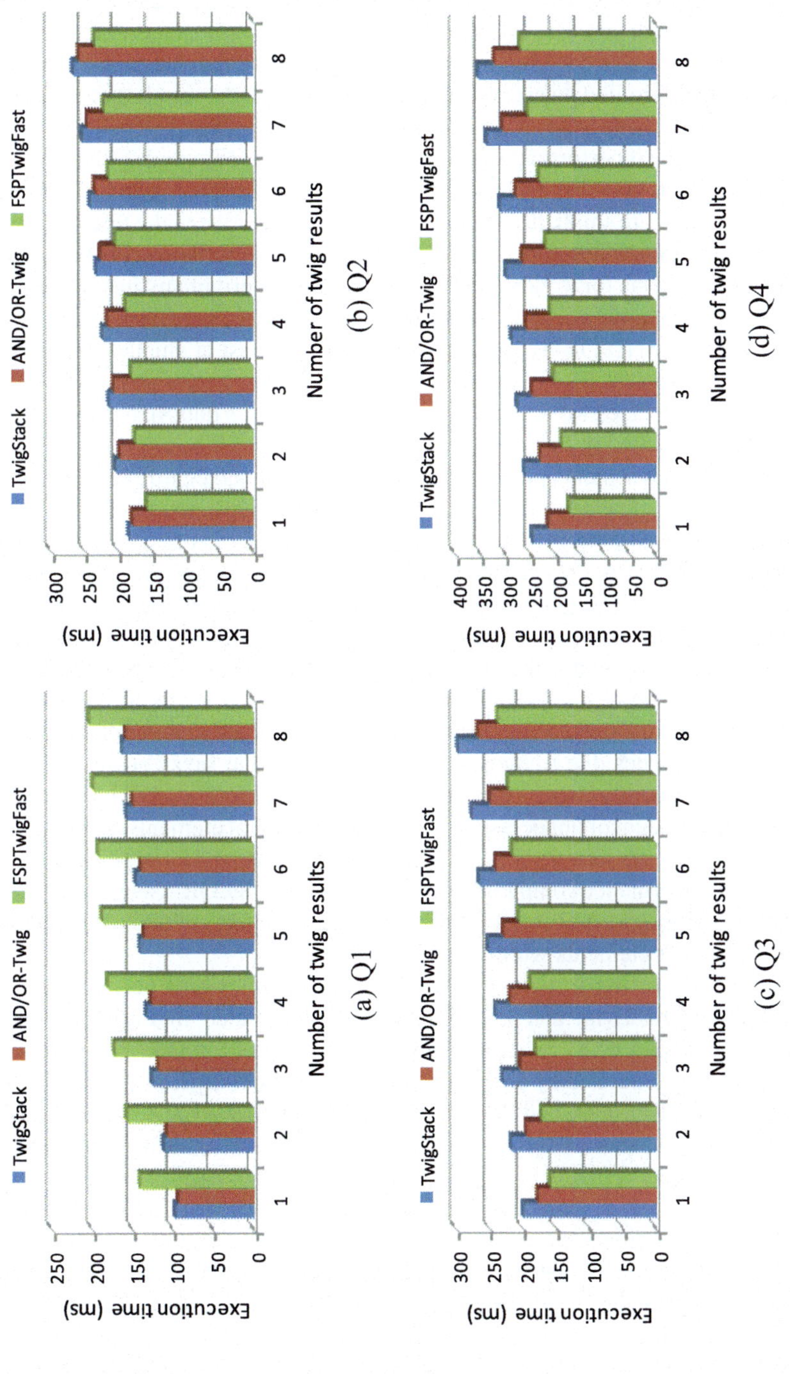

Fig. 6.8 Execution time of different number of twig results

either Q_2 or Q_3, while Q_2 and Q_3 contain more conditions than Q_1. Consequently, a greater number of conditions leads to a higher number of matched nodes, thereby necessitating additional processing time. Furthermore, the fuzzy spatiotemporal twig incurs additional time costs associated with managing data fuzziness. As depicted in Fig. 6.9, the increase in data size correlates with a rise in the number of accessed XML nodes, which in turn leads to greater time expenditure. Notably, FSPTwigFast demonstrates improved performance as data size increases. Although FSPTwigFast initially requires more time to sort leaf nodes compared to TwigStack and AND/OR-Twig when data sizes are small, this time cost diminishes relative to the time required for processing intermediate solutions as data size grows. Importantly, FSPTwigFast does not incur time costs for irrelevant nodes that do not contribute to the final result, as it does not generate intermediate solutions. A clear observation from Fig. 6.9a is that TwigStack and AND/OR-Twig outperform FSPTwigFast at smaller data sizes. However, in Fig. 6.9b–d, FSPTwigFast surpasses both TwigStack and AND/OR-Twig. The execution times for queries Q_1, Q_2, Q_3, and Q_4, as data size varies, align with the results presented in Fig. 6.8.

The third set of experiments presents the empirical findings regarding the number of access elements associated with the utilization of TwigStack, AND/OR-Twig, and FSPTwigFast. As illustrated in Fig. 6.10, the number of access elements required by FSPTwigFast for query Q_1 exceeds that of both TwigStack and AND/OR-Twig. Conversely, for queries Q_2, Q_3, and Q_4, the number of access elements needed by FSPTwigFast is significantly lower than that required by the other two methods. This discrepancy can be attributed to the presence of numerous temporal, fuzzy spatial, or fuzzy spatiotemporal nodes in queries Q_2, Q_3, and Q_4, in contrast to the exclusively crisp nodes present in Q_1. Specifically, FSPTwigFast necessitates the reading of a greater number of nodes due to the inherent structure of the fuzzy spatiotemporal data model. Consequently, when the query pertains solely to crisp data, FSPTwigFast incurs a higher count of unnecessary fuzzy spatiotemporal elements. However, in scenarios where the query involves fuzzy spatiotemporal data, FSPTwigFast demonstrates superior performance compared to TwigStack and AND/OR-Twig, as the latter two methods tend to return all relevant answers, which often include a substantial amount of redundant elements within the twigs.

In the fourth set of experiments, we analyze the memory cost performance of TwigStack, AND/OR-Twig, and FSPTwigFast. As illustrated in Fig. 6.11, both TwigStack and AND/OR-Twig demonstrate superior performance compared to FSPTwigFast when generated by query Q_1. Conversely, FSPTwigFast exhibits better performance than TwigStack and AND/OR-Twig for queries Q_2, Q_3, and Q_4. The memory cost is influenced by the number of accessed elements, which aligns with the findings presented in Fig. 6.10.

In the final series of experiments, we assess the accuracy of the methodologies employed. The evaluations encompass questions Q_1, Q_2, Q_3, and Q_4, as illustrated in Fig. 6.12, with anticipated outcomes aligned with the desired results. In Q_1, where only crisp data is present, the precision scores for TwigStack, AND/OR-Twig, and FSPTwigFast all achieve a value of 1.0. Conversely, in Q_2 and Q_3, the precision of TwigStack and AND/OR-Twig is lower than that of FSPTwigFast. Notably, the

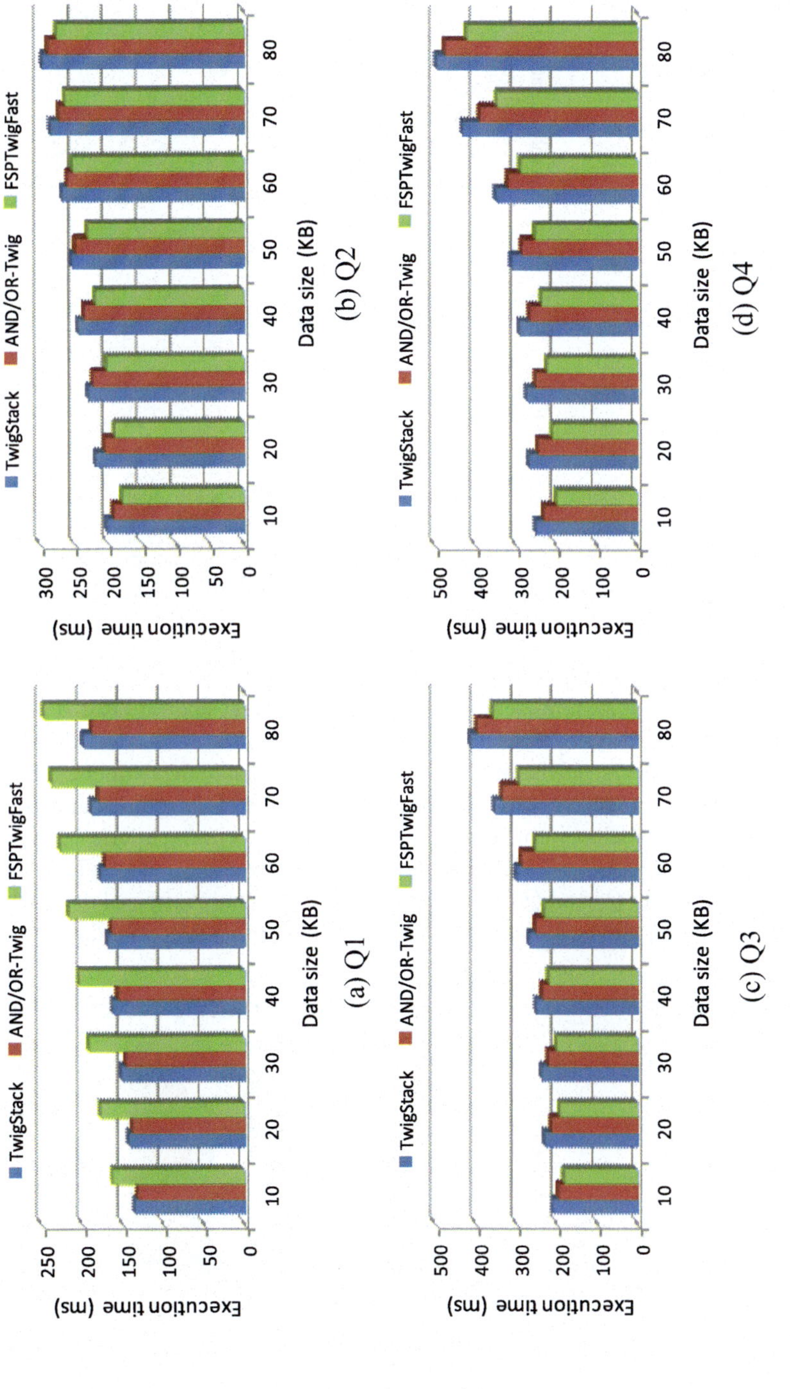

Fig. 6.9 Execution time of different data size

Fig. 6.10 Number of access elements

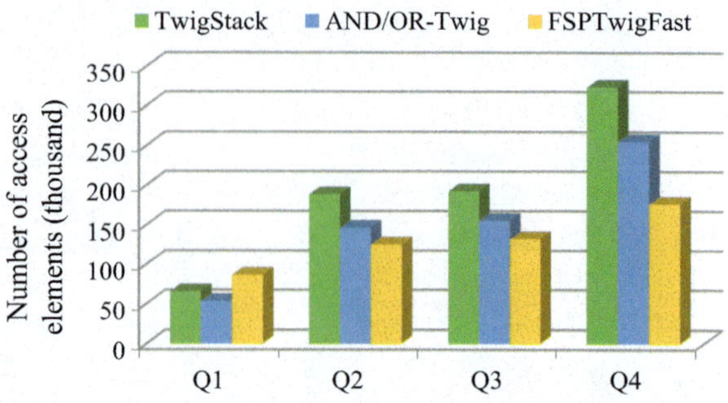

Fig. 6.11 Memory cost of queries

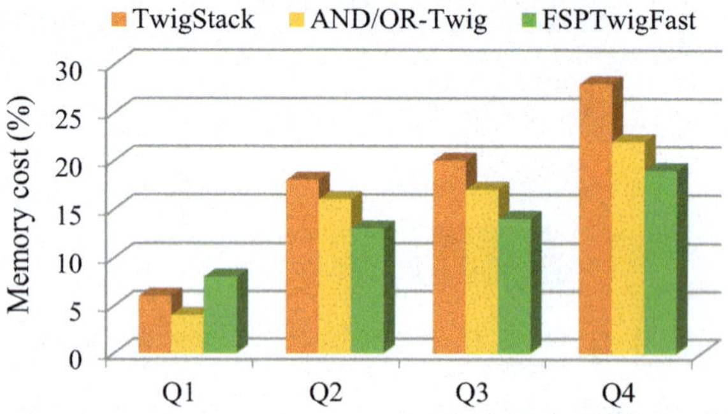

precision of TwigStack remains relatively constant across Q_2 and Q_3, as it is unable to accommodate the querying semantics. In contrast, the precision of AND/OR-Twig in Q_3 surpasses that in Q_2, attributable to its capability to handle fuzzy queries. In Q_4, FSPTwigFast attains a precision of 1.0, which exceeds that of TwigStack and AND/OR-Twig, due to the inclusion of fuzzy spatiotemporal data in the queries. Furthermore, the precision scores for TwigStack and AND/OR-Twig in Q_4 are slightly lower than those observed in Q_2 and Q_3, as Q_4 addresses fuzzy spatiotemporal data, while Q_2 focuses solely on temporal data and Q_3 exclusively on fuzzy spatial data.

Fig. 6.12 Precisions of queries

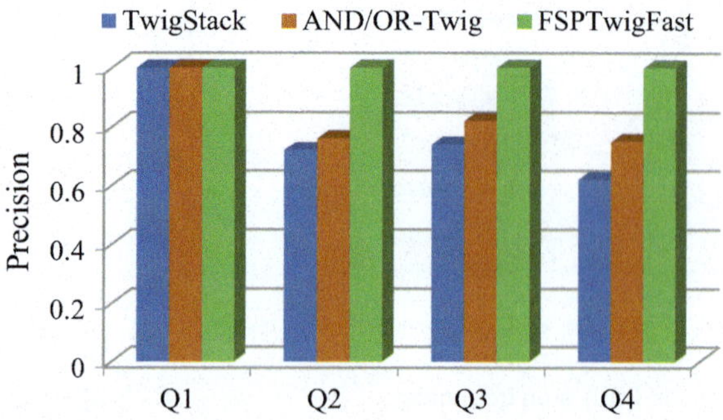

The results from the five experimental groups indicate that, in scenarios where the twig query is composed solely of crisp data, both TwigStack and AND/OR-Twig demonstrate superior performance relative to FSPTwigFast. Conversely, when the twig query incorporates fuzzy spatiotemporal data, FSPTwigFast exhibits a marked advantage over the existing methodologies.

6.6 Summary

In this chapter, we present a methodology for representing fuzzy spatiotemporal data by incorporating temporal and spatial attributes that correspond to the inherent fuzziness present in crisp data. Furthermore, we enhance the Dewey code to annotate fuzzy spatiotemporal data for specialized processing and to ascertain the structural relationships among fuzzy spatiotemporal nodes within XML documents. Our approach employs streams to store leaf nodes in the XML document that correspond to leaf query nodes, which are subsequently filtered to eliminate any unmatched nodes. Following this filtering process, output lists are generated for each matched leaf node.

References

1. J. Tekli, G. Tekli, R. Chbeir, Combining offline and on-the-fly disambiguation to perform semantic-aware XML querying. Comput. Sci. Inf. Syst. **20**(1), 423–457 (2023)
2. P. Lukás, R. Baca, M. Krátký, T.W. Ling, Demythization of structural XML query processing: comparison of holistic and binary approaches. IEEE Trans. Knowl. Data Eng. **33**(4), 1439–1452 (2021)
3. N. Papadakis, S. Kartakis, K. Papadakis, E. Papadaki, A tool for spatial reasoning in XML documents. Int. J. Seman. Comput. **9**(1), 67–104 (2015)
4. Z. Brahmia, H. Hamrouni, R. Bouaziz, TempoX: a disciplined approach for data management in multi-temporal and multi-schema-version XML databases. J. King Saud Univ. Comput. Inf. Sci. **34**(1), 1472–1488 (2022)
5. L. Bai, Z. Cui, X. Duan, H. Fu, Keyword coupling query of spatiotemporal data based on XML. J. Intell. Fuzzy Syst. **42**(3), 2219–2228 (2022)
6. Y. Chen, P. Revesz, Querying spatiotemporal XML using DataFox, in *Proceedings of the 2nd ACM International Conference on Web Intelligence* (2003), pp. 301–309
7. Z. Ma, T. Li, Y. Yan, An approach of top-k keyword querying for fuzzy XML. Computing **100**(3), 303–330 (2018)
8. T. Li, Z. Ma, A structure-based approach of keyword querying for fuzzy XML data. Int. J. Knowl. Based Intell. Eng. Syst. **22**(2), 125–140 (2018)
9. L. Zhu, J. Wang, L. Bai, A general characterization of integrating and querying heterogeneous fuzzy spatiotemporal XML data. Earth Sci. Inf. **16**(4), 3303–3321 (2023)
10. L. Bao, J. Yang, C.Q. Wu, H. Qi, X. Zhang, S. Cai, XML2HBase: storing and querying large collections of XML documents using a NoSQL database system. J. Parallel Distrib. Comput. **161**, 83–99 (2022)
11. J. E. Córcoles, P. González, Using RDF to query spatial XML, in *Proceedings of the International Conference on Web Engineering* (2004), pp. 316–329

12. Y.S. Chang, H.D. Park, XML web service-based development model for internet GIS applications. Int. J. Geogr. Inf. Sci. **20**(4), 371–399 (2006)
13. K. Nørvåg, Temporal query operators in XML databases, in *Proceedings of the 2002 ACM Symposium on Applied Computing* (2002), pp. 402–406
14. F. Rizzolo, A.A. Vaisman, Temporal XML: modeling, indexing, and query processing. Very Large Data Bases **17**(5), 1179–1212 (2008)
15. L. Bai, L. Yan, Z. Ma, Determining topological relationship of fuzzy spatiotemporal data integrated with XML twig pattern. Appl. Intell. **39**(1), 75–100 (2013)
16. N. Bruno, N. Koudas, D. Srivastava, Holistic twig joins: optimal XML pattern matching, in *Proceedings of the 28th ACM SIGMOD International Conference on Management of Data* (2002), pp. 310–321
17. J. Lu, T. W. Ling, C. Y. Chan, T. Chen, From region encoding to extended Dewey: on efficient processing of XML twig pattern matching, in *Proceedings of the 31st International Conference on Very Large Data Bases* (2005), pp. 193–204
18. J. Lu, T. Chen, T.W. Ling, Efficient processing of XML twig patterns with parent child edges: a look-ahead approach, *in: Proceedings of the 13th ACM CIKM International Conference on Information and Knowledge Management*, pp. 533–542 (2004)
19. L. Bai, Y. Li, J. Liu, FSPTwigFast: holistic twig query on fuzzy spatiotemporal XML data. Appl. Intell. **47**(4), 1224–1239 (2017)
20. I. Tatarinov, S. Viglas, K.S. Beyer, J. Shanmugasundaram, E.J. Shekita, C. Zhang, Storing and querying ordered XML using a relational database system, in *Proceedings of the 2002 ACM SIGMOD International Conference on Management of Data* (2002), pp. 204–215
21. J. Liu, Z. Ma, Y. Yan, Matching twigs in fuzzy XML. Inf. Sci. **181**(1), 184–200 (2011)
22. Hurricane forecast of Atlantic (2014). https://www.wunderground.com/?MR=1
23. J. Liu, Z. Ma, Q. Qv, Dynamically querying possibilistic XML data. Inf. Sci. **261**(4), 70–88 (2014)

Chapter 7
Query Relaxation and Top-*k* Result Sorting of Fuzzy Spatiotemporal XML Data

7.1 Introduction

Given its importance for practical spatiotemporal applications, the querying of spatiotemporal information has emerged as a critical area of focus within the domain of spatiotemporal data management [1–5]. Nevertheless, in real-world spatiotemporal applications, certain users may struggle to articulate their intentions clearly or may lack explicit query objectives, particularly within ambiguous contexts. For instance, in the case of a travel website, a user might request information on all cities in Hebei province that feature a river or sea and experience sunny weather. Upon processing this query, the travel website may return a list of cities that possess a river or sea and are forecasted to have sunny weather during the upcoming weekend. However, it is important to note that, irrespective of the user's preferences, cities with a river or sea that are expected to have cloudy weather that weekend, cities with sunny weather that lack a river or sea, or cities with a river or sea that are predicted to have sunny weather the following week could also fulfill the user's requirements. Consequently, it becomes essential to relax the original query parameters in order to obtain a broader range of relevant results.

Query relaxation is frequently employed to increase the volume of query results. Notable advancements have been made in the query relaxation of spatiotemporal data, with the fundamental premise of these methodologies centered on the reduction or reconfiguration of the constraints imposed by the original query to broaden its reach. Liu and Yan [6] have formulated a set of schema-aware relaxation rules and developed a corresponding penalty model predicated on weight modification. Building upon this foundation, they propose an adaptive query relaxation technique that exclusively returns results pertinent to the specified query conditions. Fazzinga et al. [7] introduce a method for XPath query relaxation, which facilitates XML query relaxation processing through the application of rewriting rules. Yang et al. [8] present a heuristic probabilistic relaxation approach for road network matching, aimed at integrating the most current information from multiple sources of geospatial

L. Bai and L. Zhu, *Fuzzy Spatiotemporal XML Data Management*, Studies in Computational Intelligence 1183, https://doi.org/10.1007/978-3-031-81033-6_7

data. Ford and Kench [9] investigate the effects of typhoons and associated relaxation strategies. Additionally, Mitra and Pallickara [10] design a scalable, query-oriented distributed data integration framework that dynamically generates precise interpolations for targeted spatiotemporal scopes, while also estimating the uncertainty associated with such interpolations in instances of spatiotemporal misalignment of data points.

The increasing prevalence of XML has led to heightened interest in the representation and analysis of XML data. Numerous initiatives have been undertaken to model XML data effectively. Ma and Yan [11] explore the multi-granularity of data fuzziness in XML through the lens of possibility theory, resulting in the development of a fuzzy XML data model that addresses various types of fuzziness. In a related vein, Bai et al. propose a series of methodologies for modeling fuzzy spatiotemporal data based on XML, integrating the labeling and ordering of XML documents with the characteristics of fuzzy spatiotemporal data [12–15]. Additionally, with respect to querying XML data, Xu et al. [15] introduce a contextual preference query method alongside content scoring mechanisms. Li et al. [16] examine probabilistic threshold keyword queries (PrTKQ) over XML data, proposing two querying algorithms that leverage probability density functions. Furthermore, Campi et al. [17] present a flexible framework for querying semi-structured XML data grounded in fuzzy set theory, which includes a novel function designed to assess the degree of similarity between two XML trees. Lastly, Damiani et al. [18] discuss a graph-based approach for executing blind queries on unvalidated XML information.

Following the relaxation of queries, a subsequent challenge arises in the form of "information overload," characterized by an excessive number of query results. To address this issue, numerous studies have concentrated on the sorting of results. Jones et al. [19] integrate the retrieval probability model with experimental data from the TREC planning materials, proposing a ranking algorithm known as PIR. Agrawal et al. [20] adopt a democratic approach to reconcile preferences gathered from diverse sources, utilizing these preferences to establish representative rankings that facilitate the prompt delivery of ranked responses. Marian et al. [21] introduce an adaptive top-k query-processing strategy for XML, capable of evaluating both exact and approximate matches, with approximation defined through the relaxation of XPath axes. Weigel et al. [22] investigate the interplay between scoring methodologies and XML indices to enhance ranking efficiency, proposing IR-CADG, an extension of data guides that incorporates keyword considerations and integrates ranking based on both structural and content elements. Yan et al. [23] present a preference-based ranking model designed to manage approximate queries in XML, which relies on a user-defined interest score to generate the top-k results, thereby increasing the query burden on users. Additionally, Li et al. [24], Sun et al. [25], Watson et al. [26], and Bai et al. [27] also explore top-k retrieval methods that yield the optimal top-k results for users, organized according to the established ranking criteria.

In this chapter, we propose a query relaxation approach that incorporates the weighting of each designated attribute. Furthermore, the query outcomes are categorized into accurate results and fuzzy results. We present a top-k sorting approach for the query results, which integrates path probability into the ranking process based on contextual preferences, thereby differentiating between accurate and fuzzy results.

7.2 Preliminaries

In this subsection, we will first elucidate the p-document model employed for the representation of XML data. Subsequently, we will present the encoding methodology utilized for XML, which underpins our research. Lastly, we will introduce the notion of query relaxation.

7.2.1 Probabilistic XML Model

In this chapter, we examine a widely recognized probabilistic XML model known as $PrXML^{\{ind,mux\}}$ [28], which is initially introduced in [29]. Within this framework, a probabilistic XML document (referred to as a p-document) is conceptualized as a labeled tree structure. This tree comprises two distinct categories of nodes: *ordinary* nodes, which represent the actual data, and *distributional* nodes, which denote the probability distributions associated with their child nodes. Distributional nodes are further classified into two types: IND and MUX. An IND node features children that are *independent* of one another, whereas the children of a MUX node are *mutually-exclusive*, meaning that only one child can be present in any given random instance of the document, referred to as a possible world. Each edge in the XML tree is assigned a real number within the interval (0,1], which signifies the conditional probability of the child node's occurrence under the parent node, contingent upon the parent node's existence. An illustrative example of a p-document is presented in Fig. 7.1. Edges that are unweighted default to a conditional probability of 1. In the p-document depicted in Fig. 7.1, *ordinary* nodes are identified by their tag names, such as A, B1, B2, C1, and C2. The two types of *distributional* nodes are represented visually, with MUX nodes illustrated as rectangular boxes with rounded corners and IND nodes as circles. For instance, the IND2 node has two children, C1 and B2, with respective existence probabilities of 0.6 and 0.5. Consequently, the probability of neither C1 nor B2 appearing is calculated as $(1-0.6)\times(1-0.5)= 0.2$. Similarly, the MUX2 node encompasses three children: D1, E2, and the IND3 node, with existence probabilities of 0.5, 0.3, and 0.1, respectively. Therefore, the probability of none of these children appearing is also determined as $1-0.5-0.3-0.1=0.1$.

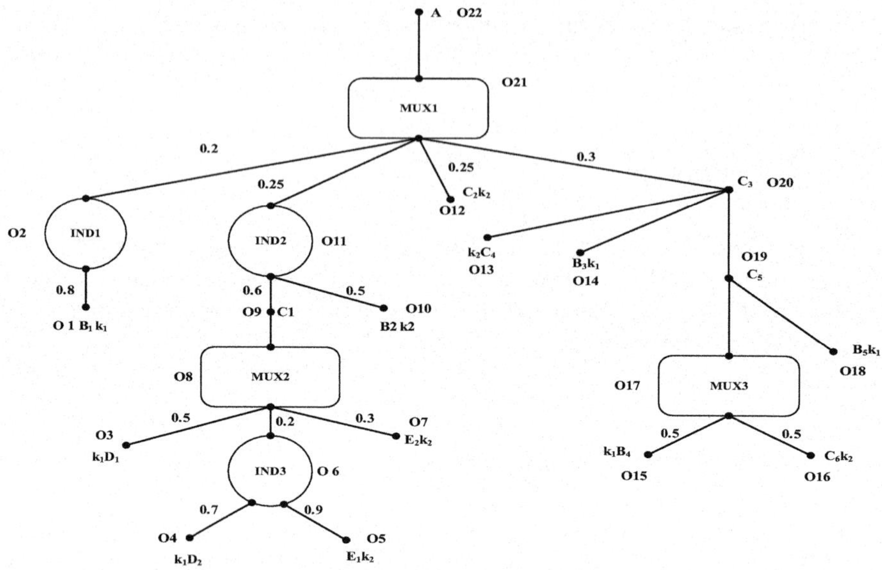

Fig. 7.1 A p-document

7.2.2 Encoding Probabilistic XML Data

The Dewey encoding scheme has been validated and utilized as a proficient and effective method in prior keyword search methodologies [28, 30]. This scheme originates from the necessity to indicate the position of an element's occurrence within the context of information retrieval. In this chapter, we extend upon the Dewey encoding scheme to accommodate probabilistic XML documents. To achieve this, we assign a numerical representation to each node that corresponds to its Dewey number, while appending the character "M" or "I" preceding the number to denote MUX or IND nodes, respectively.

Figure 7.2 illustrates the outcomes of encoding the nodes depicted in Fig. 7.1. Furthermore, each leaf node retains a list of conditional probabilities that span from the root to the respective leaf node. These documented conditional probabilities can be utilized to compute the probability of the SLCA results. The encoding results are presented in Fig. 7.2. Initially, the code for the root node A is assigned the value of 1. Subsequently, the nature of its child nodes is taken into account. If the child node of A is an original node, its code is prefixed with 1.*x*, where *x* begins at 1 and increases sequentially from left to right. Conversely, if the child node of A is of the IND or MUX type, its code is denoted as 1.Ix or 1.Mx, respectively. Following this coding scheme, the code for MUX1 is designated as 1.M1. MUX1 has four children, which, from left to right, are IND1, IND2, C2, and C3. The character "I" is appended when coding IND1 and IND2, resulting in their codes being 1.M1.I1 and 1.M1.I2,

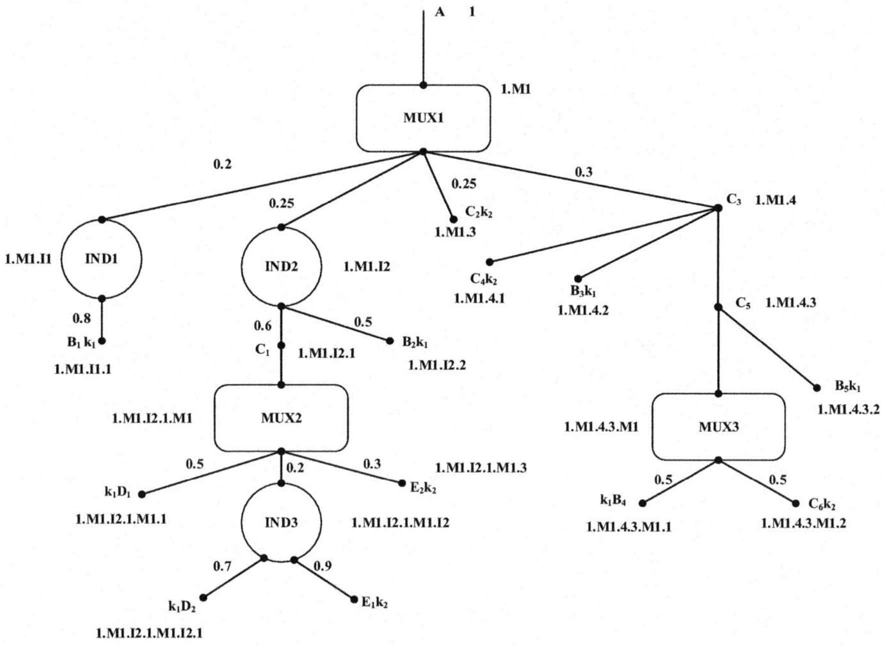

Fig. 7.2 An encoded p-document

respectively. For the original nodes C2 and C3, the codes are assigned as 1.M1.3 and 1.M1.4, respectively.

7.2.3 *Query Relaxation*

To more effectively address the user's initial query intent, it is necessary to implement a relaxation of the original query. Initially, relevant definitions are introduced, including a specific definition of query relaxation, as noted by [31].

Definition 7.1 (*Fuzzy temporal attribute*) The fuzzy temporal attribute is denoted as $FTA = \{l_t, \tau\}$. $\mathrm{l_t} = [\mathrm{t_s}, \mathrm{t_e}]$ distinctly characterizes the value of the temporal attribute, with t_s and t_e representing the starting and ending time, respectively. $\tau \in (0, 1]$ serves as the membership function that encapsulates the inherent ambiguity associated with the attribute.

Definition 7.2 (*Fuzzy spatial attribute*) A fuzzy spatial attribute is denoted as $FSA = \{l_s, \tau\}$. $l_s = \{x_a, x_b, y_a, y_b\}$ distinctly characterizes a specific region within a two-dimensional coordinate system, where (x_a, y_a) (x_b, y_b) denote points in this space. $\tau \in (0, 1]$ signifies the degree of ambiguity associated with the attribute, specifically the likelihood that the attribute value is accurate.

The membership function is determined through the principles of fuzzy statistics. In this context, $\tau_A(x_0) = \lim_{n \to \infty} \frac{times\ of\ x_0 \in A^*}{n}$, where x_0 represents an element within the universe X. A^* denotes a set that varies randomly, while A is characterized as a fuzzy set with A^* serving as its flexible boundary. Additionally, n refers to the number of fuzzy statistical experiments conducted.

Definition 7.3 (*Fuzzy spatiotemporal data model*) Fuzzy spatiotemporal data model (*FSTD*) is characterized by a 4-tuple, denoted as (*OID, GEA, FTA, FSA*), where

- *OID*: the identification of fuzzy spatiotemporal data, which depicts the evolving states;
- *GEA*: the name of general attribute set associated with fuzzy spatiotemporal data, which signifies the general attribute;
- *FTA*: the name of temporal attribute set associated with fuzzy spatiotemporal data, which signifies the temporal attribute;
- *FSA*: the name of spatial attribute set associated with fuzzy spatiotemporal data, which signifies the spatial attribute.

Definition 7.4 (*Query relaxation*) Utilize the R to model a relationship within a spatiotemporal framework characterized by n tuples $\{t_1, \ldots, t_n\}$ and m attributes $\{A_1, A_2, \ldots, A_m\}$. The query Q is formulated on R, incorporating conjunctive conditions of the form $Q = \sigma \wedge C_i$, $i \in (1, k)$, where $2 \leq k \leq m$, $C_i = A_i \theta a_i$, $\theta \in \{>, <, =, \geq, \leq, \text{between}\}$. In this context, each A_i represents an attribute of the dataset D, while a_i signifies a specific value or range within its corresponding domain. By modifying the original query Q, one can derive a relaxed query Q, which aims to identify all tuples within D that exhibit a degree of similarity to the original query, surpassing a predetermined similarity threshold $\lambda \in (0, 1)$. This process can be calculated as follows:

$$Q(D) = \{t | t \in D, Sim(Q, t) \geq \lambda\} \tag{7.1}$$

In Definition 7.4, $Q(D)$ denotes a set of query results that meets the criteria of a relaxed query, while λ represents the threshold established by either the system or the user.

7.3 Fuzzy Spatiotemporal Query Relaxation

The primary concept underlying fuzzy spatiotemporal query relaxation involves adjusting the parameters of each attribute in accordance with its assigned weight. In this subsection, we first introduce a method for calculating the weight of attributes, followed by a formula for determining the query relaxation threshold based on these weights. Subsequently, we present distinct methods for relaxing queries pertaining to temporal and spatial attributes. Finally, we propose an algorithm for implementing fuzzy spatiotemporal query relaxation.

7.3.1 Attribute Weight Distribution

The initial phase of query relaxation involves the distribution of attribute weights, which establishes the foundation for the relaxation range. Given that each user possesses unique preferences, it is essential to assign varying weights to attributes according to individual users. The allocation of attribute weights is conducted using Inverse Document Frequency (IDF), which takes into account the distribution of designated attributes alongside the user's query history. The weight assigned to a specific attribute (w_i) is computed as follows:

$$w_i = \frac{IDF_i(x)}{\sum_{j=1}^{k} IDF_j(x)} \tag{7.2}$$

where k represents the quantity of attributes delineated by the query, while $IDF_i(x)$ denotes the IDF value of the attribute A_i.

The sub-threshold for each designated attribute can be determined based on the attribute weights and the relaxation threshold λ associated with the query. For a given query Q, the sub-threshold φ_i for each specified attribute within Q can be calculated using Formulas (7.3) and (7.4).

$$\frac{\mathrm{w_i}}{\varphi_\mathrm{i}} = \cdots = \frac{\mathrm{w_n}}{\varphi_\mathrm{n}} \tag{7.3}$$

$$\lambda = \sum_{\mathrm{i}=1}^{\mathrm{n}} w_i \times \varphi_i \tag{7.4}$$

where n represents the number of specified attributes, w_i denotes the weight of the attribute A_i, and λ indicates the given relaxation threshold.

The relaxation threshold is primarily employed to broaden the scope of numerical attributes. For instance, when the variable q in the original query condition represents a numerical interval $[q_a, q_b]$, the relaxed query interval is:

$$\left[q_a - h_i\sqrt{\frac{1-\varphi_i}{\varphi_i}},\ q_b - h_i\sqrt{\frac{1-\varphi_i}{\varphi_i}}\right] \tag{7.5}$$

where $h_i = 1.06\sigma_i n^{-1/5}$ (σ_i represents the standard deviation of all values on the numeric attribute A_i, n denotes the total count of tuples in the database).

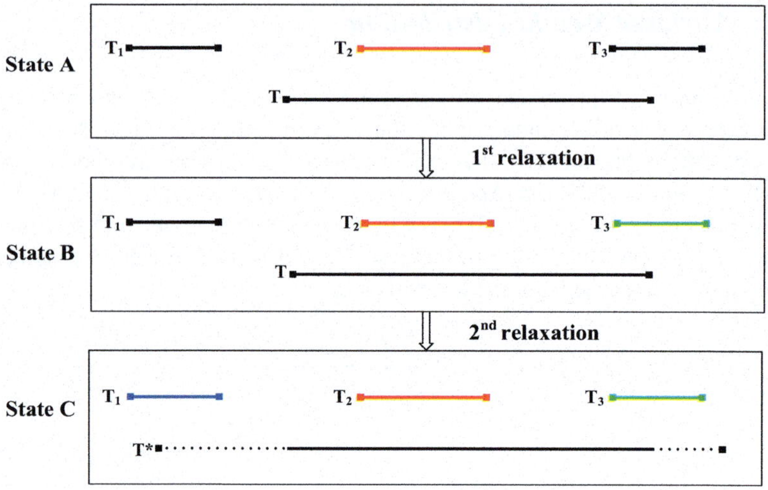

Fig. 7.3 The relaxation of temporal attributes

7.3.2 *Temporal Attribute Relaxation Query Method*

The process of query relaxation concerning temporal attributes is comprised of two distinct phases. Initially, one must identify the time intervals that intersect with the specified query time range and incorporate these into the result set. Subsequently, the query interval is expanded in accordance with the established relaxation threshold, and any time intervals that intersect with this newly defined query range are then added to the result set.

In Fig. 7.3 (State A), the time interval T_2 falls within the initial query range T, thereby qualifying T_2 as the precise result for the query. In Fig. 7.3 (State B), a portion of the time interval T_3 overlaps with T, satisfying the query criteria; consequently, T_3 is included in the result set. In Fig. 7.3 (State C), the extension of T to T^* in pursuit of additional results, results in an intersection between T_1 and T^*, leading to the inclusion of T_1 in the result set.

7.3.3 *Spatial Attribute Relaxation Query Method*

The process of query relaxation concerning spatial attributes is delineated in two distinct phases. Initially, it is necessary to identify the nodes that intersect with the original spatial query parameters, which are then incorporated into the result set. Subsequently, the query's scope is broadened in accordance with the established relaxation threshold, allowing for the inclusion of nodes that intersect with the newly defined query range into the result set. The methodology for relaxing spatial attributes is illustrated in Fig. 7.4.

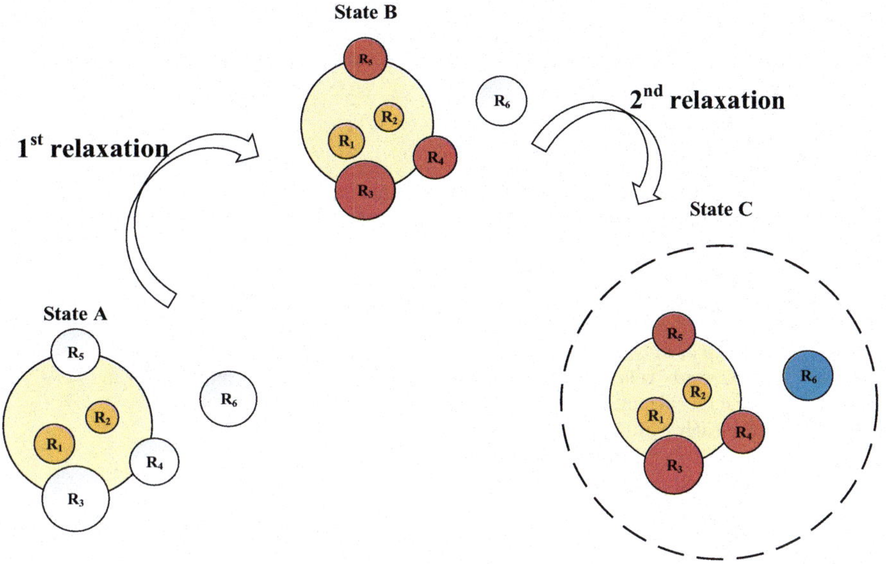

Fig. 7.4 The relaxation of spatial attribute

The yellowish circle represents the spatial range C_Q specified in the initial query conditions. In the initial state, depicted in Fig. 7.4 (State A), only the information nodes R_1 and R_2 satisfy the criteria of the initial query C_Q. In Fig. 7.4 (State B), to enhance the results, nodes R_3, R_4 and R_6, which intersect with C_Q, are incorporated into the result set. If the query results remain insufficient, a new query range, denoted as C_{Q*}, is illustrated by a dashed circle, which is determined based on the relaxation threshold. As indicated in Fig. 7.4 (State C), node R_6 is included within the boundaries of C_{Q*}.

7.3.4 *Fuzzy Spatiotemporal Query Relaxation Algorithm*

This subsection proposes a novel algorithm known as SPQR (spatiotemporal query relaxation), which incorporates the relaxation of general attributes, as well as fuzzy temporal and fuzzy spatial attributes. The comprehensive procedure for implementing SPQR is delineated in Algorithm 7.1.

Algorithm 7.1 SPQR

Input: initial query condition Q, specified number of results u, relaxation threshold λ, the weight of all attribute W $(w_1, w_2, \ldots, w_i)$
Output: set 1, set 2
01 **initialize** (set 1)
02 **initialize** (set 2)
03 **query**(Q) // The function of the query is to execute the query
04 **if num**(set 1) >u **then** //num function is used to calculate the number of results in the result set
05 **return** set 1
06 **else**
07 **do** {
08 **for each** c_i **in** Q
09 $\varphi_i \leftarrow$ **sub_relaxationthreshold**(λ, w_i)// according to Formulas (7.3) and (7.4)
10 **if** general attributes == c_i **then**
11 **for each** v **in** D(c_i)
12 **if sim**(v, a_i) **then** add v to set 2
13 **if** fuzzy temporal attributes == c_i **then**
14 $[t_s, t_e] \leftarrow \left[t_s - h_T\sqrt{\frac{1-\varphi_T}{\varphi_T}}, t_e + h_T\sqrt{\frac{1-\varphi_T}{\varphi_T}}\right]$ // according to Formulas (7.3), (7.4), and (7.5)
15 **if** fuzzy spatial attributes == c_i **then**
16 $\{x_a, x_b, y_a, y_b\} \leftarrow$
17 $\left\{x_a - h_L\sqrt{\frac{1-\varphi_L}{\varphi_L}}, x_b + h_L\sqrt{\frac{1-\varphi}{\varphi_L}}, y_a - h_L\sqrt{\frac{1-\varphi_L}{\varphi_L}}, y_b + h_L\sqrt{\frac{1-\varphi_L}{\varphi_L}}\right\}$ //according to Formulas (7.3), (7.4), and (7.5)
18 $\lambda \leftarrow \lambda - 0.2$
19 **query**($Q^{\sim}$) // $Q^{\sim}$ is the query after relaxation
20 } **while num**(set 1 ∪ set 2) < u
21 **endif**
22 **return** set 2

In the query relaxation process, initial empty sets are established to accommodate the relaxed results (Lines 1–2). Subsequently, the primary query Q is executed, and the results are stored in set 1 (Line 3). If the number of results in set 1 exceeds the user-specified threshold u, set 1 is returned as the output (Lines 4–5). To obtain additional results, the sub-relaxation threshold φ_i for each attribute c_i is computed based on the attribute weight w_i and the relaxation threshold λ (Lines 6–9). For general attributes, a flexible query approach is employed for relaxation (Lines 11–12). In the case of fuzzy temporal attributes, the range is adjusted according to the specified Formulas (7.3), (7.4), and (7.5) (Lines 13–14). Similarly, for fuzzy spatial attributes, the new scope is calculated using the same formulas (Lines 15–17). The relaxation threshold λ is subsequently updated to decrease at a gradient of 0.2 (Line 18), and a new query is executed based on the revised query conditions (Line 19). This process continues until the number of results in both set 1 and set 2 surpasses the threshold u (Lines 20–22).

7.4 Sorting Method of Multiple Results

The outcomes within the set are presented in an unordered manner following the relaxation of queries, which fails to satisfy user requirements. In this subsection, we first introduce a method for calculating result scores that takes into account both path probability and attribute satisfaction. Subsequently, we propose an algorithm named SPQRR (spatiotemporal query relaxation and results ranking) designed to rank multiple query results based on the computed result scores.

7.4.1 Result Score Calculation

Diverse result sets employ distinct scoring methodologies. In the case of set 1, the scores are denoted as P_{path}, which signifies the path probability value of the elements within the XML document tree, derived from the probability sequence. Conversely, for set 2, the scoring formula is represented as $G_s \times (T_s \times W_t) \times (L_s \times W_l) \times P_{path}$, where G_s, T_s, and L_s correspond to the satisfaction levels of general attributes, temporal attributes, and spatial attributes, respectively. Additionally, W_t and W_l denote the weights assigned to the temporal and spatial attributes.

To assess the satisfaction of general attributes G_s, we employ a methodology grounded in semantic similarity for computation, as represented by the following formula:

$$G_s\left(A_k, Q_{A_k}\right) = \sum_{i=1}^{n} W_i \times sim(A_i, Q_{A_i}) \tag{7.6}$$

where n represents the number of general attributes present in the initial query condition Q, and W_i denotes the weight assigned to attribute A_i in the data set. The values of T_s and L_s are determined using the formulas outlined in Table 7.1.

Table 7.1 Satisfaction calculation of temporal and spatial attributes

	Temporal attributes	Spatial attributes
Exact query	$T_s = 1$	$L_s = 1$
First round of relaxation	$T_s = \frac{[t_{n1}, t_{n2}] \cap [t_a, t_b]}{[t_{n1}, t_{n2}]}$	$L_s = \frac{C_n \cap C_Q}{C_n}$
Second round of relaxation	$T_s = \frac{t_b - t_a}{t_b - t_a + 2\delta} \times \frac{[t_{n1},\ t_{n2}] \cap [t_a - \delta,\ t_b + \delta]}{[t_{n1},\ t_{n2}]}$	$L_s = \frac{C_Q}{C_{Q*}} \times \frac{C_n \cap C_{Q*}}{C_n}$

7.4.2 Sorting Algorithm

In contrast to previous efforts [20], our approach incorporates the generation and clustering of tuple arrangements during the online processing phase, thereby enhancing query efficiency. The SPQRR algorithm is structured into two distinct phases. Initially, each result is evaluated based on path probability and attribute satisfaction. Subsequently, the Top-k results are selected and presented based on their scores. The SPQRR algorithm is delineated as follows.

Algorithm 7.2 SPQRR

Input: set 1, set 2
Output: ordered sequence of results in set 1 and set2
01 *score[2] [n]* ← 0 // set the score array to zero, where n is the maximum number of results in set 1 and set 2
02 **if** set 1 **then**
03 **for each** $result_i$ **in** set 1
04 $score[0][result_i] \leftarrow \boldsymbol{P_{path}}$ // P_{path} is the function used to calculate the P_{path} value of all general attributes in the set 1
05 **endif**
06 **if** set 2 **then**
07 **for each** $result_j$ **in** set 2
08 $score[1]\,[result_j] \leftarrow G_s \times (T_s \times W_t) \times (L_s \times W_l) \times P_{path}$ //according to Formula (7.2), (7.6) and Table 7.1
09 **endif**
10 **sort** (score, set 1, set 2)
11 **return** set 1, set 2

In the results sorting procedure, an array is initially established to accommodate the scores (Line 1). The scores for the results across various sets are computed sequentially. Specifically, the results in set 1 are evaluated using the P_{path} scoring method (Lines 2–5), where the P_{path} values for all attributes of each result are derived from the XML document tree. Conversely, the scores for the results in set 2 are represented by $G_s \times (T_s \times W_t) \times (L_s \times W_l) \times P_{path}$ (Lines 6–9). The variable G_s denotes the satisfaction score for general attributes, which is calculated in accordance with Formulas (7.2) and (7.6). Additionally, T_s represents the satisfaction score for fuzzy temporal attributes, while *ls* signifies the satisfaction score for fuzzy spatial attributes; the respective calculation formulas for these scores are provided in Table 7.1. Following the computation of the scores, the elements within the sets are sorted and subsequently returned (Lines 10–11).

7.5 Experiments

7.5.1 Experimental Setup

The experiments were conducted on a computer equipped with Windows 10, featuring an Intel Core i5 processor operating at 1.9 GHz and 8 GB of RAM. All algorithms were implemented using Python version 3.7.

In the course of our evaluations, we established a typhoon database sourced from the World-Wide-Weather Site. This database comprises 100,000 tuples, encompassing both general attributes (including wind speed and type) and spatiotemporal attributes (such as time and location). We implemented a fuzzy processing methodology to enhance the data's suitability for experimental inquiries into querying fuzzy spatiotemporal XML data.

To assess the efficacy of our methodology, we conduct three distinct sets of experiments. The initial set involves a comparative analysis of the SPQR algorithm against the AIMQ and AQR algorithms, focusing on metrics such as Recall, Precision, and F-Score to determine the effectiveness of the SPQR algorithm. The second set of experiments evaluates the efficiency of the SPQR algorithm by examining execution time and memory usage. Finally, the third set compares the SPQRR algorithm with the PIR and CPR algorithms to evaluate the performance of the SPQRR.

7.5.2 Experimental Results

The initial set of experiments is designed to assess the effectiveness of the SPQR algorithm. To measure effectiveness, we utilize Recall, Precision, and F-Score metrics. Due to the varying methodologies employed in the relaxation of attributes, the experiments are conducted independently. We begin with the query relaxation experiments focused on temporal attributes, as illustrated in Table 7.2, which presents examples of simple temporal queries. Subsequently, we perform experiments on spatial attributes, utilizing the queries outlined in Table 7.3. Finally, we examine spatiotemporal queries, with corresponding examples provided in Table 7.4. The experiments are executed using three distinct query relaxation algorithms: SPQR, AQR [32] and AIMQ [33].

Figures 7.5, 7.6 and 7.7 present the outcomes of the query relaxation experiment conducted using the typhoon dataset. The results indicate that the SPQR algorithm demonstrates a significant advantage over the AMIQ and AQR algorithms in terms of recall, precision, and F-Score. The metrics for recall and precision consistently meet our expectations, achieving high accuracy levels (exceeding 70% on average) and acceptable recall rates (averaging over 50%). Two primary factors contribute to these results. Firstly, both the AIMQ and AQR algorithms execute relaxation processing in a single step, lacking the granularity necessary for effective relaxation. Secondly,

Table 7.2 Examples of simple temporal query

User query	Types	Practical query example	
Q I	Simple temporal query	Q1	Time ∈ [2017-03-18, 2017-04-02]
		Q2	Time ∈ [2017-03-18, 2017-04-02] ∩ Wind Type = 1
		Q3	Time ∈ [2017-03-18, 2017-04-02] ∩ Wind Type = 1 ∩ Wind Speed = 120
		Q4	Time ∈ [2017-03-18, 2017-04-02] ∩ Wind Type = 1 ∩ Wind Speed = 120 ∩ Pressure = 330

Table 7.3 Examples of simple spatial query

User query	Types	Practical query example	
QII	Simple spatial query	Q5	Longitude ∈ [46, 49] ∩ Latitude ∈ [− 20, − 15] ∩ Wind Type = 1
		Q6	Longitude ∈ [37, 40] ∩ Latitude ∈ [23, 26] ∩ Wind Type = 1
		Q7	Longitude ∈ [37, 40] ∩ Latitude ∈ [23, 26] ∩ Wind Type = 1 ∩ Wind Speed = 120
		Q8	Longitude ∈ [37, 40] ∩ Latitude ∈ [23, 26] ∩ Wind Type = 1 ∩ Wind Speed = 120 ∩ Pressure = 330

Table 7.4 Examples of spatiotemporal query

User query	Types	Practical query example	
QIII	Spatiotemporal query	Q9	Time∈ [2017-03-18, 2017-04-02] ∩ Longitude∈ [37, 40] ∩ Latitude∈[23, 26]
		Q10	Time∈ [2017-03-18, 2017-04-02] ∩ Longitude∈ [37, 40] ∩ Latitude∈[23, 26] ∩ Wind Type = 1
		Q11	Time∈ [2017-03-18, 2017-04-02] ∩ Longitude∈ [37, 40] ∩ Latitude ∈ [23, 26] ∩ Wind Type = 1 ∩ Wind Speed = 120
		Q12	Time∈ [2017-03-18, 2017-04-02] ∩ Longitude∈ [37, 40] ∩ Latitude ∈ [23, 26] ∩ Wind Type = 1 ∩ Wind Speed = 120 ∩ Pressure = 330

the SPQR algorithm implements multiple relaxation iterations based on the established relaxation threshold, thereby enhancing the accuracy of the results. In practical scenarios, when users query spatial or temporal data, they typically seek related information within a specified time interval or spatial location, making precision particularly critical. Additionally, the similarity calculations for spatiotemporal attributes are performed offline and stored in an auxiliary structure, which significantly reduces the time required for online queries when users engage in spatiotemporal searches.

Figures 7.8, 7.9, and 7.10 illustrate the performance of three algorithms (SPQR, AIMQ, and AQR) regarding Recall across various relaxation thresholds. It is observed that a lower relaxation threshold corresponds to a higher recall rate for all three algorithms. Notably, the SPQR algorithm consistently outperforms both AIMQ and AQR in terms of recall across the different thresholds examined. At a

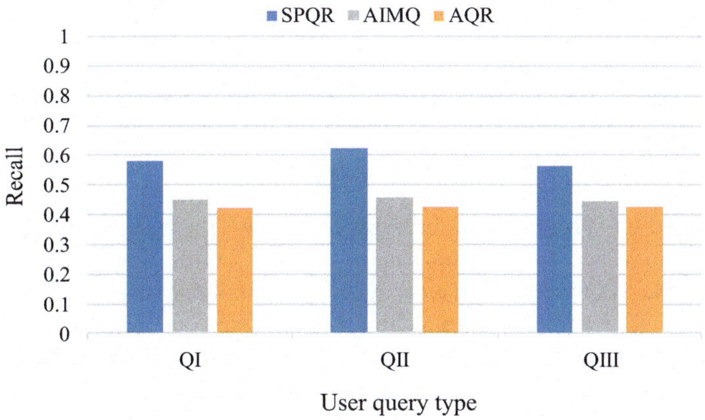

Fig. 7.5 Recall of the queries relaxation in typhoon DB

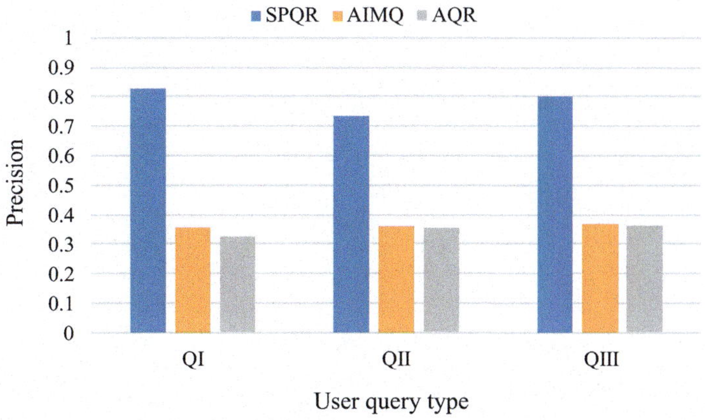

Fig. 7.6 Precision of the queries relaxation in typhoon DB

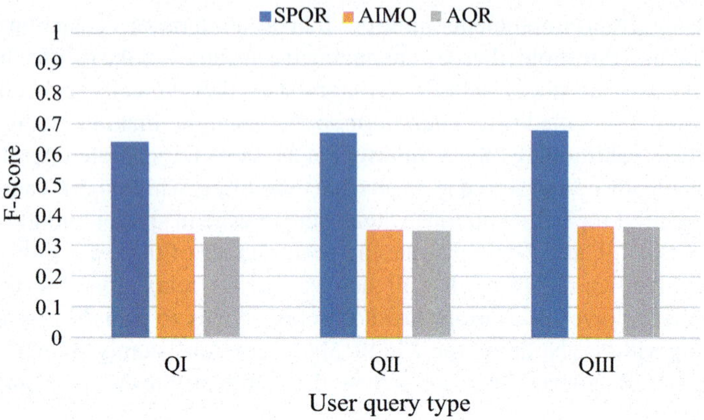

Fig. 7.7 F-Score of the queries relaxation in typhoon DB

relaxation threshold of 0.4, the overall recall rate of the SPQR algorithm stabilizes, achieving an average recall rate exceeding 0.9. This finding may serve as a valuable reference for determining optimal relaxation thresholds in future relaxation query operations.

The second set of experiments is designed to assess the performance of the SPQR algorithm with respect to execution time and memory utilization. Unlike other query algorithms, the SPQR algorithm employs a T-list auxiliary structure to facilitate the rapid identification of potential matching objects. We conduct comparative experiments to illustrate its efficiency. For these experiments, datasets of sizes 1000, 5000, and 8000 are utilized, and three distinct types of queries are performed on these datasets.

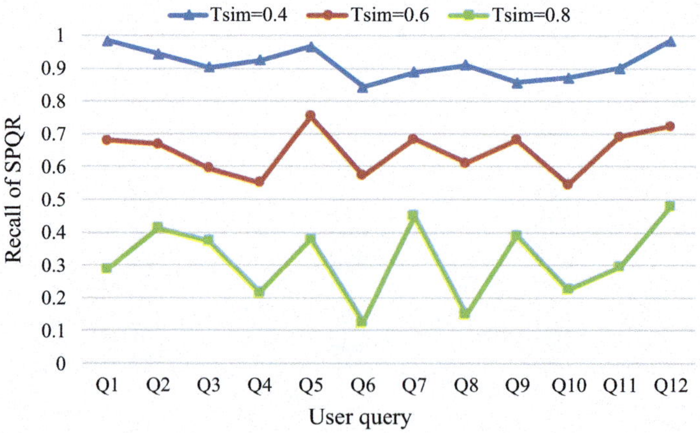

Fig. 7.8 Recall of SPQR under different T_{sim}

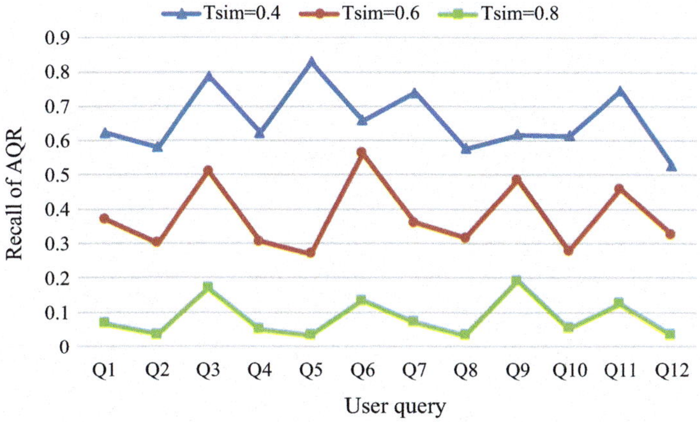

Fig. 7.9 Recall of AQR under different T_{sim}

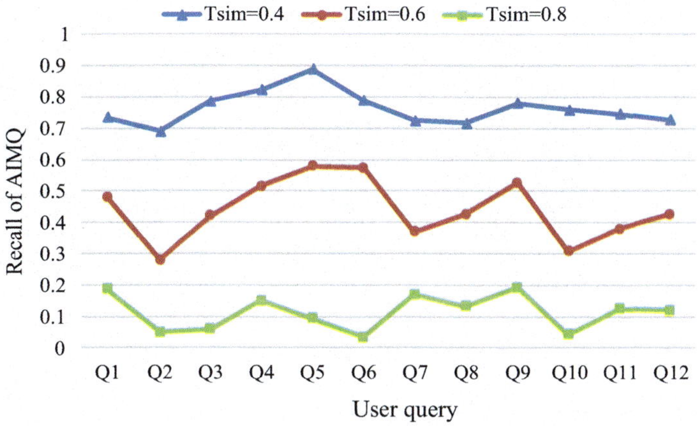

Fig. 7.10 Recall of AIMQ under different T_{sim}

With regard to execution time, as illustrated in Figs. 7.11, 7.12, and 7.13, the auxiliary structures demonstrate a distinct advantage. This benefit is not readily apparent with smaller datasets; however, as the dataset size increases, the execution time associated with auxiliary structures is significantly lower compared to scenarios without such structures. Conversely, in terms of memory consumption, as depicted in Figs. 7.14, 7.15, and 7.16, auxiliary structures exhibit a higher demand for memory resources. This is particularly pronounced when complex queries are executed on large datasets, resulting in substantial memory usage. Taking both factors into account, it is justifiable to prioritize reduced execution time over increased memory consumption,

The third set of experiments is designed to assess the impact of the SPQRR algorithm. We present the accuracy of the top-10 results within the dataset.

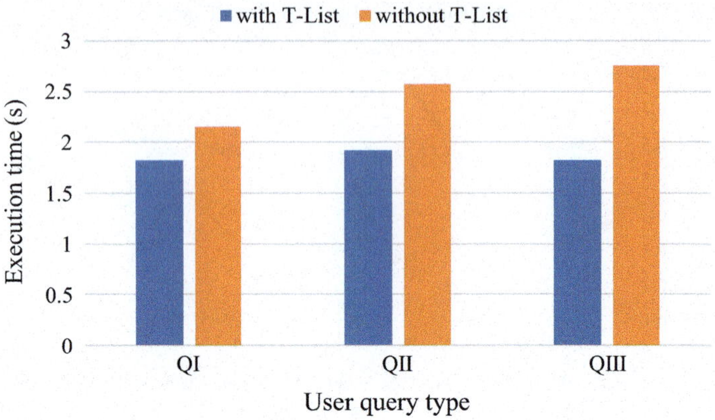

Fig. 7.11 Execution time for dataset size of 1000

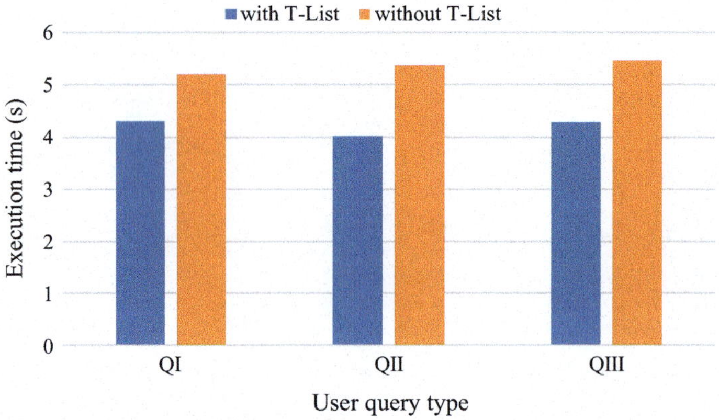

Fig. 7.12 Execution time for dataset size of 5000

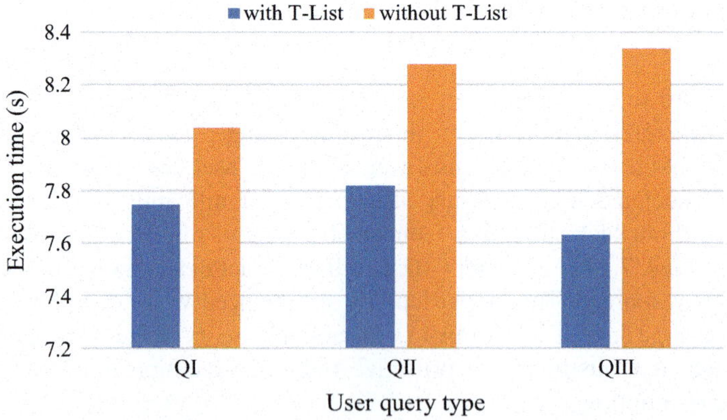

Fig. 7.13 Execution time for dataset size of 8000

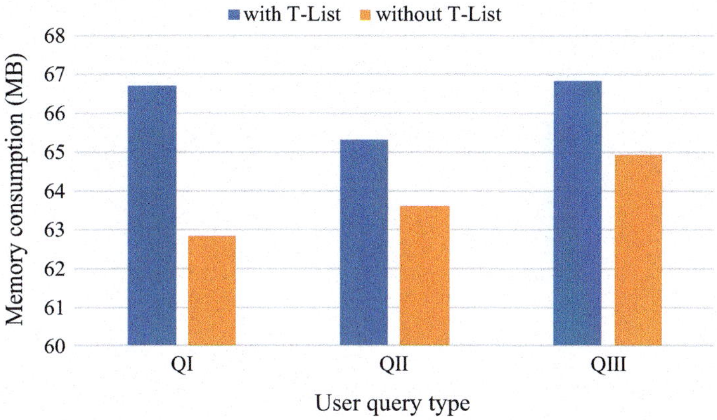

Fig. 7.14 Memory consumption for dataset size of 1000

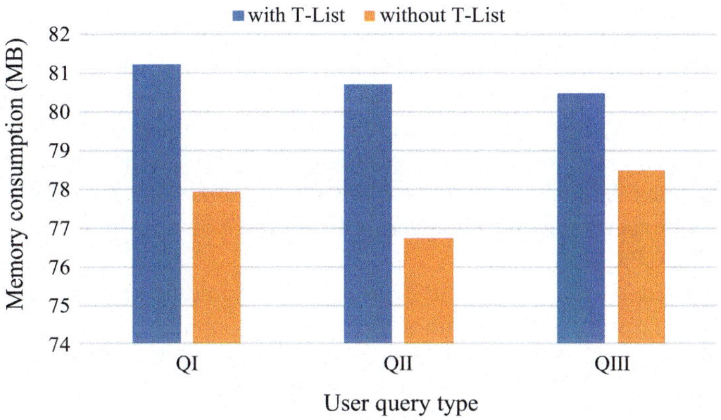

Fig. 7.15 Memory consumption for dataset size of 5000

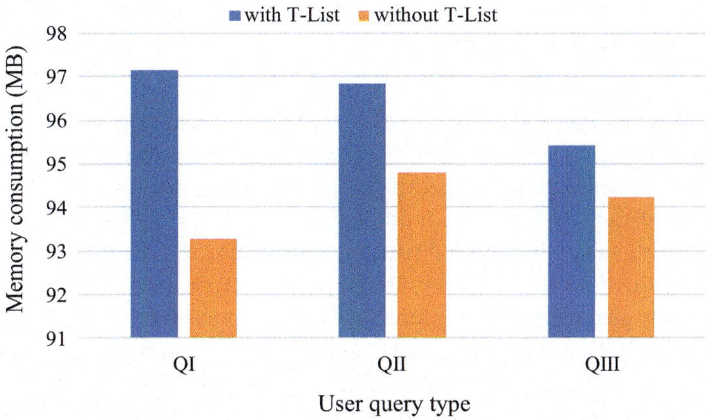

Fig. 7.16 Memory consumption for dataset size of 8000

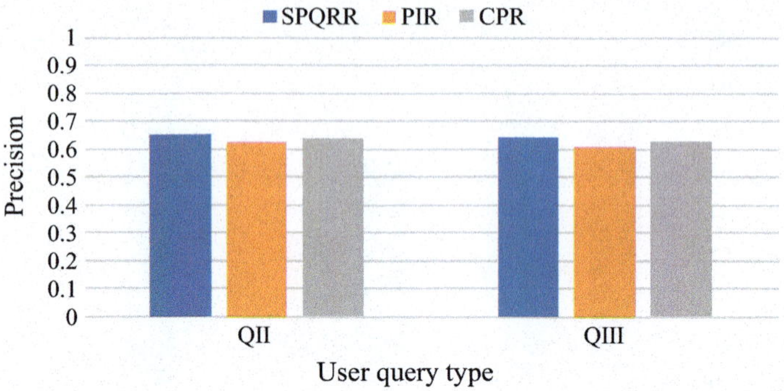

Fig. 7.17 Precision of SPQRR, PIR and CPR under Q II and Q III conditions

$$Precision = \frac{\{top-10\ results \cap relevant\ results\}}{10} \tag{7.7}$$

In Formula (7.7), $\{top-10\ results \cap relevant\ results\}$ denotes the quantity of tuples within the dataset that are identified as relevant results. To guarantee the reliability of the experimental outcomes, we initially formulate ten queries for each category of query and subsequently compute the *Precision* associated with these query conditions. Ultimately, we derive the average *Precision* for each query type. This series of experiments is conducted utilizing three distinct algorithms: SPQRR, PIR [19], and CPR [20].

As illustrated in Fig. 7.17, the SPQRR algorithm demonstrates superior precision in the sorting of results. This algorithm takes into account both general and spatiotemporal attributes during the scoring process, in contrast to the PIR and CPR algorithms, which focus solely on a single attribute. Consequently, the SPQRR algorithm is more adept at ranking fuzzy spatiotemporal results.

7.5.3 Comparative Study

Table 7.5 presents a comparative analysis of query relaxation in XML data. It is evident that the majority of existing research on query relaxation primarily focuses on general data, neglecting the incorporation of spatiotemporal features. Additionally, while investigations into the modeling and querying of XML data do address spatial, temporal, or spatiotemporal features, there is a lack of further exploration into their implications for query relaxation and result sorting. In contrast, our approach integrates fuzzy general attributes alongside spatiotemporal attributes, utilizing XML for data representation and examining query relaxation within this framework.

Table 7.5 Comparative study on query relaxation of XML data

	Spatial features	Temporal features	Spatiotemporal features	Fuzzy features	Query relaxation	XML based
Muslea and Lee [34]					✓	
Chen and Li [35]					✓	
Bosc et al. [36]				✓	✓	
Mottin et al. [37]					✓	
Martinenghi and Torlone [38]				✓	✓	
Li et al. [16]				✓		✓
Ford and Kench [9]	✓	✓	✓		✓	
Meng et al. [31]					✓	
Zhang et al. [39]				✓		✓
Ours	✓	✓	✓	✓	✓	✓

7.5.4 Discussion

This subsection evaluates the effectiveness and efficiency of the SPQR algorithm through a series of experiments. A total of twelve query examples, categorized into three distinct query types, were tested across three algorithms: SPQR, AIMQ, and AQR. The experimental findings indicate that the SPQR algorithm outperforms the other two algorithms. Specifically, the average recall and precision of the SPQR algorithm are 10% and 20% higher, respectively, than those of the other algorithms. Notably, in the QIII experiment, which focuses on spatiotemporal queries, the SPQR algorithm achieved a recall of 56.5% and a precision of 80.3%. The F-Score recorded was 68.3%, nearly double that of the AQR algorithm. Furthermore, to determine the optimal parameters for the algorithm, multiple experimental sets were conducted under varying relaxation thresholds. The results indicate that the SPQR algorithm performs optimally at a relaxation threshold of 0.4.

The query relaxation methodology that we have proposed is applicable across various domains that involve temporal or spatial data. For instance, a travel website may need to identify beaches suitable for beachcombing within the urban area of Qinhuangdao, Hebei Province, in the near future. Utilizing the SPQR algorithm, the initial temporal constraint is established to be within one week, while the spatial constraint is confined to the urban area of Qinhuangdao, supplemented by tidal information, including the timings of high and low tides. However, it may be determined that no beaches satisfy these specified criteria. Consequently, the temporal range can be extended to two weeks, and the spatial range can be broadened from the urban area to encompass the entire region. Under these relaxed query parameters, beaches such as Beidaihe and Nandaihe can be identified. To yield additional results, the term "beaches" may be substituted with "locations," which would also return sites such as Dongshan Bathing Beach and Pigeon Nest Park. Ultimately, the results are organized in descending order based on the extent of query relaxation.

Despite the aforementioned advantages, several limitations persist. First, the methodology should aim to minimize memory usage associated with the auxiliary structure while maintaining query efficiency, potentially through the implementation of compressed storage structures. Second, incorporating users' historical query preferences could enhance precision further. Lastly, the approach could be refined by integrating emerging semantic technologies to facilitate approximate querying across both structured and unstructured data, as well as linked data.

7.6 Summary

In this chapter, we investigate a query relaxation algorithm designed to address the issue of small or empty result sets encountered during query operations in fuzzy spatiotemporal XML databases. Our findings indicate that the average recall and precision of our algorithm surpass those of the AIMQ algorithm by 10% and 20%, respectively. Furthermore, we establish two distinct sets to store the outcomes of both accurate and fuzzy queries. This classification of result sets facilitates the subsequent selection of the query result that most effectively aligns with user requirements amidst a multitude of results. Additionally, following the query relaxation process, we propose an efficient algorithm for sorting the results. Experimental results validate that our approach to spatiotemporal query relaxation, combined with multiple result sorting, effectively addresses the challenge of insufficient answers while achieving a higher precision rate of 80.3% for SPQR and 64.4% for SPQRR.

References

1. C. Li, G. Baciu, Y. Wang, J. Chen, C. Wang, DDLVis: real-time visual query of spatiotemporal data distribution via density dictionary learning. IEEE Trans. Visual Comput. Graphics **28**(1), 1062–1072 (2022)
2. A. Sözer, A. Yazici, H. Oguztüzün, Indexing fuzzy spatiotemporal data for efficient querying: a meteorological application. IEEE Trans. Fuzzy Syst. **23**(5), 1399–1413 (2015)
3. A. Sözer, A. Yazici, H. Oguztüzün, O. Tas, Modeling and querying fuzzy spatiotemporal databases. Inf. Sci. **178**(19), 3665–3682 (2008)
4. L. Zhu, J. Wang, L. Bai, A general characterization of integrating and querying heterogeneous fuzzy spatiotemporal XML data. Earth Sci. Inf. **16**(4), 3303–3321 (2023)
5. L. Bai, Z. Cui, X. Duan, H. Fu, Keyword coupling query of spatiotemporal data based on XML. J. Intell. Fuzzy Syst. **42**(3), 2219–2228 (2022)

6. J. Liu, D.L. Yan, Answering approximate queries over XML data. IEEE Trans. Fuzzy Syst. **24**(2), 288–305 (2016)
7. B. Fazzinga, S. Flesca, F.F. Furfaro, XPath query relaxation through rewriting rules. IEEE Trans. Knowl. Data Eng. **23**(10), 1583–1600 (2011)
8. B. Yang, Y. Zhang, X. Luan, A probabilistic relaxation approach for matching road networks. Int. J. Geogr. Inf. Sci. **27**(2), 319–338 (2013)
9. M.R. Ford, P.S. Kench, "Spatiotemporal variability of typhoon impacts and relaxation intervals on Jaluit Atoll, Marshall Islands. Geology **44**(2), 159–162 (2016)
10. S. Mitra, S.L. Pallickara, Confluence: adaptive spatiotemporal data integration using distributed query relaxation over heterogeneous observational datasets, in *Proceedings of the 11th IEEE/ ACM International Conference on Utility and Cloud Computing* (2018), pp. 184–193
11. Z. Ma, L. Yan, Fuzzy XML data modeling with the UML and relational data models. Data Knowl. Eng. **63**(3), 972–996 (2007)
12. L. Bai, L. Yan, Z. Ma, Fuzzy spatiotemporal data modeling and operations in XML. Appl. Artif. Intell. **29**(3), 259–282 (2015)
13. L. Bai, Y. Li, J. Liu, FSPTwigFast: holistic twig query on fuzzy spatiotemporal XML data. Appl. Intell. **47**(4), 1224–1239 (2017)
14. L. Bai, L. Zhu, An algebra for fuzzy spatiotemporal data in XML. IEEE Access **7**(1), 22914–22926 (2019)
15. Y. Xu, W. Qu, Z. Li, G. Min, K. Li, Z. Liu, Efficient *k*-means++ approximation with MapReduce. IEEE Trans. Parallel Distrib. Syst. **25**(12), 3135–3144 (2014)
16. J. Li, C. Liu, R. Zhou, J.X. Yu, Quasi-SLCA based keyword query processing over probabilistic XML data. IEEE Trans. Knowl. Data Eng. **26**(4), 957–969 (2014)
17. A. Campi, E. Damiani, S. Guinea, S. Marrara, G. Pasi, P. Spoletini, A fuzzy extension of the XPath query language. J. Intell. Inf. Syst. **33**(3), 285–305 (2009)
18. E. Damiani, L. Tanca, F.A. Fontana, Fuzzy XML queries via context-based choice of aggregations. Kybernetika **36**(6), 635–655 (2000)
19. K.S. Jones, S. Walker, S.E. Robertson, A probabilistic model of information retrieval: development and comparative experiments: Part 2. Inf. Process. Manage. **36**(6), 809–840 (2000)
20. R. Agrawal, R. Rantzau, E. Terzi, Context-sensitive ranking, in *Proceedings of the ACM SIGMOD International Conference on Management of Data* (2006), pp. 383–394
21. A. Marian, S. Amer-Yahia, N. Koudas, D. Srivastava, Adaptive processing of top-*k* queries in XML, in *Proceedings of the 21st International Conference on Data Engineering* (2005), pp. 162–173
22. F. Weigel, H. Meuss, K.U. Schulz, F. Bry, Content and structure in indexing and ranking XML, in *Proceedings of the 7th International Workshop on the Web and Databases* (2004), pp. 67–72
23. W. Yan, Z. Ma, F. Zhang, X. Meng, Relaxing queries based on XML structure and content preference, in *Proceeding of the Web Information Systems Engineering* (2010), pp. 44–57
24. J. Li, C. Liu, R. Zhou, W. Wei, Top-*k* keyword search over probabilistic XML data, in *Proceedings of the 27th International Conference on Data Engineering* (2011), pp. 673–684
25. C. Sun, C. Chan, A. Goenka, Multiway SLCA-based keyword search in XML data, in *Proceedings of the 16th International Conference on World Wide Web* (2007), pp. 1043–1052
26. G. Watson, An algorithm for the single facility location problem using the Jaccard metric. SIAM J. Sci. Stat. Comput. **4**(4), 748–756 (1993)
27. L. Bai, X. Duan, B. Qin, Adaptive query relaxation and top-*k* result sorting of fuzzy spatiotemporal data based on XML. Int. J. Intell. Syst. **37**(3), 2502–2520 (2022)
28. Y. Xu, Y. Papakonstantinou, Efficient keyword search for smallest LCAs in XML databases, in *Proceedings of the 2005 ACM SIGMOD International Conference on Management of Data* (2005), pp. 537–538
29. A. Nierman, H. Jagadish, ProTDB: probabilistic data in XML, in *Proceedings of 28th International Conference on Very Large Data Bases*. Hong Kong SAR (2002), pp. 646–657
30. L. Guo, F. Shao, C. Botev, J. Shanmugasundaram, XRANK: ranked keyword search over XML documents, in *Proceedings of the 2003 ACM SIGMOD International Conference on Management of Data* (2003), pp. 16–27

31. X. Meng, X. Zhang, Y. Tang, C. Bi, Adaptive query relaxation and top-*k* result ranking over autonomous web databases. Knowl. Inf. Syst. **51**(2), 395–433 (2017)
32. X. Zhang, X. Meng, Y. Tang, C. Bi, Adaptive query relaxation and result categorization based on data distribution and query context. J. Inf. Sci. Eng. **33**(5), 1375–1396 (2017)
33. U. Nambiar, S. Kambhampati, Answering imprecise queries over autonomous web database, in *Proceedings of the 22nd International Conference on Data Engineering* (2006), pp. 45–45
34. I. Muslea, T.J. Lee, Online query relaxation via Bayesian causal structures discovery, in *Proceedings of the 20th National Conference on Artificial Intelligence* (2005), pp. 831–836
35. Z. Chen, T. Li, Addressing diverse user preferences in SQL-query-result navigation, in *Proceedings of the International Conference on Management of Data* (2007), pp. 641–652
36. P. Bosc, A. Hadjali, O. Pivert, Empty versus overabundant answers to flexible relational queries. Fuzzy Sets Syst. **159**(12), 1450–1467 (2008)
37. D. Mottin, A. Marascu, S. Basu-Roy, G. Das, T. Palpanas, Y. Velegrakis, IQR: an interactive query relaxation system for the empty-answer problem, in *Proceedings of the International Conference on Management of Data* (2014), pp. 1095–1098
38. D. Martinenghi, R. Torlone, Taxonomy-based relaxation of query answering in relational databases. VLDB J. **23**(5), 747–769 (2014)
39. Z. Zhao, Z. Ma, L. Yan, An efficient classification of fuzzy XML documents based on kernel ELM. Inf. Syst. Front. **23**(1), 515–530 (2019)

Chapter 8
Adaptive Query Relaxation and Result Categorization of Fuzzy Spatiotemporal XML Data

8.1 Introduction

As discussed in Chap. 7, query relaxation is frequently employed to increase the volume of query results. The effective implementation of query relaxation has garnered considerable interest from both academic researchers and industry practitioners. Notable advancements have been made in the realm of query relaxation for both general and spatiotemporal data. The fundamental premise of these methodologies involves the reduction or reconfiguration of the constraints imposed on the original query to broaden its applicability. In the context of query relaxation for general data, Muslea and Lee [1] introduce the TOQR algorithm, which utilizes a limited dataset to uncover implicit relationships among domain attributes, subsequently leveraging this domain knowledge to relax unsuccessful queries. Mottin et al. [2] present the IQR, which dynamically proposes a single relaxation of the original query conditions at a given time based on specific optimization objectives. Martinenghi and Torlone [3] propose a logical model and a class of abstract query languages designed for querying relational datasets characterized by vague schemas, contingent upon the availability of taxonomies. Zhang et al. [4] introduce a framework that facilitates automatic query relaxation by taking into account the distribution of query values within the database. Meng et al. [5] offer a model that incorporates user preferences during the query relaxation process, thereby enhancing the relevance and reliability of the returned items. Bosc et al. [6] present a solution that adjusts the fuzzy conditions present in the user query through the application of suitable transformations. Conversely, with respect to query relaxation for spatiotemporal data, Yang et al. [7] propose a heuristic probabilistic relaxation method for road network matching, aimed at integrating the most current information from multiple sources of geospatial data. Ford and Kench [8] investigate the effects and relaxation strategies following typhoon events. Mitra and Pallickara [9] design a scalable, query-oriented distributed data integration framework that dynamically generates precise interpolations for specified spatiotemporal scopes, while also estimating the uncertainty

L. Bai and L. Zhu, *Fuzzy Spatiotemporal XML Data Management*, Studies in Computational Intelligence 1183, https://doi.org/10.1007/978-3-031-81033-6_8

associated with such interpolations in instances of spatiotemporal misalignment of data points.

Following the relaxation of queries, a subsequent challenge arises, namely the phenomenon of "information overload," characterized by an excessive volume of query results. This issue has catalyzed significant research interest in the classification of results, leading to the development of various solutions [4, 10, 11, 12]. For instance, Chen and Li [10] introduce a cost-based algorithm designed to construct a navigational tree that incorporates user query history. Similarly, Chakrabarti et al. [11] automatically generate a labeled hierarchical category structure based on the content of the tuples within the answer set. Afif et al. [12] present a hybrid genetic algorithm that employs rule-based categorization to improve the precision of categorization rules and yield accurate classifiers. Tang et al. [13] utilize interactions ranging from two to five variables for feature selection, applying their methodology to text categorization. Additionally, Zhang et al. [4] propose an innovative algorithm for categorizing query results that takes into account user contextual preferences and data distribution. Building upon this foundation, they introduce a new algorithm for generating a navigational tree aimed at the categorization of query results. While these approaches offer diverse methodologies for result classification, they exhibit limitations in their applicability to spatiotemporal data. This is primarily due to their tendency to treat spatiotemporal features as general attributes and their reliance on deterministic attribute values, which neglects the inherent fuzziness associated with such data.

The increasing adoption of XML for data representation has spurred significant interest in the study of XML data. XML data are frequently structured in a tree model, which allows for flexible organization of nodes at various levels and the representation of node fuzziness along the path. Consequently, numerous research initiatives have been directed towards the modeling and querying of fuzzy XML data [14], as well as fuzzy spatiotemporal XML data [15–17]. In the context of querying XML data, Amer-Yahia et al. [18] formalized the concept of query relaxation and established a ranking scheme for XML queries. Xu et al. [19] introduced a contextual preference query method alongside content scoring, while Li et al. [20] explored probabilistic threshold keyword queries (PrTKQ) over XML data, proposing two querying algorithms that utilize probability density functions. Additionally, Sözer et al. [21] presented an innovative approach for querying spatiotemporal data associated with fuzzy spatial and complex objects, as well as spatial relations. Bai et al. [22] proposed an adaptive query relaxation and result categorization method for fuzzy spatiotemporal data based on XML, ensuring compatibility with spatiotemporal features during the execution of related queries. Although these efforts primarily concentrate on querying XML data, they also contribute foundational methodologies for fuzzy spatiotemporal query relaxation and result categorization within the XML framework.

Table 8.1 presents a comparative analysis of query relaxation and result categorization pertaining to XML data. It is evident that the majority of existing research on query relaxation and/or result categorization predominantly focuses on general data, neglecting the incorporation of spatiotemporal features. Additionally, while investigations into the modeling and querying of XML data do address spatial, temporal, or spatiotemporal features, there remains a lack of exploration regarding their application in the context of query relaxation and result categorization.

Researches on fuzzy spatiotemporal query relaxation and result categorization utilizing XML continue to require further advancements. Our study does not merely extend prior investigations; rather, we introduce a novel and efficient methodology for fuzzy spatiotemporal query relaxation and result categorization based on XML. This approach effectively addresses spatiotemporal attributes in a unique manner and mitigates the influence of probability on the outcomes. We enhance the algorithm by employing clustering techniques to group spatiotemporal data, subsequently ranking these groups in descending order based on their average fuzziness. Given the rapidly increasing volume of collected fuzzy spatiotemporal data and the pressing need for effective fuzzy spatiotemporal queries, there is a critical necessity to develop a representation that facilitates the swift localization of spatiotemporal information while clearly conveying the associated fuzziness.

In this chapter, we first introduce an alternative efficient algorithm designed to address the fuzzy spatiotemporal query, specifically targeting the issue of insufficient responses in XML by incorporating considerations of probability and fuzziness. Furthermore, we propose an innovative method that takes into account the weight of each specified attribute within the query, while also analyzing the distribution of data within the database. This method aims to achieve a harmonious balance between proximity and fuzziness in the context of fuzzy spatiotemporal XML data. Secondly, we enhance the existing algorithms for generating navigation trees, thereby providing a more rational framework for categorizing the results of fuzzy spatiotemporal queries. Notably, our approach employs a clustering technique to group closely related spatiotemporal attribute values and subsequently ranks them based on their fuzziness.

8.2 Preliminary

8.2.1 Fuzzy Spatiotemporal XML Data

Fuzzy spatiotemporal data is characterized by two fundamental attributes: the spatiotemporal feature and the fuzzy feature. The spatiotemporal feature itself is comprised of spatial and temporal components. Specifically, the spatial component can be categorized into three types: point, line, and region, whereas the temporal component encompasses both time points and time intervals. Subsequently, we will delineate the various classifications of fuzzy spatiotemporal objects.

Definition 8.1 (*Types of fuzzy spatiotemporal data*) Let fuzzy temporal type be denoted as T_i and fuzzy spatial type as S_i. Then, the fuzzy spatiotemporal F_i is defined as:

Table 8.1 Comparative study on query relaxation and result categorization of (XML) data

	Spatial features	Temporal features	Spatiotemporal features	Fuzzy features	Query relaxation	Result categorization	XML based
Muslea and Lee [1]					✓		
Mottin et al. [2]					✓		
Martinenghi and Torlone [3]				✓	✓		
Zhang et al. [4]					✓	✓	
Meng et al. [5]					✓		
Bosc et al. [6]				✓	✓		
Yang et al. [7]	✓				✓		
Ford and Kench [8]	✓	✓	✓		✓		
Mitra and Pallickara [9]	✓	✓	✓		✓		
Chen and Li [10]						✓	
Chakrabarti et al. [11]						✓	
Zhao et al. [14]				✓			✓
Bai et al. [15]	✓	✓	✓				✓
Bai et al. [16]	✓	✓	✓	✓			✓
Bai and Zhu [17]	✓	✓	✓				✓
Amer-Yahia et al. [18]					✓		✓
Xu et al. [19]						✓	
Li et al. [20]				✓			✓
Sözer et al. [21]	✓	✓	✓				
Yu [23]	✓	✓	✓				
Ours	✓	✓	✓	✓	✓	✓	✓

Table 8.2 Six types of spatiotemporal data

	Spatial point	Spatial line	Spatial region
Temporal point	F_{TP}	F_{TL}	F_{TR}
Temporal interval	F_{IP}	F_{IL}	F_{IR}

$$F_i = T_i \times S_i \tag{8.1}$$

As delineated in Definition 8.1, spatiotemporal data can be categorized into six distinct types: fuzzy point at a fuzzy time point, fuzzy line at a fuzzy time point, fuzzy region at a fuzzy time point, fuzzy point within a fuzzy time interval, fuzzy line within a fuzzy time interval, and fuzzy region within a fuzzy time interval. These classifications are presented in Table 8.2.

Utilizing the six classifications of fuzzy spatiotemporal data, we develop a representation model to characterize this data.

Definition 8.2 (*Fuzzy spatiotemporal XML data model*) The fuzzy spatiotemporal object (*FSP*) can be defined as a 2-tuple $FSP = (Fuzz, ATTR)$, where:

- *Fuzz* represents the membership degree of a fuzzy spatiotemporal object exists in the current situation.
- *ATTR* represents the attributes of a fuzzy spatiotemporal object.

In Definition 8.2, both general attributes and spatiotemporal attributes are encompassed in *ATTR*. The general attributes consist of the value (*val*) and its associated fuzziness, with *val* denoting the attribute values pertinent to *FSP*. The spatiotemporal attributes are further categorized into temporal and spatial attributes. Specifically, the spatial attribute denotes the position of the *FSP*, while the temporal attribute indicates the time associated with the *FSP* in the present context. The representation of the fuzzy spatiotemporal data model is illustrated in Fig. 8.1.

XML data can be inherently depicted as ordered and directed trees, where each node signifies an attribute (or a value) and each edge denotes a relationship between attributes and values (or between attributes themselves). Subsequently, we introduce the concept of a fuzzy XML data tree as outlined in Definition 8.3.

Definition 8.3 (*Fuzzy XML data tree*) A Fuzzy XML data tree $T(V, E)$ is comprised of a node set V and an edge set E, where

- $V = \{v_1, v_2, \ldots\ldots v_n\}$ denotes a set of nodes of *FSP*
- $E = \{e_1, e_2, \ldots\ldots e_n\}$ denotes a set of edges of *FSP*
- $DE = \left(v_i, v_j\right)$ denotes a directed edge from node v_i to node v_j

In accordance with Definition 8.3, let v_1 and v_2 represent two nodes within the tree n, which is comprised of the node set $V(n)$. If $DE = (v_1, v_2) \in E(n)$ where $DE = (v_1, v_2)$ denotes the edge connecting v_1 and v_2, then the structural relationship between these nodes is classified as a parent–child relationship, with node v_1 serving as the parent of node v_2. Conversely, if there exists a path from v_1 to v_2 and

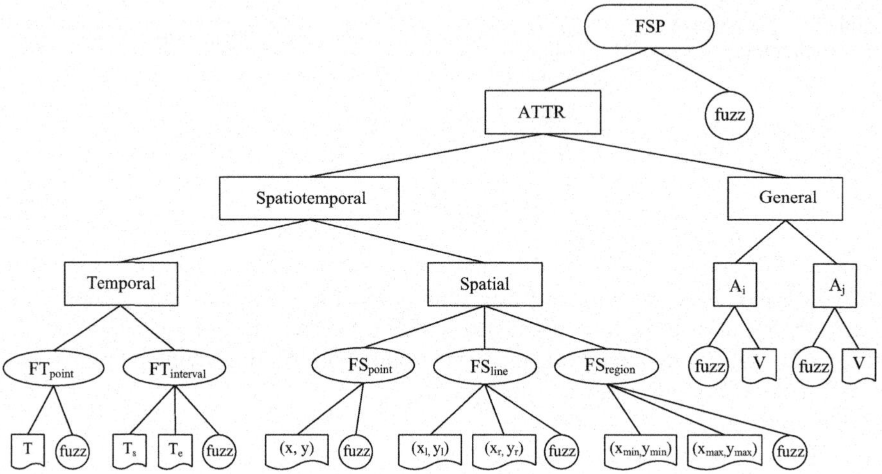

Fig. 8.1 The model of fuzzy spatiotemporal XML tree

the structural relationship between these nodes does not conform to the parent–child classification, it is then referred to as an ancestor–descendant relationship. Furthermore, if the trees s and t fulfill the conditions $V(s) \subseteq V(t)$ and $E(s) \subseteq E(t)$, it can be concluded that s is a subtree of t.

The degree of fuzziness of a child node in relation to its parent node is indicated along the connecting edge. Consequently, the computation of the fuzziness for a designated node is delineated in Definition 8.4.

Definition 8.4 (*Fuzzy XML node fuzziness representation*) Let $F(C|B), F(B|A), F(A)$. denote the degree of fuzziness associated with nodes C, B and A respectively. If $A \to B \to C$ is a chain originating from the root node A, then the fuzziness of the node can be defined as:

$$F(C) = F(C|B) \times F(B|A) \times F(A) \tag{8.2}$$

8.2.2 *Query Relaxation*

To enhance the alignment with users' initial query intentions, it is necessary to implement a relaxation of the original query. Let Q represent the original query that yields a limited number of results; upon the reduction of its constraints, it is transformed into $\tilde{Q}$. The concept of query relaxation is articulated in Definition 8.5, as noted by Meng et al. [5].

Definition 8.5 (*Query relaxation*) Utilize R to denote a relationship within a spatiotemporal database characterized by n tuples $\{t_1, t_2, \ldots\ldots, t_n\}$ and m attributes $\{A_1, A_2, \ldots\ldots, A_m\}$. The query Q is formulated on R, incorporating conjunctive conditions structured as $Q = \sigma \wedge_{i \in \{1,\ldots,k\}} C_i$ where the condition holds for $2 \leq k \leq m$, $C_i = (A_i \theta a_i)$, $\theta \in \{>, <, =, \geq, \leq, between\}$. Each attribute A_i in the foundational condition corresponds to an attribute of the dataset D, and a_i represents a value (or range) within its respective domain. By relaxing Q, a relaxed query $\tilde{Q}$ is derived, which aims to identify all tuples within D that exhibit a degree of similarity to the original qry Q, sursing a specified similarity threshold $\lambda \in (0, 1]$, as defined by:

$$\tilde{Q}(D) = \{t | t \in D, Sim(Q, t) \geq \lambda\} \tag{8.3}$$

In Definition 8.5, let $\tilde{Q}(D)$ denote a query result set that adheres tthe cteria of a relaxed query, while λ represents the threshold established by either the system or the user. The results derived from the relaxed query are those that exhibit the highest relevance to the original query. Following the acquisition of these relaxed query results, it is imperative to categorize them appropriately, and subsequently, a navigation tree should be presented to the user, taking into account the context of the query and the characteristics of the database.

8.2.3 Result Categorization

The navigation tree plays a crucial role in assisting users in making optimal selections when the query results generated by $\tilde{Q}$ are excessively numerous. Attributes deemed more significant by the user are positioned nearer to the root of the tree, while attributes at each level are organized according to their Estimated Navigation Tree Cost (*Ecost*). Consequently, the resulting navigation tree facilitates the user's decision-making process.

Definition 8.6 (*The fuzzy navigation tree*) A fuzzy navigation tree $T(V, E, L)$ is comprised of a set of nodes V, a set of edges denoted as E, and a set of labels represented as L. Each node $v \in V$ is characterized by a defined average fuzziness and a label ($label(v) \in L$) that delineates the conditions pertaining to a specific attribute.

Figure 8.2 illustrates an example of a fuzzy navigation tree representing the outcomes of query Q within a weather context. According to Definition 8.6, node v encompasses a collection of tuples $N(v)$ that fulfill all conditions imposed by its anceors, including itself. The attribute with the highest weight is positioned at the top level of the fuzzy navigation tree, while each subsequent layer is organized according to the fuzziness of the buckets, which is determined by the average fuzziness of all nodes contained within each bucket. From the user's standpoint, *Ecost* represents the cost associated with locating a satisfactory node, serving as a metric for evaluating the efficiency of the navigation tree. To elucidate the concept of *Ecost*, it is essential to first present the methodology for calculating path probability.

Definition 8.7 (*Path probability*) Let v_j. represent a leaf node within the tree T. The set $Anc(v_j)$ comprises the ancestors of the node v_j, including the node itself but excluding the root. The set $Sib(v_j)$. consists of the sibling nodes of v_i, also including the node itself. Furthermore, let K_1 denote the unit cost associated with visiting a leaf node, while K_2 signifies the unit cost for visiting an intermediate node. Consequently, the product of the probability P_j that a user explores each ancestor of the leaf node v_j (including the node itself) is defined as follows:

$$P_j = P_{(v_j)} \cdot \prod_{v_i \in Anc_{(v_j)}} P_{(v_i)} \tag{8.4}$$

In Definition 8.7, the notation P_j represents likelihood that users will exhibit interest in the sub-tree associated with the leaf node v_j. Specifically, $P_{(v_j)}$ indicates the probability that a user engages with the leaf node v_i, while $P_{(v_i)}$ signifies the probability that the user tacts with the intermediate node v_i, which serves as an ancestor to v_j. Subsequently, Definition 8.8 introduces the concept of *Ecost*.

Definition 8.8 (*Estimated navigation tree cost*) Let P_j represent the path probability, $K_1 N(v_j)$ denote the cost associated with visiting tuples at the leaf node v_j, and $K_2 \sum_{v_i \in Leaf(v_j)} |sib(v_i)|$ signify the cost incurred when visiting intermediate nodes. Consequently, *Ecost* is defined as follows:

$$Ecost(T) = \sum_{v_j \in Leaf(T)} P_j \left(K_1 N(v_j) + K_2 \sum_{v_i \in Leaf(v_j)} |sib(v_i)| \right) \tag{8.5}$$

The *Ecost* associated with a leaf node v in a classification tree is comprised of two components: the cost incurred in traversing the intermediate nodes and the cost associated with accessing all tuples contained within leaf node v. Specifically, to identify leaf node v, the user must examine the labels of each intermediate node as well as those of its siblings, necessitating visits to the intermediate nodes ($\sum_{v_i \in Leaf(v_j)} |sib(v_i)|$) from the root to the leaf node v. Conversely, it is necessary for all $N(v)$ tuples within the node v to be examined ($\sum_{v_i \in Leaf(v_j)} |sib(v_i)|$) when the user arrives at the leaf node v.

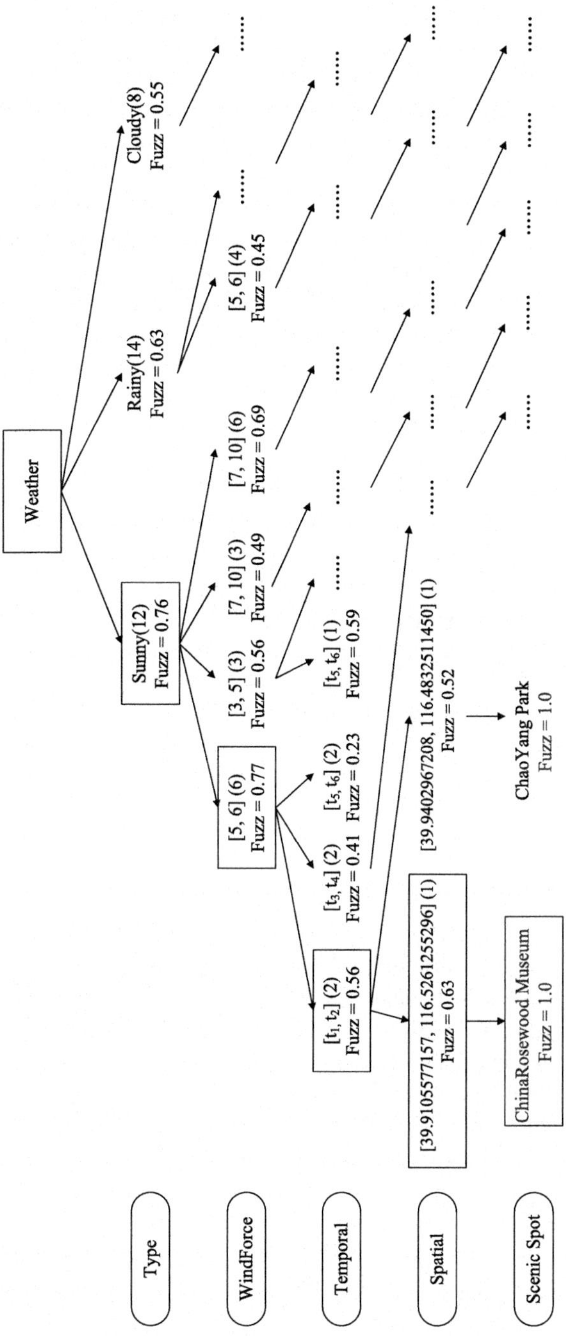

Fig. 8.2 The weather example of fuzzy navigation tree

8.3 Framework

In order to relax the initial query and yield categorized outcomes, this chapter introduces a two-step processing methodology to tackle the issue at hand. The framework for fuzzy spatiotemporal querying and result categorization is illustrated in Fig. 8.3.

The initial phase involves the relaxation of the original query during offline processing. This process entails the computation of Inverse Document Frequency (IDF) weights and the assessment of similarity for all unique attributes, alongside the documentation of data distribution. These computations are subsequently stored in auxiliary tables, which serve to significantly decrease query processing time. Upon the arrival of a new query, it is deconstructed into multiple predicates, and the IDF weights and similarity metrics for each specified query value are retrieved from the auxiliary tables. Following this, proximity is determined through fuzzy matching and similarity analysis. The query is then reformulated as a relaxed query based on the weights of the attributes and the proximity of the attribute values. Ultimately, the relevant results, which exhibit a proximity to the original query that meets or exceeds a predetermined relaxation threshold, are returned.

The subsequent step involves the classification of results derived from the relaxed query outcomes during the online phase. Based on the distribution of attribute values delineated by $\tilde{Q}$ within the database and the querying context, these attribute values are organized into several categories or "buckets." Following this categorization, the weights of the attributes and the average probabilities of each bucket are computed. In the constructed navigation tree, the attribute corresponding to each level is established based on its weight, while the arrangement of the buckets is determined by their average probabilities at each level. Ultimately, a navigation tree that is both precise and efficient is presented to the user, facilitating the selection of satisfactory results.

8.4 Fuzzy Spatiotemporal Query Relaxation

In the context of a fuzzy spatiotemporal query Q that yields limited or no results, it is important to note that a substantial number of records within the dataset may still fulfill the requirements of users. Consequently, the implementation of query relaxation becomes essential. The method for relaxing fuzzy spatiotemporal queries does not simply eliminate unsatisfactory results; rather, it assesses whether to return results based on modified criteria. Ultimately, the results provided align with the users' needs. This subsection will first elucidate the processes involved in calculating attribute weights and measuring the proximity between records, prior to introducing a viable method for fuzzy spatiotemporal query relaxation. Finally, an algorithm for fuzzy spatiotemporal query relaxation will be proposed. Table 8.3 presents a selection of datasets from our database that will be utilized in the subsequent examples.

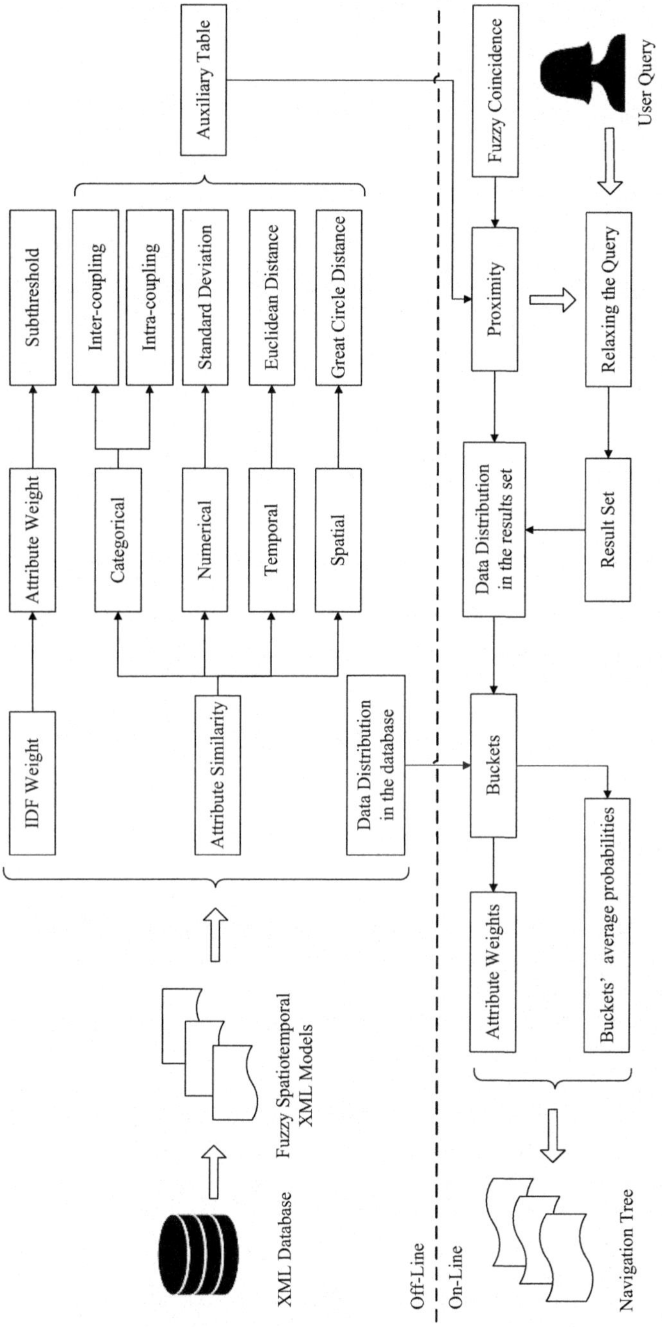

Fig. 8.3 The framework of query relaxation and result categorization of fuzzy spatiotemporal data

Table 8.3 Examples of weather forecast (City: Beijing, Date: 2018/11/19)

Type	District	Scenic spot	Wind force	Temperature (°C)	Humidity (%)	Forecast period	Location
Rainy	ChaoYang	Museum	> 5.4	4–13	83	T_{start} 08:00 T_{end} 10:00	x:39.9105577157 y:116.526125529
Rainy	ChaoYang	Park	> 6.5	4–13	76	T_{start} 08:30 T_{end} 09:00	x:39.9402967208 y:116.483251145
Rainy	DongCheng	Museum	> 6	6–11	73	T_{start} 13:00 T_{end} 16:00	x:39.9178896573 y:116.397050073
Cloudy	HaiDian	Park	> 6.5	1–18	76	T_{start} 14:00 T_{end} 16:00	x:39.9413212696 y:116.339708815
Cloudy	XiCheng	Market	> 6.3	2–10	83	T_{start} 19:00 T_{end} 21:00	x:39.9116991666 y:116.303567738
Windy	ShiJingShan	Park	> 6.3	2–10	83	T_{start} 06:00 T_{end} 12:00	x:39.9056943459 y:116.222994824
Sunny	FengTai	Park	< 5	6–13	63	T_{start} 18:00 T_{end} 22:00	x:39.8567665077 y:116.085382895
Sunny	DongCheng	Park	< 3	8–16	56	T_{start}08:00 T_{end} 11:00	x:39.9276641225 y:116.386954263
Rainy	TongZhou	Park	> 4.1	2–11	81	T_{start} 07:00 T_{end} 12:00	x:39.9074891651 y:116.688517349
Rainy	TongZhou	Temple	> 4.5	2–13	86	T_{start} 08:30 T_{end} 10:30	x:39.9192015845 y:116.655812485
Rainy	TongZhou	Market	> 4.2	2–13	86	T_{start} 09:30 T_{end} 11:30	x:39.9159190619 y:116.605782496
Rainy	TongZhou	Park	> 5.1	4–12	88	T_{start} 09:00 T_{end} 11:00	x:39.9094803705 y:116.617513273

8.4.1 Attribute Weight Assignment

The initial phase of assessing similarity involves determining the weights of attributes pertinent to a given fuzzy spatiotemporal query. Individual users possess distinct preferences, necessitating the assignment of varying weights to attributes based on user-specific criteria. The allocation of attribute weights is conducted using the IDF approach, which relies on the distribution of the relevant attributes and the user's query history. Given that the IDF weight assessment is traditionally applicable to categorical attributes, it is essential to extend its application to continuous attributes. Consequently, we propose an enhanced IDF weighting methodology grounded in the fuzzy spatiotemporal XML data model. This approach employs the categorical IDF method for independent fuzzy attributes and utilizes a fuzzy scoping IDF allocation technique for attributes characterized by fuzzy scopes.

8.4.1.1 Categorical IDF Weight Assignment

IDF serves as a metric for assessing the informativeness of data. In this chapter, we utilize the definition of IDF as proposed by Muslea and Lee [1] to evaluate the significance of numerical attribute values. Consequently, if a user places greater emphasis on the attribute node v_i compared to v_j, the degree of relaxation applied to the query criterion for the attribute node v_i should be less than that for v_j. This approach ensures that the responses generated are more pertinent to the initial query. For a specified query $A_i = v_i$, the $IDF_i(v)$ is:

$$IDF_i(v) = \log\left(\frac{n}{F_i(v)}\right) \tag{8.6}$$

where n represents the total number of sub-tree in the database, while $F_i(v)$ denotes the frequency of sub-tree in the database of $A_i = v_i$.

In particular, within a fuzzy dataset, the frequency is defined as the sum of the probabilities associated with the condition $A_i = v_i$.

As previously indicated, the significance of a categorical attribute value is regarded as equivalent to the significance of its associated attribute. For example, consider a query Q_1 at includes the condition $Humidity = 80\% \wedge Region = ChaoYang \wedge Type = Rainy \wedge City = Beijing \wedge Date = 2018/11/19$. If the frequency of $Type = Rainy$ in the Forecast dataset is 7, and the frequency of tuples in the database for the *Type* attribute is 12, we can derive the weight of the *Type* attribute by $IDF(Rainy) = \log\left({}^{12}/_{7}\right)$.

8.4.1.2 Scoping IDF Weight Assignment

In assessing the significance of scoping attribute values, it is not suitable to utilize the conventional definition of IDF due to its binary characteristics, wherein the values of u and v, while similar, remain distinct. Instead, the IDF associated with a scoping value should be contingent upon the proximity of related values. Consequently, it is necessary to broaden the scope of its application. Let $\{v\_(1), v_2, v_3 \cdots, v_n\}$ represent the set of values for attribute A_i present in the database. For a given attribute value v specified in the query, the $IDF_i(v)$ is:

$$IDF_i(v) = \log\left(\frac{n}{\sum_{j=1}^{n} e^{-\frac{1}{2}\left(\frac{v_j - v}{h}\right)^2}}\right) \tag{8.7}$$

where h represents the bandwidth parameter, $h = 1.06\,\sigma n^{-\frac{1}{5}}$, and σ denotes the standard deviation of all value nodes $\{v_1, v_2, v_3 \cdots, v_n\}$ on the $Dom(A_i)$.

In Eq. (8.7), the denominator signifies the aggregate of "contributions" to the variable v from each individual v_i within the domain Dom (A_i). These contributions are characterized by a Gaussian distribution, indicating that as the distance between v and v_i increases, the contribution from v_i diminishes correspondingly.

Additionally, there exists a generalized range query denoted as $A_i \in V_i$, where V_i represents a collection of categorical attributes or a specified range $[v_{min}, v_{max}]$ of scoping attributes. Consequently, $IDF_i(v)$ is:

$$IDF_i(v) = \max_{v \in Q_i} IDF(v) \tag{8.8}$$

Typically, the user's initial inquiry encompasses various conditions, resulting in the allocation of distinct weights to different keywords during the query relaxation process. Furthermore, it is essential to normalize these attribute weights, as this facilitates the system's ability to effectively and adaptively assess the weights of the designated attributes within the query. This normalization process involves dividing the IDF weight of each specified attribute by the total IDF weight of all specified attributes. Consequently, the weight w_i associated with the attribute A_i identified in the query can be computed as follows:

$$w_i = \frac{IDF_i(v)}{\sum_{j=1}^{k} IDF_j(v)} \tag{8.9}$$

where k represents the number of attributes specified by the query, and $IDF_i(v)$ denotes the IDF value of the attribute A_i.

From Eqs. (8.7) to (8.9), it is observed that an attribute exhibiting a higher frequency of occurrence is associated with a lower IDF value, which in turn results in a correspondingly reduced weight for that attribute. Consequently, when the original query is relaxed, it is the attributes with lower weight values that require relaxation. Subsequently, the sub-threshold for each designated attribute can be determined based on the attribute weights and the relaxation threshold, denoted as λ, for the query.

For a conjunctive query Q defined over a dataset D, the sub-threshold φ_i for each designated attribute within Q can be determined using Eqs. (8.10) and (8.11) where

$$\frac{w_1}{\varphi_1} = \frac{w_2}{\varphi_2} = \ldots = \frac{w_k}{\varphi_k} \tag{8.10}$$

$$\lambda = \sum_{i=1}^{k} w_i \times \varphi_i \tag{8.11}$$

where k. represents the number of specified attributes, w_i. denotes the weighofhe attribute A_i. specified by Q, and λ indicates the given relaxation threshold.

To enhance the efficiency of query processing, it is essential to compute the proximity between fuzzy attribute values in advance, prior to execution. Consequently, the subsequent subsection will focus on the methodologies for measuring proximity among fuzzy attribute values, encompassing both fuzzy general attribute values and fuzzy spatiotemporal attribute values. Notably, general attributes are categorized into two distinct types: categorical attributes and scoping attributes.

8.4.2 Fuzzy Spatiotemporal Attribute Proximity Assignment

In light of the unique characteristics of fuzzy spatiotemporal objects, this subsection is organized into two distinct segments. The initial segment addresses the proximity of fuzzy general attribute values, while the subsequent segment focuses on the proximity of fuzzy spatiotemporal attribute values.

8.4.2.1 Proximity of Fuzzy General Attribute Values

This subsection begins by delineating the fuzzy general attribute into two distinct categories: fuzzy categorical attributes and fuzzy scoping attributes. Subsequently, it presents methodologies for assessing the proximity of values associated with fuzzy general categorical attributes, followed by an introduction to the proximity calculation techniques pertinent to general fuzzy scoping attribute values.

The proximity coefficient serves as a quantitative metric for assessing the similarity between two entities, a single entity and a group of entities, or between two groups of entities. A fundamental criterion for evaluating proximity is the similarity coefficient. To derive the relative similarity coefficient between two values of a categorical attribute, an adaptive coupling relationship measurement method is employed [4]. This model incorporates two key parameters: Intra-coupling relationship (*IaR*) and Inter-coupling relationship (*IeR*). *IaR* pertains to the frequency of occurrences of the values within their shared attribute, while *IeR* refers to the interactions of other attributes associated with this attribute. Additionally, the Intra-coupling and Inter-coupling relationships of a pair of values are integrated to form their coupling relationship, thereby reflecting their relative semantic similarity.

In Table 8.3, two values of the attribute *District* are considered similar when they occur with comparable frequencies, exemplified by the pairs *HaiDian* with *ChaoYang* and *ShiJingShan* with *FengTai*. These pairs exhibit a degree of intra-coupling similarity; however, it is important to note that the former pair (*HaiDian* with *ChaoYang*) demonstrates a greater degree of similarity than the latter (*ShiJing-Shan* with *FengTai*). This is attributable to the fact that the former pair shares a higher number of common values across other attributes. Specifically, the values for the attributes *Humidity* and *WindForce* associated with *HaiDian* and *ChaoYang* are identical (76% and > 6.5, respectively), and the values for the attribute *Price* are nearly equivalent. Consequently, the assessment of similarity between values of the attribute *District* should also take into account the interdependencies with other attributes, such as *Humidity*, *WindForce*, and *Weather type*, in order to accurately determine

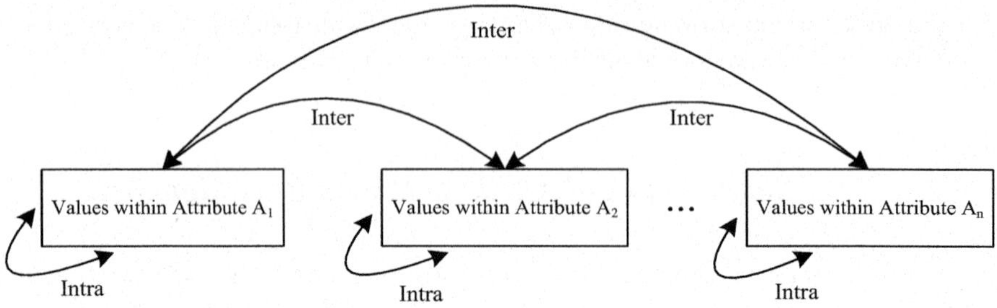

Fig. 8.4 The explanation of *IaR* and *IeR*

inter-coupling similarity. In conclusion, the relationships of intra-coupling and inter-coupling serve to reflect their relative semantic similarity. A concise overview of *IaR* and *IeR* is illustrated in Fig. 8.4.

The variation in the frequency of attribute value occurrences serves as an indicator of value similarity within the context of frequency distribution. Consequently, the *IaR* between two values can be assessed by utilizing their current frequency within the corresponding attribute domain. Specifically, the *IaR* between a pair of values (x and y) associated with a categorical attribute node (A_j) in Tree T can be determined as follows:

$$\delta_{A_j}^{IaR} = \frac{N_{A_j(x)} \times N_{A_j(y)}}{N_{A_j(x)} + N_{A_j(y)} + N_{A_j(x)} \times N_{A_j(y)}} \tag{8.12}$$

where $N_{A_j}(x)$ represents the cumulative possibility of sub-trees that include the value node x associated with the attribute node A_j, while $N_{A_j}(y)$ signifies the cumulative possibility of sub-trees that include the value node y linked to the attribute nodeA_j.

As indicated by Eq. (8.12), an increase in the frequency of attribute values within a pair results in a closer *IaR*. For instance, as presented in Table 8.3, the attribute value *ChaoYang* occurs twice ($N_{A_j}(ChaoYang) = 2$), DongCheng also occurs twice ($N_{A_j}(DongCheng)$), while FengTai appears only once ($N_{A_j}(FengTai) = 1$) in Dom (*Region*). Consequently, their intra-coupling similarity is $\delta_{Region}^{IaR}(ChaoYangandDongCheng) = {}^4\!/_8$ and $\delta_{Region}^{IaR}(ChaoYangandFengTai) = {}^2\!/_5$. This observation is justifiable, given that *ChaoYang* and *DongCheng* are situated at the same latitude, with *ChaoYang* being in closer proximity to *DongCheng*. Therefore, the similarity between *ChaoYang* and *DongCheng* is greater than that between *ChaoYang* and *FengTai*.

In addition to the intra-coupling similarity relationship among values, interactions arising from other attributes also influence their proximity. For instance, consider the values *ChaoYang* and *HaiDian* within the attribute *District*, as presented in Table 8.3. The set of values associated with *ChaoYang* under the attribute *ScenicSpot* is {*Museum*, *Park*}, while that for *HaiDian* is {*Park*}. It is evident that the intersection set U of *ChaoYang* and *HaiDian* concerning the attribute *ScenicSpot* is $\{Park\}$, indicating that *Park* is a common value shared by both *ChaoYang* and *HaiDian*. To assess the inter-coupling *IeR* between attribute value nodes, it is essential to introduce the concept of ICP Information Conditional Probability [5]. The $ICP(A|B)$ denotes the probability of the occurrence of event A given that another event B has already transpired. Consequently, the $ICP(u|x)$ represents the likelihood of u co-occurring with x when x is present in the relationship.

$$P_{A_k|A_j(u|x)} = \frac{\left|T_{A_k(u)} \cap T_{A_j(x)}\right|}{\left|T_{A_j(x)}\right|} \tag{8.13}$$

where $T_{A_j(x)}$ denotes the collection of tuples that include the value x for the attribute node A_j, while $T_{A_k(u)}$ denotes the collection of tuples that include the value u for the attribute node A_k.

Utilizing *ICP*, we can delineate the inter-coupling relationship between two variables (x and y) concerning attribute A_j. through their shared values on attribute A_k. as follows:

$$P^{Is}_{A_j|A_k}(x, y) = \sum_{u \in U} \min\left\{P_{A_k|A_j(u|x)}, P_{A_k|A_j(u|y)}\right\} \tag{8.14}$$

where U denotes the intersection set pertaining to attribute node Ak for the variables x and y, u represents an element of the set U, $P_{A_k|A_j(u|x)}$. signifies the ICP of u in relatto x, and $P_{A_k|A_j(u|y)}$ denotes the ICP of u in relation to y.

Subsequently, the inter-coupling relationship ($\delta^{IeR}_{A_j}(x, y)$) between value nodes x and y with respect to attribute A_j. can be determined as follows:

$$\delta^{IeR}_{A_j}(x, y) = \sum_{k=1,k\neq j}^{m} w_k \times P^{Is}_{A_j|A_k(x,y)} \tag{8.15}$$

where $\sum_{k=1}^{m} w_k = 1., w_k \in [0, 1]$

Utilizing the concepts of Intra-coupling ($\delta^{IaR}_{A_j}$) and Inter-coupling ($\delta^{IeR}_{A_j}$), the proximity relationship I_{proxA_j} between a pair of value nodes (x, y) that are associated with the same attribute A_j. is calculated as follows:

$$I_{proxA_j}(x, y) = \alpha \cdot \delta^{IaR}_{A_j}(x, y) + (1 - \alpha) \cdot \delta^{IeR}_{A_j}(x, y) \tag{8.16}$$

where $\delta^{IaR}_{A_j}(x, y)$ represents the Intra-coupling, and $\delta^{IeR}_{A_j}(x, y)$ denotes the Inter-coupling.

Additionally, fuzzy scoping attributes represent the second category of fzgal attributes. The methodology for assessing the proximity between two values of scoping fuzzy attributes is grounded in fuzzy set theory. The fuzzy relationship denoted as "close to" serves as a metric for estimating the similarity between two scoping values [6]. The membership function associated with the concept of "close to" is presented as follows:

$$\mu_{close\ to\ Y}(u) = \frac{1}{1 + \left(\frac{u-Y}{\beta}\right)^2} \tag{8.17}$$

According to Eq. (8.17), an increase in the value of β results in a broader curve. Consequently, for a specified set of scoping value nodes $\{v_1, v_2, v_3 \ldots v_n\}$ associated with attribute A_k, the proximity coefficient N_{proxA_k}. between nodes v_i. and v_j can be determined using Eq. (8.18).

$$N_{proxA_k}(v_i, v_j) = \frac{1}{1 + \left(\frac{v_i - v_j}{h}\right)^2} \tag{8.18}$$

In the context of scoping fuzzy general attribute value proximity, when a condition "$A_k = q$" is applied to query Q, φ_k represents the sub-threshold for A_k. The relaxation range of the fuzzy scoping query criterion associated with the attribute A_k is articulated in Eq. (8.19). Additionally, if the condition imposed on a scoping attribute is expressed as an interval ($A_k \in [q_{min}, q_{max}]$), the initial fuzzy condition can be adjusted as indicated in Eq. (8.20).

$$\left[q - h\sqrt{\frac{1-\phi_k}{\phi_k}}, q + h\sqrt{\frac{1-\phi_k}{\phi_k}}\right] \tag{8.19}$$

$$\left[q_{\min} - h\sqrt{\frac{1-\phi_k}{\phi_k}}, q_{\max} + h\sqrt{\frac{1-\phi_k}{\phi_k}}\right] \tag{8.20}$$

Following the determination of the proximity of fuzzy general attributes, it is essential to address the measurement of spatiotemporal attributes. Given that spatiotemporal data is defined by its continuous alteration in spatial position over time, the methodologies applicable to fuzzy general attributes are insufficient for addressing fuzzy spatiotemporal attributes. Consequently, a novel approach for measuring fuzzy spatiotemporal distance is proposed to effectively address this issue.

8.4.2.2 Proximity of Fuzzy Spatiotemporal Attribute Values

A fuzzy spatiotemporal object can be defined as a fuzzy spatial object that undergoes changes over time, wherein each spatial object possesses a distinct value characterized by fuzziness within the spatial domain at any given moment. For instance, the location of a fuzzy spatiotemporal object is represented by its latitude(x) and longitude(y). The interrelationship between temporal and spatial attributes is illustrated in Fig. 8.5a and b. The analysis of temporal and spatial attributes will be conducted independently prior to their integration. The computation of fuzzy spatiotemporal proximity is articulated in Eq. (8.21).

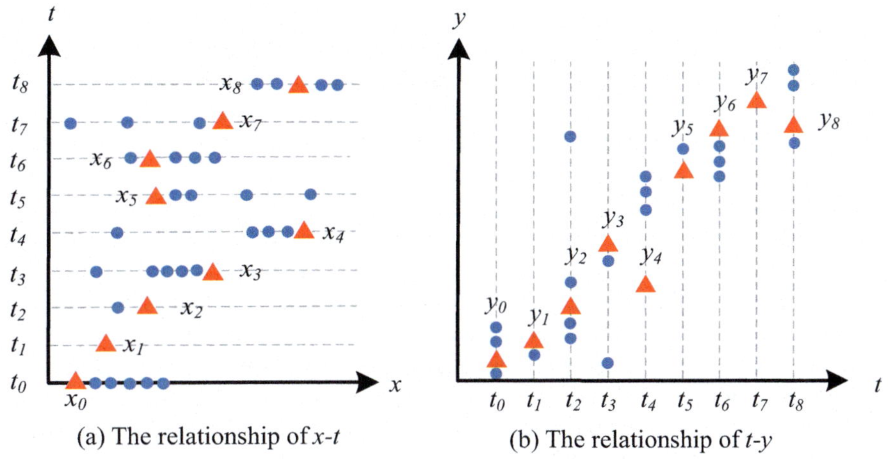

Fig. 8.5 The relationship of temporal attribute and spatial attribute

$$ST_{prox} = \alpha \cdot S_{prox}(S_i, S_j) + (1 - \alpha) \cdot T_{prox}(T_i, T_j) \tag{8.21}$$

where $S_{prox}(S_i, S_j)$ represents the proximity of spatial attribute, $T_{prox}(T_i, T_j)$ denotes the proximity of temporal attribute, and $\alpha \in [0, 1]$ indicates an adjustable parameter utilized to ascertain the weights of temporal and spatial attributes, thereby facilitating the efficient and effective resolution of the spatiotemporal relationships among spatiotemporal objects.

The closeness of fuzzy spatiotemporal attribute values is contingent upon both similarity and the fuzzy coefficient. Consequently, a methodology is proposed that assesses similarity through distance measurement while also accounting for the fuzzy coefficient. This approach effectively addresses the challenges associated with quantifying the similarity of spatiotemporal attributes and gives due consideration to the inherent fuzziness arising from uncertainties in data boundaries or inaccuracies in temporal (or spatial) measurements.

Considering the similarity between two temporal attributes, denoted as ($T\_$ i, T_j), along with their fuzzy coincidence, and two spatial attributes, represented as (S_i, S_j), along with their fuzzy coincidence, the proximity of the temporal and spatial attributes can be determined through the following calculation:

$$T_{prox}(T_i, T_j) = FC \times T_{sim}(T_i, T_j) \tag{8.22}$$

$$S_{prox}(S_i, S_j) = FC \times S_{sim}(S_i, S_j) \tag{8.23}$$

Subsequently, the methodology for calculating the similarity (T_{sim} and S_{sim}), as well as the fuzzy coefficient (FC) pertaining to both temporal and spatial characteristics, will be presented.

The Euclidean distance is widely recognized as the predominant method for assessing the distance between two objects in existing research literature. As a traditional metric, it provides a definitive evaluation of the distance between objects. However, when considering the complexities of spatiotemporal features, the application of various distance measures to spatial and temporal attributes is necessitated by the concept of relative distance. For instance, while the Euclidean distance may yield small values for latitude and longitude, these values may correspond to significantly larger distances in reality, indicating a greater relative distance. Conversely, the values associated with time points and intervals, when measured using Euclidean distance, tend to reflect their actual magnitudes, resulting in a relatively small relative distance. To enhance representation and calculation accuracy, we have employed distinct measures. In this chapter, we evaluate the similarity of temporal attribute values using an improved version of the Euclidean distance, taking into account the specific characteristics of spatiotemporal data. Nevertheless, the Euclidean distance model proves inadequate for spatial data applications. This inadequacy arises from the limited scale of calculations, where the differences in longitude and latitude yield small values that do not accurately represent the larger distances they may signify in reality. This issue can be effectively addressed through the implementation of the Great-Circle Distance model. Consequently, the Great-Circle Distance model is utilized to measure the distances of spatial attribute values, thereby rectifying accuracy discrepancies.

In the context of temporal attributes that can be categorized into discrete time points and continuous time intervals, a standardized methodology is employed to address the various mapping relationships between these points and intervals. The initiation and conclusion of a time period are represented as a specific point within a two-dimensional coordinate system (illustrated in Fig. 8.6). Subsequently, an enhanced version of the Euclidean distance is utilized to precisely assess the similarity between two temporal attribute values. A smaller distance between the two points indicates a higher degree of similarity. A notable enhancement in this approach is the stipulation that the distance $T_{dist}(T_i, T_j)$ should be set to zero when two time intervals (T_i and T_j) completely overlap. Based on the temporal information ($[T_{i_{start}}, T_{i_{end}}]$ *and* $[T_{j_{start}}, T_{j_{end}}]$) of the two objects, he similarity of fuzzy temporal attribute values is articulated in Eq. (8.24), while the corresponding distance is computed using Eq. (8.25).

$$T_{sim}(T_i, T_j) = 1 - \frac{T_{dist}(T_i, T_j)}{Maxdist} \tag{8.24}$$

$$T_{dist}(T_i, T_j) = \sqrt{\left(T_{i_{start}} - T_{j_{start}}\right)^2 + \left(T_{i_{end}} - T_{j_{end}}\right)^2} \tag{8.25}$$

- Time point $T_{i_{start}} = T_{i_{end}}, T_{j_{start}} = T_{j_{end}}$.
- Time interval $T_{i_{start}} < T_{i_{end}}, T_{j_{start}} < T_{j_{end}}$.

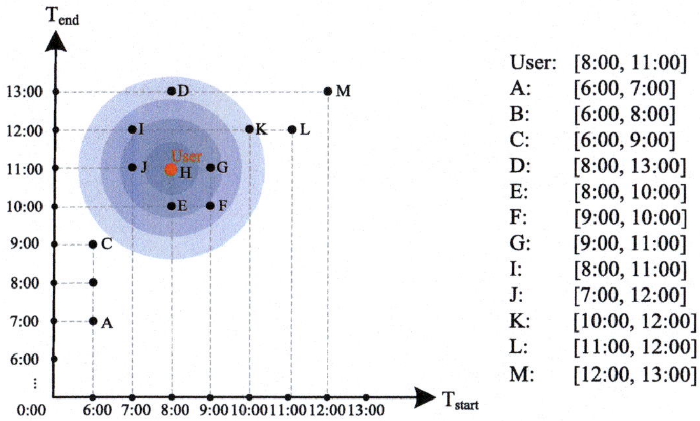

Fig. 8.6 The Euclidean distance in temporal attribute values

where *Maxdist* denotes the maximum distance among all temporal attribute values in the relevant spatiotemporal database.

For instance, if we consider $Maxdist = 5$ as the maximum duration of continuous rainfall within the pertinent database, and two values, $T_{ChaoYangPark}[8:00, 10:00]$ and $T_{MusicPark}[9:00, 11:00]$, from Table 8.3 that pertain to the temporal attribute, we can subsequently compute the temporal distance and their corresponding temporal similarity as $T_{dist}\left(T_{ChaoYangPark}, T_{MusicPark}\right) = \sqrt{2}$ and $T_{sim}\left(T_{ChaoYangPark}, T_{MusicPark}\right) = 1 - {}^{T_{dist}}\!/_{Maxdist} = 1 - {}^{\sqrt{2}}\!/_{5} = 0.7172$.

There exist two specific scenarios that are fully overlapping ($\left[T_{j_{start}}, T_{j_{end}}\right] \subseteq [T_i, T_i]$) and intersecting ($\left[T_{j_{start}}, T_{j_{end}}\right] \supseteq [T_i, T_i]$). These scenarios are directly incorporated into the results generated when a user submits a query ($[T_i, T_i]$). For instance, a traveler in Beijing may wish to explore options between 8:00 and 10:00 in the morning, necessitating an assessment of which scenic locations are not experiencing rain during that timeframe. It is evident that the user seeks information regarding both the anticipated and current rainfall conditions within this period. Consequently, it is imperative to integrate these two specific scenarios into the results. For example, if the user queries the rainy period from 9:00 to 11:00, it is essential to also include overlapping intervals such as [9:30, 11:30] or [8:30, 11:30].

In the context of spatial attributes, the similarity of spatial attribute values ($S_i and S_j$) is assessed through their spatial distance. The earth is modeled as a sphere, with each point on its surface represented by a set of coordinates (longitude and latitude). The geographic distance on this spherical surface is approximately equivalent to the spatial distance. Various methodologies exist for quantifying spatial distance [24–26]. In this chapter, we have selected the Great-Circle Distance method as our approach for measuring spatial distance. Subsequently, we will provide a detailed calculation of the Great-Circle Distance.

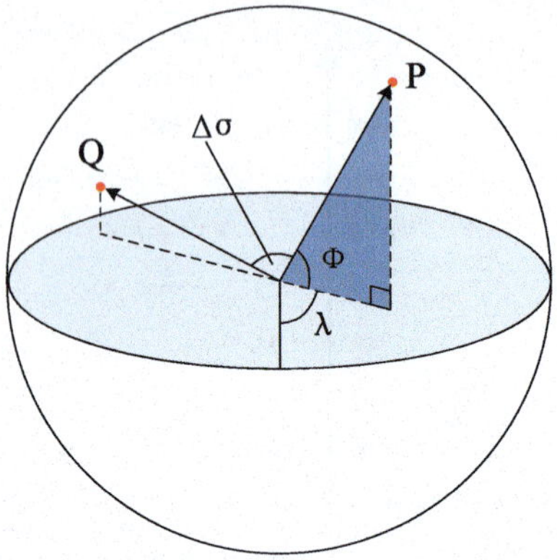

Fig. 8.7 The measurement of the spatial distance between two points

As illustrated in Fig. 8.7, the central angle $\Delta\sigma$ between two spatial points P (latitude (Φ_1) and longitude (λ_1)) and Q (latitude (Φ_2) and longitude (λ_2)) can be determined using the formula presented in Eq. (8.26).

$$\Delta\sigma = \arccos(\sin\Phi_1 \cdot \sin\Phi_2 + \cos\Phi_1 \cdot \cos\Phi_2 \cdot \cos(\Delta\lambda)) \tag{8.26}$$

According to Eq. (8.26), the minimum spatial distance between points P and Q is denoted as d, which is calculated as the product of the sphere's radius (sphere (r)) and the central angle ($\Delta\sigma$). Subsequently, the similarity of fuzzy spatial attribute values can be derived as indicated in Eq. (8.27). Notably, *Maxdist* represents the maximum geographical distance threshold, which serves the purpose of normalizing spatial similarity.

$$S\_sim(S_i, S_j) = 1 - \frac{S_{dist}(S_i, S_j)}{Maxdist} \tag{8.27}$$

where $S_{dist}(S_i, S_j)$ denotes the great circle distance of two objects.

For instance, considering two geographical coordinates (116.68°E, 39.91°N) and (116.61°E, 39.90°N), the great circle distance between these points is calculated to be 101,356 m. Given that the maximum distance threshold is set at 20 km, the spatial similarity between these two points can be obtained as $\left(1 - {}^{101356\text{m}}/_{200000m}\right) =$ 0.4931.

In the context of measurement and data recording, certain errors are unavoidable due to the inherent challenges associated with accurately capturing spatiotemporal information. Additionally, each spatiotemporal attribute value within the database possesses a degree of fuzziness. Historically, the approach to addressing this fuzziness has involved the multiplication of fuzziness and similarity; however, this method is not universally applicable. For instance, a scenario may arise where two attribute values exhibit a high degree of similarity, yet their respective fuzziness levels are low, leading to a diminished proximity when calculated as the product of similarity and fuzziness. Consequently, one of the values may be excluded from the results due to its low fuzzy degree, despite the user's satisfaction with both values. To mitigate the impact of fuzzy degree on the results, we propose the introduction of an association weight factor to assess fuzzy coincidence. Given two attribute values along with their corresponding fuzziness (δ_i, δ_j) and a defined fuzzy threshold γ, the calculation of fuzzy coincidence can be articulated as follows:

$$FC = \frac{min(\delta_i, \delta_j, \gamma)}{\gamma} \tag{8.28}$$

Following the introduction of the proximity measurement methodology across various attribute types, the proximity for each data point within the dataset is computed. Data points exhibiting proximity values that exceed a predetermined threshold will be integrated into the final results. However, the initial phase of calculating proximity involves assessing the similarity between designated attribute values, a process that can be particularly time-intensive for the system, especially when dealing with large datasets. Consequently, it is essential to develop auxiliary information for each attribute type during the offline phase and to store this information in an Auxiliary Table. Subsequently, we will outline the algorithm for fuzzy spatiotemporal query relaxation.

8.4.3 Fuzzy Spatiotemporal Query Relaxation Algorithm

The query rewriting approach introduced by Zhang et al. [4] is deemed inadequate for fuzzy spatiotemporal XML data due to its inefficiency in managing spatiotemporal information. Additionally, the relaxed queries generated by this method tend to yield numerous results with low relevance, which is not acceptable. Consequently, the suitability of each data point is assessed dynamically through the application of a proximity measurement metric.

Given a query $Q = \{q_1 \wedge q_2 \wedge \ldots \wedge q_k\}$, a relaxation threshold λ, and the attribute weights $W = \{w_1, w_2, \ldots, w_k\}$, the query relaxing algorithm can be depicted as follows.

Algorithm 8.1 The query relaxing algorithm

Input: query $Q = \{q_1 \wedge q_2 \wedge \ldots \wedge q_k\}$, relaxed threshold λ, attribute weights $W = \{w_1, w_2, \ldots, w_k\}$, the Auxiliary table G, the XML dataset $D = \{d_1, d_2, \ldots, d_n\}$

Output: the result dataset R, compute the sub-threshold φ_i for each attribute A_i specified by query Q

01 $R \leftarrow$ empty set
02 **for each** datum d in dataset D:
03 **for each** predicate q_i with original limited range (A_i in L):
04 $v \leftarrow$ the value of d on attribute A_i
05 **if** A_i is a scoping attribute
06 $\text{h} \leftarrow G(A_i)$
07 replace L with $\left[q_{min} - h\sqrt{\frac{1-\varphi_i}{\varphi_i}}, q_{max} + h\sqrt{\frac{1-\varphi_i}{\varphi_i}}\right]$ to be L'
08 **end if**
09 **if** A_i is a spatiotemporal attribute
10 $FC \leftarrow \frac{min(\delta_d, \gamma)}{\gamma}$, δ_d is the possibility of value v
11 $proximity \leftarrow FC \cdot \max(G(A_i, v, l)), l \in L$
12 **end if**
13 **if** v not in L' or $FC < \varphi_i$:
14 continue
15 **else**
16 **if** $proximity < \varphi_i$:
17 continue
18 **end if**
19 **add** v into R
20 **Return** R

In the query relaxation process, an initial empty set (R) is established to hold the relaxed results (Line 01). Subsequently, the query relaxation algorithm examines the dataset, where each predicate attribute is associated with its respective value and fuzzy coincidence (Lines 02–04). For the scoping attribute, the original constrained range is adjusted (Lines 05–08) in accordance with Eq. (8.20). Regarding the spatiotemporal attribute (Lines 09–12), its corresponding proximity is calculated based on similarity and fuzzy coincidence (FC). The decision to include a value in the results set is made based on the proximity and the attribute sub-threshold (Lines 13–20). Additionally, similarities are recorded in the Auxiliary Table during the offline phase, significantly reducing the need for real-time calculations of similarities during online processing. This approach also leads to a substantial decrease in response time (RT) during online operations.

Upon relaxing the initial query and acquiring the resultant dataset, it is imperative to classify the results to enhance user satisfaction and reduce the time required for information retrieval. Subsequently, we will rank the query results based on the context of the query and the distribution of the data.

8.5 Result Categorization of Fuzzy Spatiotemporal Data

The relaxed query frequently yields a substantial volume of results. To effectively address user requirements, it is imperative to optimize these results. Consequently, this subsection will categorize the result set based on users' contextual preferences and the distribution of data within the database. Subsequently, a navigation tree, organized by conditional significance, will be presented to the user. Finally, the algorithm utilized for generation will be proposed.

8.5.1 User Context Preference and Data Distribution

In order to construct an optimal tree that minimizes information overload for the user, it is essential to identify the suitable attributes for each layer within the XML data structure. Additionally, it is important to recognize that users may exhibit varying preferences depending on the context; therefore, we incorporate users' contextual preferences into our analysis.

Numerous methodologies exist for ascertaining a user's contextual preferences, one of which involves analyzing query history to infer user preferences [5]. However, these approaches often overlook scenarios in which query histories are generated by multiple individuals or in the absence of user feedback. Furthermore, they primarily focus on general data without accounting for spatiotemporal features. To achieve a more nuanced and precise understanding of user preferences, we propose a method that considers the user's current query conditions as the contextual framework. This approach entails the assignment of attribute weights in conjunction with data distribution. Notably, KL-distance is a widely utilized technique for quantifying distributional differences. As highlighted by Zhang et al. [4], the relationship between an attribute A_i and a query Q can be assessed by examining the disparity between the distribution of values for attribute A_i in the query results and their distribution within the database. A greater disparity suggests a stronger correlation between A_i and the query. The definition of KL-Distance can be summarized as follows.

Definition 8.9 (*KL-distance*) Let *A_j* denote a categorical attribute within the database *D*, with *Dom*(*A_j*) representing the set of values $\{a_{j1}, a_{j2} \ldots\ldots a_{jk}\}$. The result set corresponding to query Q is denoted as R. The KL-distance of the attribute A_j from the database D to the result set R is defined in Eq. (8.29).

$$D_{KL}(D||R) = \sum_{i=1}^{k} P\big(A_j = a_{ji}|D\big) \log \frac{P\big(A_j = a_{ji}|D\big)}{P\big(A_j = a_{ji}|R\big)} \tag{8.29}$$

where $P\big(A_j = a_{ji}|D\big)$ represents to the probability that $A_j = a_{ji}$ in database D, and $P\big(A_j = a_{ji}|R\big)$ denotes to the probability that $A_j = a_{ji}$ in result set R.

The method for calculating KL-distance is applicable to categorical attributes; however, it is essential to also take into account scoping attributes and spatiotemporal attributes. To tackle this issue, the attributes are organized into several buckets within the database D and the resultant set R.

8.5.2 Bucket Construction

The bucket method is designed to segment data distributions. It categorizes various types of attributes by dividing the complete range of values into a series of intervals, referred to as "buckets," and subsequently counts the number of values that reside within each interval. The primary objective of constructing these buckets is to convert all types of attributes into categorical attributes. Furthermore, the process of bucket construction varies depending on the specific type of attribute being analyzed.

In the context of fuzzy categorical attributes, a bucket is defined as a distinct value a_{ji} of the fuzzy categorical attribute A_j. For fuzzy scoping attributes, the values are initially categorized into multiple buckets, followed by the application of a histogram construction algorithm as proposed by Zhang et al. [4]. The challenge of constructing appropriate buckets for fuzzy spatiotemporal attributes arises from the highly discrete distribution of these values, which is exacerbated by the indistinct boundary lines between fuzzy scoping and fuzzy categorical attributes. Consequently, there is a pressing need for a clustering method that can treat all fuzzy spatiotemporal objects as points within a multidimensional vector space. To address this issue, we employ the k-means clustering method [27], which partitions a vector space into distinct segments while maximizing the similarity of points within each segment. This method operates through an iterative process of partitioning and reassigning points, as illustrated in Fig. 8.8. However, it is important to note that the classical k-means algorithm does not guarantee convergence to a global optimum, as its outcomes are influenced by the initial selection of clusters. In cases where the initial cluster centers are poorly chosen, the resulting clusters may be suboptimal. To mitigate this issue, we utilize the improved k-means algorithm (k-means++) [19, 28], which effectively addresses the initial cluster selection problem by ensuring that the initial cluster centers are positioned at significant distances from one another. Subsequently, the bucket construction algorithm is delineated in Algorithm 8.2.

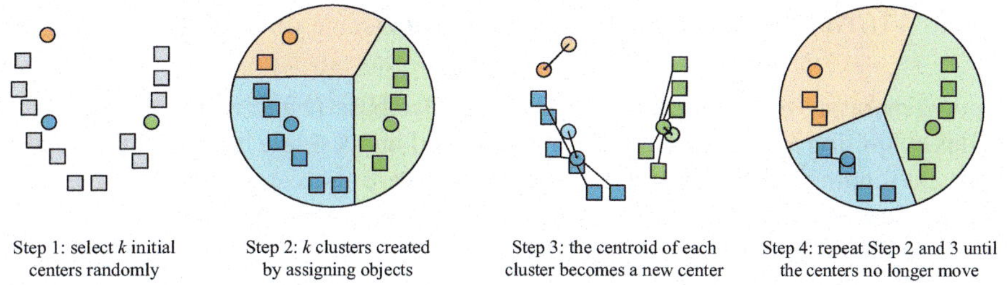

Fig. 8.8 Demonstration of the classical k-means algorithm

Algorithm 8.2 Bucket construction algorithm

Input: Attribute A_j and $Dom(A_j)$, total number of sub-trees $|T|$, minimum bucket number k
Output: Bucket B_{Dj} for attribute A_j

01 **if** A_j is a *fuzzy categorical attribute* **then**
02   **for each** distinct value a_{ji} of A_j
03   **use** query condition "$A_j = a_{ji}$" to get the number n of result tuples c from D
04   **add** a bucket (a_{ji}, n) into H_{Dj}
05 **if** A_j is a *fuzzy scoping attribute* with domain range $[a_{low}, a_{up}]$ **then**
06   **set** $\lambda = |T|/k$, $low = a_{low}$, $up = a_{up}$
07   **do**
08   **use** query condition "$low \le A_j \le a_{up}$" get the tuple number n
09   **if** $n \le \lambda$ **then**
10   **add** a bucket (low, up, n) into H_{Dj}
11   **set** $low = up$, $up = a_{up}$
12   **else**
13                 $up \leftarrow low + (up - low)/2$
14   **while** $low \le a_{up}$
15 **if** A_j is a *fuzzy spatiotemporal attribute* **then**
16   **build** k clusters $c_1, c_2, \ldots, c_k$ by using the k-means $++$ clustering algorithm
17   **for each** cluster c
18             $n \leftarrow$ the number of result tuples from cluster c
19   **add** a bucket(c, n) into B_{Dj}
20   **Return** B_{Dj}

The values of attributes are organized into distinct categories through a bucket construction algorithm. For categorical attributes, each unique value is designated as a separate category (Lines 01–04). In the case of scoping attributes, these are divided into smaller intervals by modifying the upper and lower boundaries (Lines 05–14). Spatiotemporal attributes are clustered utilizing the k-means algorithm (Lines 15–20). Following the establishment of the buckets, all scoping and spatiotemporal attributes are converted into categorical attributes. This transformation enables the computation of data distribution and the determination of attribute weights for the purpose of categorization.

8.5.3 Attribute Weight for Categorization

Upon generating the histogram of A_j utilizing database D and result R, categorical nodes exhibiting identical values are consolidated into a single bucket. Additionally, adjacent scoping nodes with closely related values are also categorized into the same bucket. Subsequently, the frequency of occurrences within each bucket is employed to derive the probability distribution.

Definition 8.10 (*Probability distribution*) P_{D_jk} is characterized as the probability distribution associated with the k-th bucket of A_j for the database D, while P_{R_jk} also represents the probability distribution of the k-th bucket of A_j for the result R.

$$P_{D_jk} = \frac{c_{D_jk}}{|D|} \tag{8.30}$$

$$P_{R_jk} = \frac{c_{R_jk}}{|R|} \tag{8.31}$$

where c_{D_jk} represents the frequency of the k-th bucket in H_{D_j}, and c_{R_jk} denotes the frequency of the k-th bucket in H_{R_j}.

Next, for the j-th attribute A_j, its weight w_j can be assigned as:

$$w_j = \frac{D_{KL}\left(P_{D_j}, P_{R_j}\right)}{\sum_{i=1}^{m} D_{KL}\left(P_{D_i}, P_{R_i}\right)} \tag{8.32}$$

As delineated in Eq. (8.32), each categorical attribute is associated with a specific weight. However, the selection of a categorical attribute, denoted as A, for level l in the construction of a navigation tree necessitates careful consideration. To resolve this issue, a single-valued method is employed to partition the values of $Dom(A)$ into distinct groups for each intermediate node at level $(l - 1)$. The single-value partitioning approach assesses the number of unique values that the categorical attribute possesses and the extent to which it can be categorized. Specifically, the scoping attribute is converted into a categorical attribute through the application of a histogram, while the spatiotemporal attribute is transformed using the k-means++ clustering technique. Subsequently, the single-value partition method is applied to these transformed attributes. Ultimately, this process facilitates the determination of each level within the navigation tree. It can be intuitively inferred that attributes deemed more significant by the user will be positioned closer to the root node of the tree.

Furthermore, to enhance user accuracy in selecting and swiftly locating the required information, the subcategories at each hierarchical level are organized in descending order based on the average probability of all tuples within a designated bucket. This methodology facilitates the establishment of the navigation tree.

8.5.4 Navigation Tree Generating Algorithm

The process of constructing the optimal navigation tree can be delineated into two primary phases. The first phase involves identifying the relevant attributes for each level of the navigation tree based on assigned weights. The second phase entails establishing the sequence of buckets according to the average probability associated with each bucket at every level.

Algorithm 8.3 Navigation tree generating an algorithm

Input: Database D, query results R, categorizing attributes weights, Bucket B_{D_j}
Output: A navigation tree with the minimum navigation cost
Create a root node (level $= 0$) and add it to T
$l = 1$;
while there exists at least one category v at level l-1 with $|N(v)| > M$
    select A_j with the maximum weight in attributes retained and has not used as an attribute for the level l
    If A_j is a *fuzzy categorical attribute* **then**
        for each value A_i of A_j in $|N(v)|$
            create a bucket under v
            add those tuples with $A_j = v_j$ to that bucket
        end for
        compute the average probability of each bucket of v
        sort buckets of A_j in descending order according to the buckets' average probabilities
If A_j is a *fuzzy scoping attribute* **then**
    for each interval $\left[low_j, up_j\right]$ of B_{D_j} for A_j in $|N(v)|$
        create a bucket under interval $\left[low_j, up_j\right]$
        add tuples with $low_i < A_j < up_i$ to that bucket
    end for
    compute the average probability of each bucket of $low_i < A_j < up_i$
    sort buckets of A_j in descending order according to the buckets' average probabilities
If A_j is a *fuzzy spatiotemporal attribute* **then**
    for each cluster c of B_{D_j} for A_j in $|N(v)|$
        create a bucket under v
        add tuples whose value in cluster c to that bucket
        end for
    compute the average probability of each bucket of cluster c
    sort buckets of A_j in descending order according to the buckets' average probabilities
    $l = l + 1$;
end while
Return T

This algorithm is designed to construct a navigation tree that incorporates attribute categorization weights and buckets. The process initiates with the establishment of a root node at the first level ($l = 0$) of tree T. Subsequently, for each subsequent level ($l = k$), subcategories are generated from the preceding level ($l = k - 1$) and are incorporated as nodes into tree T.

Furthermore, the prerequisite for the generation of a new level is the presence of at least one category containing more than M tuples at the current level; otherwise, the algorithm will terminate (Line 03). For the subsequent level, the attribute selected, denoted as A_j, is the one with the highest weight among the existing attributes that has not yet been utilized for level l (Line 04). The attribute with the maximum weight for each level is identified, after which the arrangement of buckets is established based on the various types of attributes. In the case of the fuzzy categorical attribute A_j, a bucket is created, and the relevant tuples $(A_j = v_j)$ are incorporated into this bucket. Subsequently, the average probability for each bucket of v is calculated, and the buckets of A_j are sorted in descending order according to their average fuzziness (Lines 05–11). For the fuzzy scoping attribute A_j, a bucket is established within the interval $\left[low_j, up_j\right]$, and the pertinent attribute values are added as child nodes, which are then sorted based on the average fuzziness of the bucket (Lines 12–18). Similarly, for the fuzzy spatiotemporal attribute A_j, a bucket is created under v, and tuples corresponding to the value in cluster c are added to this bucket. These tuples are then ranked in descending order according to their fuzziness (Lines 19–25). Ultimately, a navigation tree is constructed.

Within the navigation tree, the attribute positioned at the first level ought to represent the most significant attribute for the user within the specified context. Concurrently, the preceding category at the same level should align more closely with the user's intent.

8.6 Experiments

8.6.1 Experimental Setup

The experiments were performed on a computer operating with OSX 10.12, equipped with an Intel Core i5 1.6 GHz dual-core CPU and 8 GB of RAM. The relational database management system utilized was MySQL version 5.5 (server version), while the XML document database employed was BaseX. All algorithms were implemented in Python version 2.7.10, with connectivity to the relational database established via the SQL Alchemy library, and the BaseX Client was used to interface with the document database.

8.6.1.1 Datasets

For the purposes of our evaluations, we established a weather forecast database (ForecastDB) sourced from the World-Wide-Weather Site. This database comprises 100,000 tuples, which incorporate partial features that effectively represent the inherent fuzziness of spatiotemporal XML data. We implemented a fuzzy processing methodology [29] to enhance the data's suitability for experimental inquiries into querying fuzzy spatiotemporal XML data. The attributes of the final evaluated tuples encompass fuzzy categorical attributes (such as *Type*, *ScenicSpot*, *Humidity* and so on), fuzzy scoping attributes (including *WindForce*, *Temperature* and so on), as well as fuzzy spatiotemporal attributes (*time* and *position*).

8.6.1.2 Evaluating Methodology

To assess the efficacy of our methodology, we conduct three distinct sets of experiments. The initial set focuses on spatiotemporal data proximity measurement, aimed at evaluating the precision of both the spatiotemporal and general attribute value proximity measurement techniques. The second set involves query relaxation experiments designed to validate the impact of the proposed query relaxation strategy. Finally, the third set encompasses query result categorization experiments, which are conducted to evaluate the performance of the proposed result classification method.

8.6.2 Experimental Results

8.6.2.1 Spatiotemporal Data Proximity Measuring Experiments

The objective of this series of experiments is to assess the precision of spatiotemporal and general attribute value proximity measurement techniques within the context of fuzzy queries. Given that the parameter requirements for general attributes differ from those for spatiotemporal attributes in proximity measurement, we have chosen to conduct the experiments independently. Initially, we will perform proximity measurement experiments focused on general attributes.

Table 8.4 presents the various proximity values obtained under the query condition "type = sunny", calculated across datasets of varying sizes. The parameter α is established at 0.5, while γ is set at 0.7.

Table 8.4 illustrates that the weather type *Sunny* exhibits a close proximity to *Windy* and *Cloudy*, while demonstrating a significant distance from *Rainy* and *Snowy*. This observation aligns with real-world meteorological patterns, where sunny days are typically associated with either windy or cloudy conditions, with occurrences of rain or snow being relatively infrequent. This finding supports the notion that

Table 8.4 The proximity results on fuzzy categorical attribute

Fuzzy categorical value	Values	2,000 tuples	5,000 tuples	10,000 tuples
Type = Sunny	Windy	0.644	0.646	0.648
	Cloudy	0.606	0.601	0.603
	Rainy	0.485	0.489	0.493
	Snowy	0.389	0.382	0.391

the proximity calculations for fuzzy categorical values are both rational and justifiable. Consequently, when a user queries for sunny locations and receives limited results, the system is designed to automatically expand the search parameters to include *Cloudy* or *Windy* conditions in order to yield more pertinent results. Furthermore, Table 8.4 indicates that the proximity measures for other relevant values remain consistent across datasets of varying sizes, and the relative order of proximity among adjacent attribute values is preserved regardless of dataset size. This consistency suggests that our proximity measurement method is stable for evaluating fuzzy categorical attributes across different dataset sizes. Additionally, these operations are conducted offline, significantly minimizing the time expenditure associated with online processing.

In relation to spatiotemporal attributes, Table 8.5 presents the proximity results concerning temporal attributes, while Table 8.6 delineates the results pertaining to spatial attributes. The proximity of fuzzy spatiotemporal attribute values is contingent upon both similarity and the fuzzy coefficient, with the latter being influenced by the degree of possibility. In this context, the parameter *MaxDist* for the temporal attribute is set at 24, whereas for the spatial attribute, it is established at 10,000 m.

As indicated in Tables 8.5 and 8.6, certain data points exhibit proximity, as evidenced by significant overlaps in time intervals and the adjacency of spatial distances. However, the observed proximities are unexpectedly minimal due to their low probabilities. This observation aligns with the premise that, within a fuzzy dataset, the probability (or accuracy of the forecast) holds considerable significance for users. Specifically, information that demonstrates both appropriate similarity and acceptable probability tends to yield higher proximity, as it is more likely to align with the user's query intentions. Conversely, the data presented in Tables 8.5 and 8.6 reveal that certain locations, such as *Shi Sha Lake* and *Bei Hai Park*, despite their geographical closeness, exhibit low proximities, suggesting a reduced likelihood of favorable weather conditions in the future. In contrast, other locations, including *The Lama Temple* and *Temple of Confucius*, achieve higher proximities due to their greater probabilities of experiencing sunny weather. This finding is logical, as users may prefer to select a destination with favorable weather, even if it is situated at a greater distance, rather than opting for a nearby location with poor weather conditions. Furthermore, this fuzzy coincidence allows for the establishment of an anticipated equilibrium between similarity and probability, thereby facilitating the determination of appropriate proximity for each attribute value. It is important to note that the

Table 8.5 The proximity results on fuzzy temporal attribute

Fuzzy temporal value	Values	Possibility	FC ($\lambda = 0.7$)	T_{sim}	Proximity
Type = Sunny 10:00–14:00	10:00–14:00	0.32	0.45	1.0	0.45
	10:00–16:00	0.28	0.40	1.0	0.40
	08:00–12:00	0.64	0.91	0.88	0.81
	12:00–16:00	0.71	1.0	0.88	0.88

Table 8.6 The proximity results on fuzzy spatial attribute

Fuzzy spatial value	Values	Distance (m)	Possibility	FC ($\lambda = 0.7$)	S_{sim}	Proximity
Type = Sunny Position = BeiHaiBei (39.83N, 116.39E)	Position = ShiShaLake (39.94N, 116.38E)	895	0.23	0.34	0.91	0.3089
	Position = BeiHaiPark (39.92N, 116.38E)	586	0.19	0.28	0.94	0.2608
	Position = ConfuciusTemple (39.95N, 116.41E)	2842	0.72	1.0	0.71	0.7158
	Postion = LamaTemple (39.95N, 116.42E)	3058	0.56	0.80	0.69	0.5519

results presented in Table 8.6 represent one of several plausible outcomes; alternative reasonable results may emerge, indicating that users might favor locations in proximity to attractions when proximity values are computed using varying values of FC and T_{sim}, in accordance with user preferences.

8.6.2.2 Query Relaxation Experiments

The objective of this series of experiments is to assess the impact of the query relaxation algorithm on fuzzy spatiotemporal datasets. To ensure the comprehensiveness and adequacy of our algorithm, we categorize user queries into five distinct types that encompass the full range of potential user inquiries. Subsequently, we generate a substantial number of queries corresponding to these five categories (as detailed in Table 8.7). The outcomes of these queries are either empty or yield minimal results, not exceeding twelve tuples. We then apply our relaxation method to the original queries, resulting in satisfactory answers.

To assess the efficiency and accuracy of our methodology, referred to as spatiotemporal query relaxation (SPQR), in the context of user query relaxation techniques, we engaged five participants (designated as A, B, C, D, and E). Each participant was tasked with formulating five queries across five distinct query categories. For each query, denoted as Q*i*, a corresponding result set was generated. This result set comprised a balanced assortment of relevant and irrelevant tuples, derived from the top-20 result tuples produced by three different query relaxation algorithms: SPQR, AQR (which assigns weights to each property based on the data distribution within the database) [4], and TOQR [5]. The methodologies employed by AQR and TOQR share similarities with SPQR, as they facilitate the automatic relaxation of initial queries pertaining to fuzzy spatiotemporal data. In contrast, these algorithms differ significantly from IQR (an interactive query relaxation system) [2] and taxonomy-based relaxation [3], both of which necessitate user interaction and domain-specific knowledge. Consequently, we selected SPQR, AQR, and TOQR for the generation of the result sets. Subsequently, for each query, participants identified their ten most satisfactory tuples from the 60 result sets. We then compared the tuples selected by users with the 20 tuples generated by the three distinct query relaxation algorithms: SPQR, AQR, and TOQR. The effectiveness of the relaxation algorithms was evaluated based on the degree of matching. To quantify this matching, we employed *Recall*, *Precision*, and *F-Score* metrics. *Recall* [as shown in Eq. (8.33)] represents the ratio of the number of relevant tuples retrieved to the total number of relevant tuples, serving as a measure of result quality. *Precision* [as shown in Eq. (8.34)] is the ratio of the number of relevant tuples retrieved to the total number of tuples retrieved, providing an assessment of result accuracy. The *F-Score* [as shown in Eq. (8.35)] is a harmonic mean that incorporates both precision and recall, often utilized as a comprehensive performance indicator. The *Recall*, *Precision*, and *F-Score* metrics for the five query types across the three query relaxation algorithms are illustrated in Figs. 8.9, 8.10, 8.11, 8.12 and 8.13.

Table 8.7 The five types of user query

User query	Types	Practical query example	
Q1	Simple temporal query	Q1(A)	Time ∈ [10 : 00, 12 : 00]
		Q1(B)	Time ∈ [14 : 00, 16 : 00]
		Q1(C)	Time ∈ [10 : 00, 16 : 00]
		Q1(D)	Time ∈ [08 : 00, 10 : 00]
		Q1(E)	Time ∈ [10 : 00, 18 : 00]
Q2	Simple spatial query	Q2(A)	Location = (39.9413212696, 116.3397088151) ∧ ScenicSpot = Park
		Q2(B)	Location = (39.9116991666, 116.3035677389) ∧ ScenicSpot = Market
		Q2(C)	Location = (39.9056943459, 116.2229948245) ∧ ScenicSpot = Park
		Q2(D)	Location = (39.8567665077, 116.0853828956) ∧ ScenicSpot = Park
		Q2(E)	Location = (39.9276641225, 116.3869542632) ∧ ScenicSpot = Park
Q3	Spatiotemporal query	Q3(A)	Location = (39.9116991666, 116.3035677638) ∧ ScenicSpot = Market ∧ Time ∈ [19 : 30, 21 : 30]
		Q3(B)	Location = (39.9074891651, 116.688517349) ∧ ScenicSpot = Park ∧Time ∈ [13 : 30, 15 : 30)]
		Q3(C)	Location = (39.9159190619, 116.605782496) ∧ ScenicSpot = Temple ∧ Time ∈ [08 : 30, 10 : 30]
		Q3(D)	Location = (39.9094803705, 116.617513273) ∧ ScenicSpot = Park ∧Time ∈ [09 : 00, 10 : 30)]
		Q3(E)	Location = (39.9105577157, 116.526125529) ∧ ScenicSpot = Museum ∧Time ∈ [07 : 30, 09 : 00]
Q4	General query	Q4(A)	Type = Cloudy
		Q4(B)	District = ChaoYang ∧ Type = Sunny
		Q4(C)	Temperature ∈ [12, 14]
		Q4(D)	Temperature ∈ [16, 20]∧ Type = Sunny
		Q4(E)	Temperature ∈ [16, 20]∧ Type = Sunny ∧ District = DongCheng
Q5	Multiple query	Q5(A)	Time ∈ [07 : 30, 19 : 30]∧ ScenicSpot = Market ∧ type = Windy

(continued)

Table 8.7 (continued)

User query	Types	Practical query example	
		Q5(B)	Location = (39.9056943459, 116.2229948245) Location = (39.9413212696, 116.3397088151) ∧ ScenicSpot = Market ∧ Type = Cloudy
		Q5(C)	Time ∈ [08 : 00, 20 : 00]∧ Location = (39.9413212696, 116.3397088151) ∧ ScenicSpot = Park ∧ Type = Windy
		Q5(D)	Time ∈ [07 : 00, 21 : 00]∧ type = Windy ∧ Humidity ≥ 77% ∧ ScenicSpot = Museum
		Q5(E)	Time ∈ [06 : 00, 08 : 00]∧ Location = (39.9413212696, 116.3397088151) ∧ Humidity ≥ 80%

$$Precision = \frac{|\{relevant\ documents\} \cap \{retrieved\ documents\}|}{|\{retrieved\ documents\}|} \tag{8.33}$$

$$Recall = \frac{|\{relevant\ documents\} \cap \{retrieved\ documents\}|}{|\{relevant\ documents\}|} \tag{8.34}$$

$$F\text{-}Measure = \left(\frac{recall^{-1} + precision^{-1}}{2}\right)^{-1} = 2 \cdot \frac{recall \cdot precision}{recall + precision} \tag{8.35}$$

Figures 8.9, 8.10, 8.11, and 8.13 present the *Recall, Precision* and *F-Score* metrics as computed by three distinct query algorithms. The comparative analysis indicates that the SPQR algorithm exhibits a significant advantage over both the AQR and TOQR algorithms in the context of spatiotemporal query relaxation. The results obtained for recall and precision consistently meet our expectations, achieving high accuracy levels (exceeding 70% on average) and acceptable recall rates (averaging over 50%). In practical scenarios, when users conduct queries related to spatial or temporal dimensions, they typically seek information that is relevant to a specified time interval or spatial location, thereby underscoring the importance of precision in these queries. Furthermore, the computation of similarity for spatiotemporal attributes has been performed offline and stored in an auxiliary table, which significantly reduces the time required for online querying when users engage in spatiotemporal queries. In conclusion, the data illustrated in Figs. 8.9, 8.10, 8.11, and 8.13 collectively affirm that the SPQR algorithm possesses distinct advantages in the realm of spatiotemporal-related queries.

Regarding the general query, the dataset does not impose any spatial or temporal constraints. Users tend to focus more on the relationships among general attributes within the context of the given query, resulting in a less stringent requirement for precision compared to spatiotemporal queries. Additionally, the dataset contains a

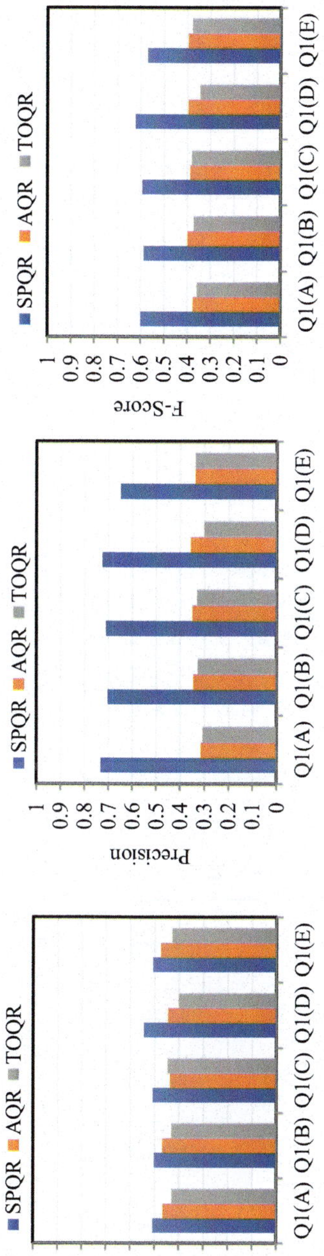

Fig. 8.9 Recall, precision, and F-score of the simple temporal queries in the Weather ForecastDB

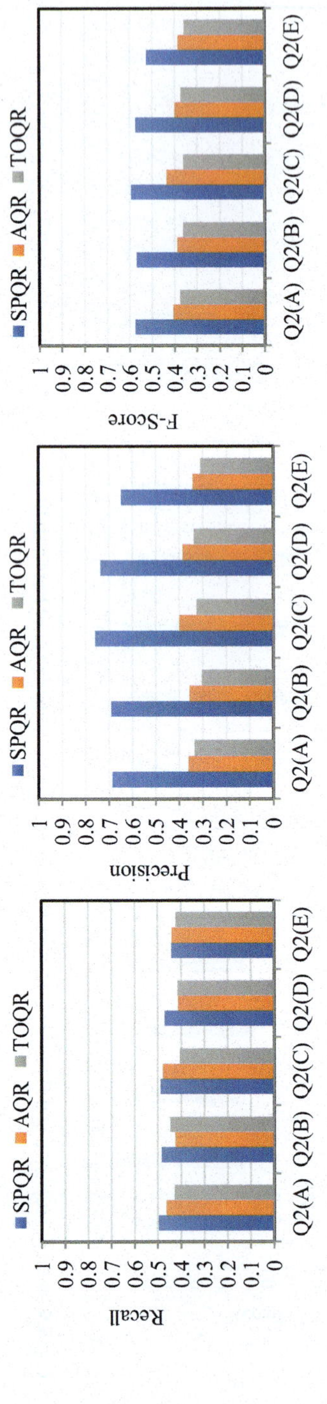

Fig. 8.10 Recall, precision, and F-score of the simple spatial queries in the Weather ForecastDB

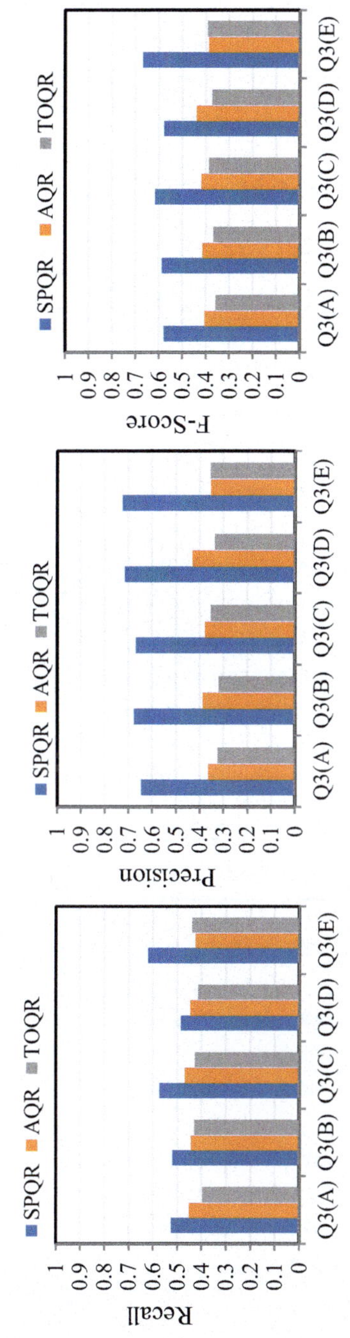

Fig. 8.11 Recall, precision, and F-score of the spatiotemporal queries in the Weather ForecastDB

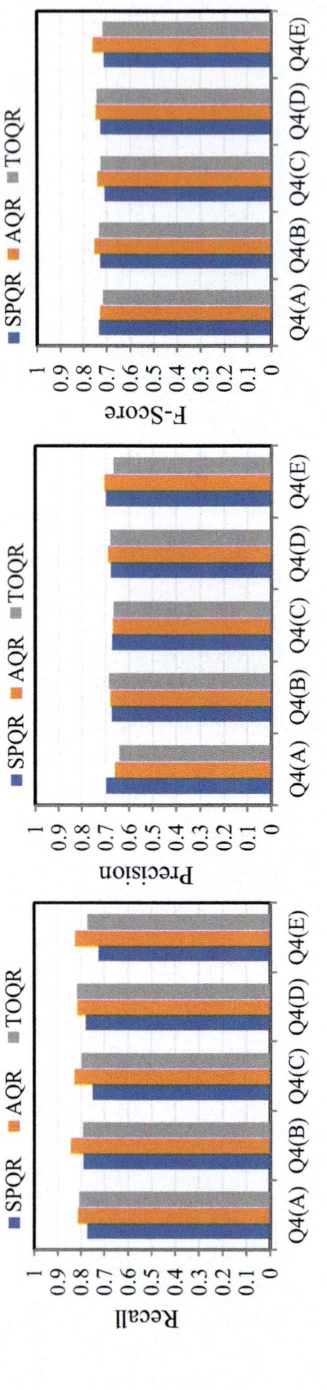

Fig. 8.12 Recall, precision, and F-score of the general queries in the Weather ForecastDB

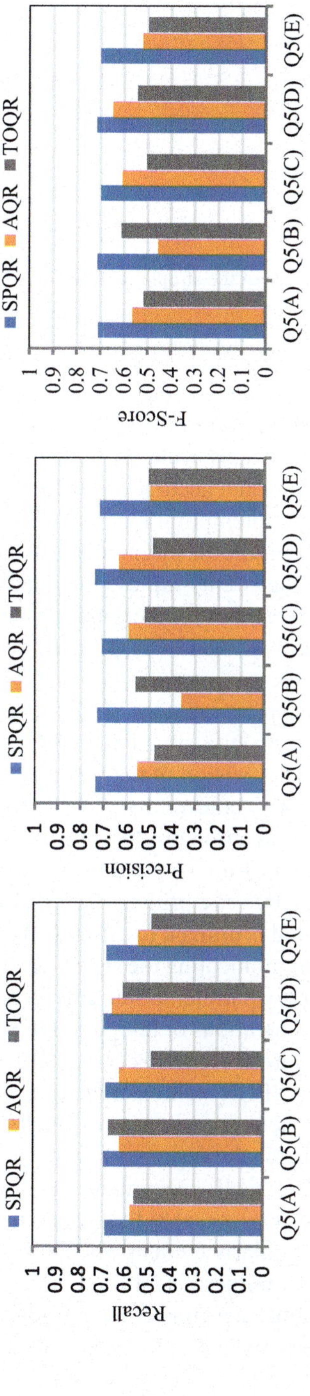

Fig. 8.13 Recall, precision, and F-score of the multiple queries in the Weather ForecastDB

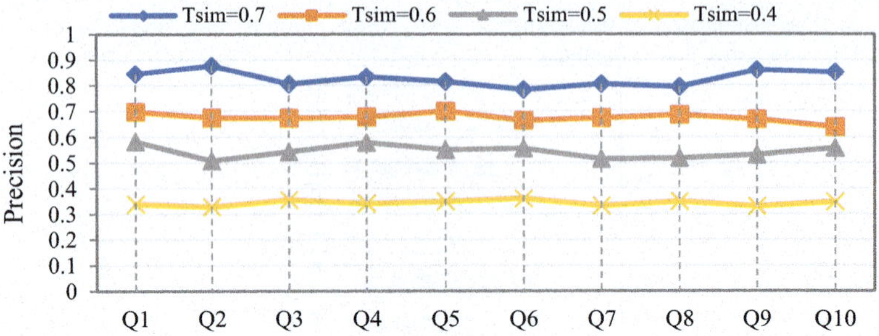

Fig. 8.14 Recall, precision, and F-score of the simple multiple query

greater variety of general attribute values than spatiotemporal attribute values. As illustrated in Fig. 8.12, the differences among the SPQR, AQR, and TOQR algorithms in relation to general queries are minimal, indicating that the SPQR algorithm performs effectively in this context as well. In conclusion, the SPQR algorithm demonstrates acceptable efficiency and accuracy in delivering the desired results for queries that are not related to spatiotemporal factors.

Furthermore, we assess the degree of alignment between the 10 tuples identified as relevant by the user and the twenty tuples generated by SPQR across various relaxation thresholds. The *Precision* metrics are employed to quantify this correspondence. Figure 8.14 illustrates the *Precision* of SPQR at different relaxation thresholds within the context of ForecastDB.

As illustrated in Fig. 8.14, the precision of the SPQR algorithm is influenced by the relaxation thresholds employed. Specifically, a decrease in the relaxation threshold correlates with a reduction in the precision of the SPQR algorithm. In the context of ForecastDB, the average precision achieved by SPQR is approximately 77%. This finding indicates that the query relaxation produced by the SPQR method is capable of effectively capturing the intent of the original query. Overall, the high precision attained by SPQR can be attributed to two primary factors. Firstly, in relation to fuzzy general attributes, the IDF method is utilized to estimate the weight of each specified attribute, thereby effectively communicating the significance of these attributes to the user. Secondly, concerning fuzzy spatiotemporal attributes, the algorithm not only assesses the proximity of spatial locations or temporal intervals but also incorporates fuzzy coincidence to adeptly address scenarios characterized by high similarity and low probability. Consequently, SPQR demonstrates the ability to effectively and accurately resolve queries that yield few or even no results in their original form.

Furthermore, utilizing XML for the representation of fuzzy values proves to be both practical and effective. The XML Document Database offers a more convenient means of expressing fuzziness compared to Relational Database Management Systems (RDBMS). This is particularly evident when dealing with large datasets, as RDBMS requires the creation of numerous tables to adequately represent complex fuzzy relations, resulting in significant time consumption. In contrast, XML documents exhibit considerable advantages due to their flexible structure, which allows

Table 8.8 The comparison of time-consumption of the similarity calculation under the different total number of records in the database between RDBMS and XML document DB

The total number of records in the database	RDBMS (ms)	XML document DB
100	8	1
500	107	8
1000	409	17
5000	8981	81
10,000	26357	269

for the direct mapping of intricate fuzzy relationships in a tree format. Additionally, the time required for constructing the offline Auxiliary Table varies between RDBMS and XML Document Databases, depending on the total number of records present, as illustrated in Table 8.8. The notable improvement highlighted in Table 8.8 is attributed to the focus on the similarity calculation step, excluding other associated costs. Moreover, in terms of similarity calculation expenses, XML's flexible structure facilitates direct traversal of the relevant nodes and values, whereas RDBMS necessitates the execution of conjunction operations.

8.6.2.3 Query Result Categorization Experiments

This subsection conducts an assessment of query result categorization experiments to evaluate the efficacy of our result classification algorithm, SPRC. The evaluation is based on two criteria: *Ecost* and the quantity of relevant tuples that correspond to high levels of user satisfaction.

The initial metric under consideration is the total navigation cost, as delineated in Definition 8.8. To streamline the analysis, we assign equal weight to the visitation of intermediate nodes and the visitation of tuples in the leaf nodes by establishing $K_1 = K_2 = 1$. Figure 8.15 illustrates the total navigation cost across ten queries of five distinct types, which are detailed in Table 8.9.

As illustrated in Fig. 8.15, the differences among the SPRC, C4.5, Cost-Based, AQR, and TOQR methods are minimal; however, SPRC demonstrates certain relative advantages. *Ecost* refers to the cost associated with navigating through nodes until the target node is reached. A lower *Ecost* signifies that users can efficiently and accurately identify a specific level within the navigation tree, as well as a particular bucket within that level. The navigation tree produced by the SPRC method effectively aligns with the user's objectives, as evidenced in Fig. 8.15. There are two primary factors contributing to the low *Ecost* achieved by SPRC. First, SPRC effectively assigns weights to attributes, ensuring that more significant attributes for the user are positioned closer to the root node. This arrangement facilitates quicker and more precise keyword identification by the user. Second, within a given level, the buckets are organized in descending order based on their fuzziness, which enhances the likelihood of meeting the user's requirements and further decreases the *Ecost*.

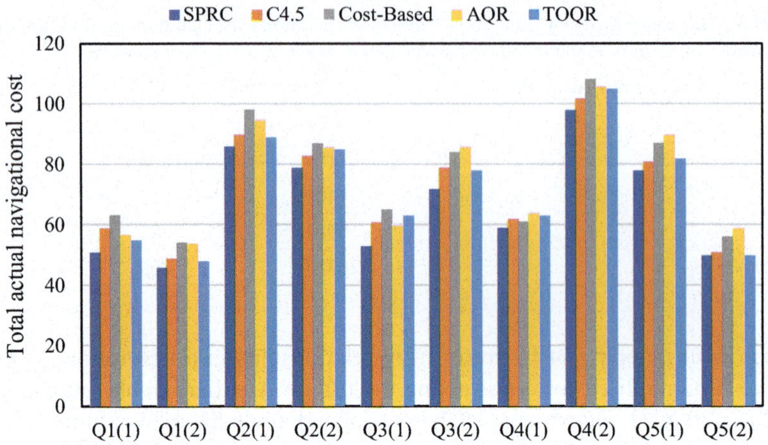

Fig. 8.15 The navigation cost under the different result categorization algorithms

Table 8.9 The given ten queries over five different types

User Query	Types	Practical query example	
Q1	Simple temporal query	Q1(1)	time ∈ [10 : 00, 12 : 00]
		Q1(2)	time ∈ [14 : 00, 16 : 00]
Q2	Simple spatial query	Q2(1)	Location = (39.9413212696, 116.3397088151) ∧ ScenicSpot = Park
		Q2(2)	Location = (39.9178896573, 116.3970500738) ∧ ScenicSpot = Museum
Q3	Spatiotemporal query	Q3(1)	Location = (39.9159190619, 116.6057824966)∧ ScenicSpot = Market ∧ time ∈ [08 : 20, 11 : 30]
		Q3(2)	Location = (39.9192015845, 116.6558124854)∧ ScenicSpot = Temple ∧time ∈ [08 : 30, 10 : 30]
Q4	General query	Q4(1)	Type = Rainy ∧ Temperature ∈ [2, 11]
		Q4(2)	Type = Rainy ∧ District = TongZhou
Q5	Multiple query	Q5(1)	Time ∈ [06 : 00, 12 : 00]∧ Location = (39.9056943459, 116.2229948245) ∧ ScenicSpot = ShiJingShan ∧ Type = Windy
		Q5(2)	Time ∈ [18 : 00, 22 : 00] ∧ Location = (39.8567665077, 16.0853828956) ∧ ScenicSpot = Fengtai ∧Temperature ∈ [6, 13]

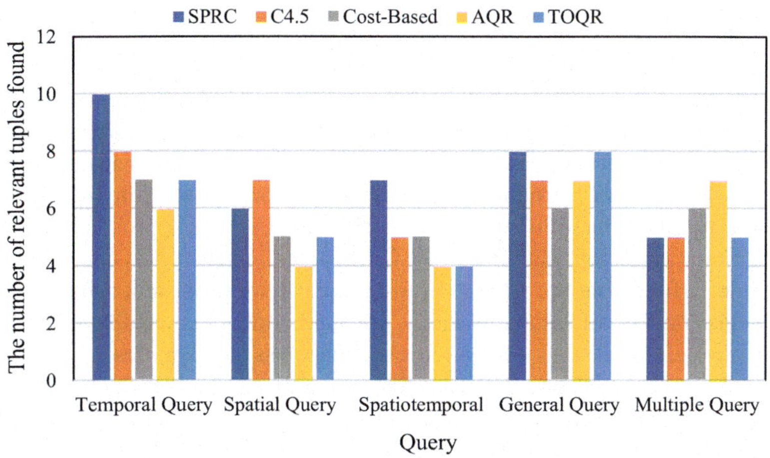

Fig. 8.16 The number of relevant tuples found by per subject with different categorization algorithms for each test query

The second metric under consideration is the quantity of relevant tuples identified. In this study, five distinct types of queries are employed, each yielding corresponding navigation trees for users tasked with locating varying numbers of relevant tuples across these different navigation structures. Users are then instructed to identify tuples that closely align with their intentions within each navigation tree. Generally, an effective categorization algorithm facilitates the user's ability to discover a greater number of relevant results. Figure 8.16 illustrates the number of relevant tuples identified by various algorithms in response to the execution of different query types, including temporal query, spatial query, spatiotemporal query, general query, and multiple query.

In Fig. 8.16, SPRC demonstrates a higher yield of relevant tuples for both temporal and spatiotemporal queries, while it produces an equivalent or marginally lower number of relevant tuples for spatial, general, and multiple queries. Consequently, our methodology exhibits superior performance in categorizing query results for temporal and spatiotemporal queries. This enhanced performance can be attributed to two primary factors. Firstly, our algorithm achieves a significant hit rate in both temporal and spatiotemporal queries. Secondly, the proximity model effectively balances similarity and fuzzy coincidence, addressing the issue of high similarity coupled with low probability, thereby providing users with a greater number of relevant tuples.

8.7 Summary

In this chapter, we examine the challenge of effectively relaxing the original fuzzy spatiotemporal query and categorizing the resultant overload of data based on XML. To enhance the retrieval of pertinent spatiotemporal information, particularly within a fuzzy framework, we introduce an innovative concept of proximity that reconciles the similarity and probability of spatiotemporal entities. This approach is anticipated to yield more satisfactory outcomes and significantly improve querying performance. Furthermore, following the relaxation of the query, we propose a robust algorithm for the classification of results. Notably, our methodology is capable of generating a result navigation tree that incorporates the fuzziness and significance of the query conditions for the user. Experimental findings indicate that our method for fuzzy spatiotemporal query relaxation and result categorization effectively addresses the issue of insufficient answers while achieving enhanced precision.

References

1. I. Muslea, T.J. Lee, Online query relaxation via Bayesian causal structures discovery, in *Proceedings of the 20th National Conference on Artificial Intelligence* (2005), pp. 831–836
2. D. Mottin, A. Marascu, S. Basu-Roy, G. Das, T. Palpanas, Y. Velegrakis, IQR: an interactive query relaxation system for the empty-answer problem, in *Proceedings of the International Conference on Management of Data* (2014), pp. 1095–1098
3. D. Martinenghi, R. Torlone, Taxonomy-based relaxation of query answering in relational databases. VLDB J. **23**(5), 747–769 (2014)
4. X. Zhang, X. Meng, Y. Tang, C. Bi, Adaptive query relaxation and result categorization based on data distribution and query context. J. Inf. Sci. Eng. **33**(5), 1375–1396 (2017)
5. X. Meng, X. Zhang, Y. Tang, C. Bi, Adaptive query relaxation and top-*k* result ranking over autonomous web databases. Knowl. Inf. Syst. **51**(2), 395–433 (2017)
6. P. Bosc, A. Hadjali, O. Pivert, Empty versus overabundant answers to flexible relational queries. Fuzzy Sets Syst. **159**(12), 1450–1467 (2008)
7. B. Yang, Y. Zhang, X. Luan, A probabilistic relaxation approach for matching road networks. Int. J. Geogr. Inf. Sci. **27**(2), 319–338 (2013)
8. M.R. Ford, P.S. Kench, Spatiotemporal variability of typhoon impacts and relaxation intervals on Jaluit Atoll, Marshall Islands. Geology **44**(2), 159–162 (2016)
9. S. Mitra, S.L. Pallickara, Confluence: adaptive spatiotemporal data integration using distributed query relaxation over heterogeneous observational datasets, in *Proceedings of the 11th IEEE/ACM International Conference on Utility and Cloud Computing* (2018), pp. 184–193
10. Z. Chen, T. Li, Addressing diverse user preferences in SQL-query-result navigation, in *Proceedings of the International Conference on Management of Data* (2007), pp. 641–652
11. K. Chakrabarti, S. Chaudhuri, S. Hwang, Automatic categorization of query results, in *Proceedings of the 2004 ACM SIGMOD International Conference on Management of Data* (2004), pp. 755–766
12. M.H. Afif, A.S. Ghareb, A. Saif, A. Bakar, O. Bazighifan, Genetic algorithm rule based categorization method for textual data mining. Decis. Sci. Lett. **9**(1), 37–50 (2020)
13. X. Tang, Y. Dai, Y. Xiang, Feature selection based on feature interactions with application to text categorization. Expert Syst. Appl. **120**, 207–216 (2019)
14. Z. Zhao, Z. Ma, L. Yan, An efficient classification of fuzzy XML documents based on kernel ELM. Inf. Syst. Front. **23**(1), 515–530 (2019)

15. L. Bai, L. Yan, Z. Ma, Fuzzy spatiotemporal data modeling and operations in XML. Appl. Artif. Intell. **29**(3), 259–282 (2015)
16. L. Bai, Y. Li, J. Liu, FSPTwigFast: holistic twig query on fuzzy spatiotemporal XML data. Appl. Intell. **47**(4), 1224–1239 (2017)
17. L. Bai, L. Zhu, An algebra for fuzzy spatiotemporal data in XML. IEEE Access **7**(1), 22914–22926 (2019)
18. S. Amer-Yahia, L.V.S. Lakshmanan, S. Pandit, FleXPath: flexible structure and full-text querying for XML, in *Proceedings of the 2004 ACM SIGMOD International Conference on Management of Data* (2004), pp. 83–94
19. Y. Xu, W. Qu, Z. Li, G. Min, K. Li, Z. Liu, Efficient k-means++ approximation with MapReduce. IEEE Trans. Parallel Distrib. Syst. **25**(12), 3135–3144 (2014)
20. J. Li, C. Liu, R. Zhou, J.X. Yu, Quasi-SLCA based keyword query processing over probabilistic XML data. IEEE Trans. Knowl. Data Eng. **26**(4), 957–969 (2014)
21. A. Sözer, A. Yazici, H. Oguztüzün, Indexing fuzzy spatiotemporal data for efficient querying: a meteorological application. IEEE Trans. Fuzzy Syst. **23**(5), 1399–1413 (2015)
22. L. Bai, A. He, M. Liu, L. Zhu, Y. Xing, Adaptive query relaxation and result categorization of fuzzy spatiotemporal data based on XML. Expert Syst. Appl. **168**, 1–19 (2021)
23. B. Yu, A query system for spatiotemporal database applications, in *Proceedings of the 18th International Conference on Systems Engineering*, pp. 28–33 (2005)
24. R. Bachi, Standard distance measures and related methods for spatial analysis. Pap. Reg. Sci. Assoc. **10**(1), 83–132 (1963)
25. H.K. Smith, G. Laporte, P.R. Harper, Locational analysis: highlights of growth to maturity. J. Oper. Res. Soc. **60**(SI), 140–148 (2009)
26. E. Porcu, M. Bevilacqua, M.G. Genton, Spatio-temporal covariance and cross-covariance functions of the great circle distance on a sphere. J. Am. Stat. Assoc. **111**(514), 888–898 (2016)
27. K.R. Žalik, An efficient k'-means clustering algorithm. Pattern Recogn. Lett. **29**(9), 1385–1391 (2008)
28. B. Bahmani, B. Moseley, A. Vattani, R. Kumar, S. Vassilvitskii, Scalable k-means++. Proc. VLDB Endow. **5**(7), 622–633 (2012)
29. J. Liu, Z. Ma, Q. Qv, "Dynamically querying possibilistic XML data. Inf. Sci. **261**(4), 70–88 (2014)